An Introduction to Metallic Corrosion

Third Edition

Ulick R. Evans

M.A., Sc.D., F.I.M., F.R.S.
Formerly Reader in the Science of Metallic Corrosion,
Cambridge University

Edward Arnold   American Society for Metals

First published 1948
by Edward Arnold (Publishers) Ltd
41 Bedford Square, London WC1B 3DQ

Reprinted 1950, 1951, 1955, 1958, 1960
Second edition 1963
First published as a paperback 1975
Third edition 1981

First US printing in hardcover 1982
published by Edward Arnold (Publishers) Ltd
and American Society for Metals
Metals Park, Ohio 44073

Library of Congress Cataloging in Publication Data

Evans, Ulick Richardson
An introduction to metallic corrosion
Bibliography: p.
Includes indexes
1. Corrosion and anti-corrosives I. Title
TA462.E83 1982 620.1′623 81–17651
ISBN 0–87170–132–4 AACR2

To R. S. H.

Printed in Great Britain

Extract from the Preface to the First Edition

I am hopeful that the book will be read, not merely by students, but by experienced scientists who have hitherto taken no interest in the subject of corrosion. It may be that the present neglect of chemical reactions involving both metals and non-metals (which, for lack of a better name, have come to be known as 'corrosion reactions'), is largely due to the non-existence of a book suitable for continuous reading. I shall be disappointed if the present volume does not do something at least to rectify the situation, and awaken interest in this *terra incognita* among those engaged in other branches of science.

Whilst this is essentially a scientific book, I have illustrated basic principles by examples taken from engineering or industrial practice, which may serve to increase the interest of the subject for engineers or industrial chemists. It would be dangerous, however, to regard so short a book as a guide to practical preventative measures, since, for reasons explained in it, protective processes, unless correctly adjusted, may often make matters worse. Those faced with such problems should consult my larger books – or better still the original papers quoted in them.

URE

Publishers' Note on the Third Edition

The second edition of this book was published in 1963 and re-issued in paperback in 1975 to meet the continuing demand. For many years Dr Evans was heavily occupied with the production of the three volumes of his major work *The Corrosion and Oxidation of Metals*, but following the publication of the third volume in 1976 he was able to prepare a revised edition of this short introductory text. The publishers would like to acknowledge the help of Dr R S Thornhill in this work. Sadly, Dr Evans died during 1980 and full responsibility for any deficiencies in proofreading must fall on the publishers.

Following the system of references used in the second edition the author has inserted numbers in the text which indicate the pages of the 1960 publication, *The Corrosion and Oxidation of Metals*, where fuller information will be found and where there are usually references to original papers embodying divergent points of view. References to material in the two supplementary volumes of *The Corrosion and Oxidation of Metals*, published in 1968 and 1976, are given in full.

Contents

1

Film Growth

General

Classification

The word 'corrosion' will be used in this book to cover all transformations in which a metal passes from the elementary to the combined condition; it will cover reactions between metals and gaseous or liquid environments. The former may result in products which can be volatile, solid or liquid depending on the temperature. Similarly, corrosive environments can consist of fluids, which may or may not be electrolytic conductors, and contain a variety of secondary constituents. The resulting attack may lead to products having solubility in the medium to varying degrees, or alternatively in the case of insoluble products to deposits on the substrate metal. Instances of corrosion may thus be divided into two classes; those which produce a solid film and those which do not.

Film formation

The present chapter is concerned with reactions which lead to thin films of product on the metal. Film-forming reactions usually slow down as the film thickens, whereas in the absence of a film the velocity of attack is unlikely to diminish, except through exhaustion of one of the participants. The fact that film-forming reactions do slow down with time accounts for the survival of metals in many potentially destructive situations. The observation that bright iron can be exposed to reasonably dry air without visible change might suggest that iron has little affinity for oxygen, but measurements of free energy show that actually the affinity is high. The true cause is different. If an oxide-free surface of iron is exposed to dry air, there is rapid combination, but as soon as the growing oxide film begins to isolate the metal from the air, oxidation becomes slow, although it does not cease entirely. Vernon[110*] found that, even after some weeks exposure to air, iron was still slowly increasing in mass.

Action of iodine on metals

The difference between filming and non-filming reactions can be illustrated by the observations of Parsons on metals introduced into solutions of iodine in various organic liquids. Whenever he used a solvent

*See Publishers' Note, p. iv.

capable of dissolving the iodide of the metal under examination, the metal passed into solution without hindrance as its iodide. But whenever the solvent was one which left a layer of solid iodide undissolved on the metallic surface, there was little attack, since the layer shut off the two reacting elements from one another. (L. B. Parsons, *J. Amer. chem. Soc.*, 1925, **47**, 1830.)

Later work showed that even where a solid layer is formed, blockage of the attack may not be immediate; film growth may begin rapidly, becoming slow as the film thickens, as found by Bannister[85] in the author's laboratory during a detailed study of film growth on silver exposed to a chloroform solution of iodine which yields silver iodide insoluble in the solvent. Some of the curves are reproduced in Fig. 1.1. Their analysis shows that, over a considerable range of thickness, the rate of thickening is inversely proportional to the thickness attained. Thus if y is the thickness reached in time t

$$dy/dt = k/y$$

where k is a constant. The equation may also be written

$$y^2 = Kt + K'$$

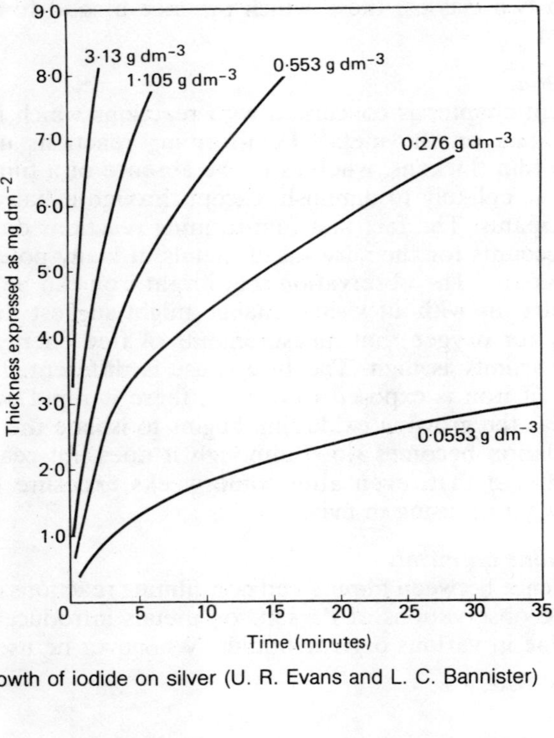

Fig. 1.1 Growth of iodide on silver (U. R. Evans and L. C. Bannister)

where K represents $2k$, and K' is the integration constant. This same 'parabolic' relationship had previously been found by Tammann[84] for the action of iodine vapour on metals.

Such an equation is in no way surprising; the rate of the passage of heat through a plate is inversely proportional to the thickness, and it is reasonable to expect the passage of matter through a film to be inversely proportional to the thickness also, whether the metal moves outwards through the film or the iodine inwards. The manner in which the curves are influenced by the iodine concentration in the liquid rather suggests that, in this case, the non-metal moves inwards; the outward movement of metal is frequently met with in high-temperature oxidation – as shown later.

If the equation $dy/dt = k/y$ were obeyed from the outset, it would imply an infinite growth rate when y was zero, that is before any film had appeared. Now chemical considerations would preclude an infinite reaction velocity; even in a liquid where no film is formed, corrosion does not proceed infinitely fast. Hence there must be a departure from the relationship in the early stages. Bannister found that the parabolic law was not obeyed at the outset, when the iodide films were very thin. Similar failure of the parabolic law in the early stages of growth will be met with later in connection with high-temperature oxidation.

Interference tints

Iodide films on silver produce a beautiful series of colours, due to interference between light reflected from the outer and inner film-surfaces. Interference may be expected whenever the effective paths travelled by the light reflected at the two surfaces differ by an odd number of half wavelengths. Consider a film being formed on a metallic surface, which is viewed by white light. As long as the film is very thin, the interference band will be in the ultraviolet (Fig. 1.2(A)) and there will be no colour. When a certain thickness is reached, depending on the film substance (usually less than 0·04 μm), the blue light reflected at the outer surface will be out of phase with that reflected at the inner surface (Fig. 1.2(B)); partial extinction of the blue waves will make the surface appear yellow. When a greater film thickness is reached, the longer green waves suffer interference (C), producing a mauve colour; a still greater thickness causes interference of yellow light, leaving the reflection blue. Since interference can again occur when the path-difference reaches 3/2, 5/2, 7/2 or 9/2 times the wavelength of the light in question, various orders of colours are possible; silver iodide films exhibit five orders, red appearing five times. The sequence of colours for films of iodide, oxide and sulphide shown in Table 1.1, is not a simple repetition, since the various absorption bands will follow one another across the spectrum at unequal intervals, as shown in Fig. 1.2. For a film of effective thickness y and refractive index n, the centres of the absorption bands will fall at wavelengths $4ny$, $4ny/3$, $4ny/5$, $4ny/7$, . . . etc. The first-order colours are produced as the first band (marked 1) passes across the visible part

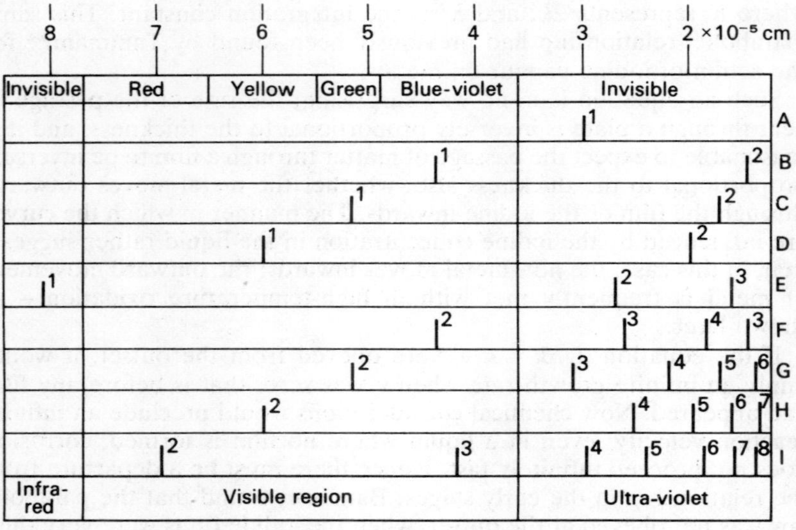

Fig. 1.2 Production of interference tints

of the spectrum (B, C and D). If the bands are narrow (as is usually the case for films on brightly reflecting metal provided the thickness is fairly uniform), the first band will have passed out into the invisible infrared before the second has entered from the invisible ultraviolet (E) thus on metals there is a silvery hiatus between the first- and second-order colours. The second-order yellow, red and blue appear when the second band is in the positions shown in F, G and H. The third band follows more closely on the second than the second follows on the first, and must enter the visible region before the second has left, so that there is always a green at the end of the second order (I). In contrast, the bands produced by air films (e.g. those responsible for Newton's rings viewed by *transmitted* light, which furnish a colour sequence similar to that given by films on metals viewed by *reflected* light) are much broader; the first band may still cause absorption in the red after the second has entered the blue, and consequently a green appears at the end of the first-order.

Oxide films produced on heating nickel sheet in air, or on exposing molten lead to air, produce almost the same colour sequence as appears on silver exposed to iodine (Table 1.1); the compositions are close to NiO and PbO. The films formed on iron heated at, say, 250°C consist of the rhombohedral α-Fe_2O_3, with magnetite underneath it; magnetite is fairly opaque and α-Fe_2O_3, which is responsible for the colours, is less transparent than NiO, so that the later colours (corresponding to the thicker films) are weak or missing. The invisible films formed on iron exposed to dry air at 18°C – or slightly higher temperatures – consist of

Table 1.1

	Colours produced by air films between glass ('Newton's rings')		Colours produced by films on metals		
Order	Viewed by reflected light	Viewed by transmitted light	Oxide films on lead or nickel; iodide-films on silver	Oxide films on iron	Oxide or sulphide films on copper
'Invisible range'	Colourless (black)	Colourless (white)	Colour of metal unchanged	Colour of iron unchanged	Colour of copper unchanged
First order	Blue	Yellow	Yellow to brown	Yellow to brown	Brown
	Faint green	Red to mauve	Rose to mauve	Mauve	Rose to mauve
	Yellow	Blue	Blue	Blue	Blue
	Red	Green	Silvery (greenish, if film is imperfect)	Silver grey	Brilliant silver (occasionally greenish)
Second order	Blue	Yellow	Yellow to brown	—	Yellow-brown
	Green	Red	Red	Pinkish blue	Red
	Yellow	Blue	Blue	Blue	Blue
	Red	Green	Green	Greenish blue	Green
Third order	Blue	Yellow	Yellow	—	Brown
	Green	Red	Red	Bluish grey (with trace of pink)	Red
	Yellow	Greyish blue	(Trace of lavender blue)	Bluish grey (specific colour)	(Trace of lavender blue)
	Red	Green	Green	Bluish grey (specific colour)	Green
Fourth order	—	Dull yellow	—	Bluish grey (specific colour)	—
	Green	Red	Red	Bluish grey (specific colour)	Dirty red
	Red	Green	Green	Bluish grey (specific colour)	Dirty green
Fifth order	Greenish	Red	Faint red passing into specific colour of film-substance	Bluish grey (specific colour)	Grey (sometimes trace of red)

cubic oxide. There is a range of solid solutions of which magnetite (Fe_3O_4) and the (cubic) γ-Fe_2O_3 are end-members; magnetite and γ-Fe_2O_3 give X-ray patterns which are very similar. The oxygen content of these thin films is not always known with certainty, but it becomes higher as the temperature becomes lower; probably the films formed at room temperature have an oxygen content approaching Fe_2O_3. On copper heated to give cuprous oxide (Cu_2O) the early colours are slightly modified by the colour of the metal, but the sequence is similar to that on nickel; the later colours may be obscured by sooty specks of cupric oxide (CuO), unless the atmosphere used has contained so little oxygen as to make the formation of CuO impossible.

If nickel oxide films are transferred from the nickel basis to a transparent support (e.g. a plate of clear plastic), the colour which any particular film produces when viewed by reflected light is complementary to that produced by the same film on the metal – as shown in Table 1.2. On iron, the thicker films have an opaque magnetite layer interposed between ferric oxide and the metal. The complementary relationship here fails. If viewed through the clear plastic, the films show the same colour as before the removal of the metallic iron, since the reflecting basis (magnetite) remains the same (Fig. 1.3). If viewed from the other side (where originally there had been metal), only the dark grey of the magnetite is seen and no colours. Below the thinner ferric oxide films, the magnetite is either too thin or too discontinuous to have any optical effect, and the complementary relationship is obtained.

A simple cell used by the author for transferring films is shown in Fig. 1.4. A later, more complicated pattern is described by the author and Tomlinson. (U. R. Evans and R. Tomlinson, *J. Appl.Chem.,* 1952, **2**, 105.) A clear plastic plate is cemented with adhesive to the film-covered face of a rectangular sheet of iron or nickel, which is then fixed in the

Table 1.2 Colours produced by oxide films (reflected light) before and after transfer (U. R. Evans)

Nickel		Iron	
Colour on metal	Colour after transfer	Colour on metal	Colour after transfer
Yellow I	Bluish white	Yellow or brown	Bluish silver
Mauve I	Whitish	Mauve or violet	Silvery grey (sometimes faintly green)
Blue I	Yellow		
'Silvery hiatus'	Red	Full blue	Golden yellow
Yellow II	Mauve to blue	Pale blue	Reddish
Red II	Green	Pinkish blue	Colours when viewed through plastic unchanged; from back, colours hidden by magnetite layer
Blue II	Yellow	Blue	(as above)
Green II	Red	Greenish blue	(as above)
Red III	Green	Bluish grey with trace of pink	(as above)

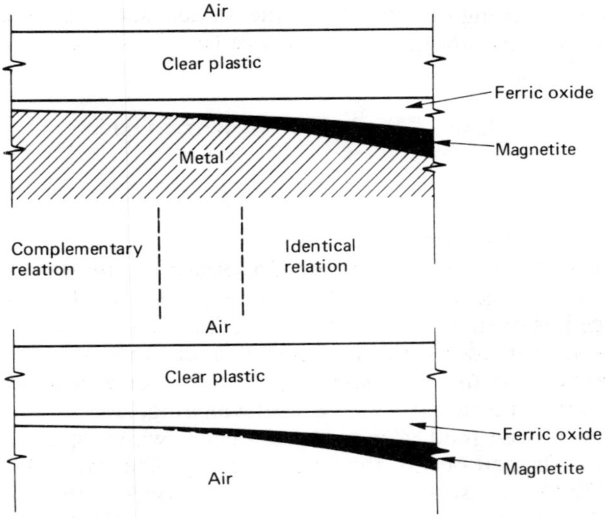

Fig. 1.3 Effect of transfer on colours shown by films

electrolytic vessel, and joined through resistances to the positive pole of a battery, so as to remove the metal by anodic corrosion, leaving the film on the clear plastic; the removal should start from the lower end of the rectangle and spread steadily upwards without leaving uncorroded islands of metal, since any portion which ceases to be connected with the external battery will fail to be removed. The solution employed contains sodium chloride along with zinc sulphate, which is added so that the cathodic reaction shall be the deposition of zinc and not the production of alkali. In the case of iron, it is necessary to ensure a high anodic

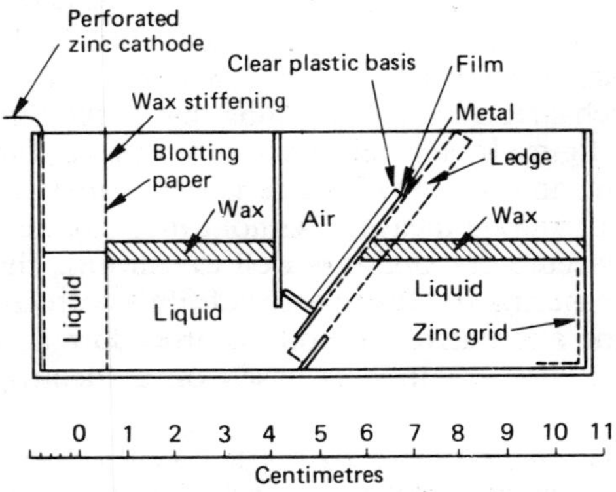

Fig. 1.4 Early form of cell for transfer of oxide films to celluloid

current density along the line where the oxide, metal and solution meet, so as to avoid the cathodic destruction of the film by means of the local cell

Iron | Solution | Ferric Oxide

(see p. 55).

Measurement of thickness of films

The simplest method (in principle) for obtaining the mean thickness of a film of oxide, sulphide or iodide is to weigh the metal before and after the surface has been converted into oxide, sulphide or iodide; the difference in mass represents the non-metal taken up.* If an oxide film is already present at the first weighing, the difference will not represent the total oxide present after the second weighing, and to avoid this error Gulbransen[782] designed a special balance in which specimens can be heated in hydrogen before the first weighing. The gravimetric method was used in the classical work of Pilling and Bedworth and of Vernon (pp. 13, 14). Bannister's curves for the growth of silver iodide films (Fig. 1.1) depended directly or indirectly on microbalance measurements.

Another method, which has given good results in the hands of Vernon, Wormwell and Nurse, is to strip oxide films from iron by dissolving the metal in a water-free solution of iodine in methanol, and then to estimate the iron in the films thus isolated.

A third method depends on the change in the ellipticity of polarized light reflected at a metallic surface – which change is modified by the presence of a film to an extent depending on the thickness. For details of this beautiful method, reference must be made to the papers of Tronstad and Winterbottom.[792] (See also Kruger, J., 1968 and 1976 vols.) In the form used by these investigators, the method is suitable for measuring invisible films too thin to give interference tints.

Tammann,[787] in his classical work on film growth, used a simpler method to obtain the thickness; he turned to the Tables showing Newton's ring colours as observed by *transmitted light* to find the thickness of the air film which gave a colour matching that of the film-covered metal; he divided this figure by the refractive index of the film substance and obtained thereby an estimate of the film on the metal. This delightfully simple plan of obtaining measure without measurement involves, however, certain inaccuracies, and it is well to calibrate the colour scale of thickness by measuring the thicknesses of films corresponding to different tints by some alternative method – a procedure adopted by Bannister. Another excellent method is based on a spectrophotometer. (A.

* If the increase in mass on area A is m', the film thickness is $m'M/(M - m)DA$ where D is the density of the film substance, M its relative molecular mass and m the mass of metal in the molecule.

Charlesby and J. J. Polling, *Proc. roy. Soc. (A)*, 1955, **227**, 434.) Vernon pointed out that the colour produced by an oxide film is not always a simple function of the oxygen uptake per unit area; a given uptake produced by a long exposure to oxygen at low temperature may develop a different tint to that arising from the same uptake obtained by a short exposure at a higher temperature. (W. H. J. Vernon, *Trans. Faraday Soc.*, 1935, **31**, 1674.) This is not surprising; oxide produced by inward penetration (e.g. along grain boundaries) will contribute nothing to the colour. If that is the essential cause of the discrepancies which have sometimes occasioned mistrust in conclusions based on observation of interference colours, the discrepancies should be avoided if the film is produced, not at the expense of the basis metal, but by deposition of material from an external source.

To test this point, research was carried out at Cambridge on colours produced by deposition of a molybdenum oxide (essentially dioxide with rather less molybdenum than the formula MoO_2 would predict) obtained by cathodic treatment in ammonium molybdate. The colour sequence, extending over several orders, was found to be the same as that obtained for iodide films on silver, as shown in Table 1.1. The coulombs per unit area needed to produce a standard tint varied slightly with the metal on which deposition was taking place, but for a given basis metal the same number of coulombs was needed to produce that tint whether provided by a weak current continuing for a long time or by a stronger current for a shorter time. Details will be found in the 1976 vol., pp. 379–99.

The electrometric method for measuring film thickness, used in a simple form at Cambridge by Bannister for studying iodide films, was developed for oxide films by Miley.[774] The circuitry is shown in Fig. 1.5. Most later investigators have used closed cells and de-aerated liquid. In many cases different solutions have been adopted. For iron Eurof Davies[773] retained ammonium chloride (the solution used by Miley) and

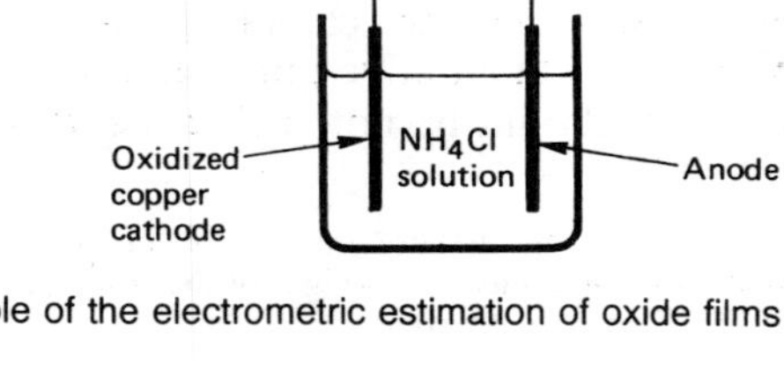

Fig. 1.5 Principle of the electrometric estimation of oxide films

obtained results in agreement with other methods. Hancock and Mayne[776] used a de-aerated potassium chloride solution. Others have adopted buffered solutions, since accumulations of alkali may introduce complications; a phosphate buffer was used for copper by Mills[776] and for silver by Eurof Davies[774]; a borate buffer is preferred in Cohen's laboratory.[776]

If a voltmeter or potentiometer is joined in parallel with the reduction cell, the reading will show a movement when reduction of oxide is complete, since a larger potential difference is needed to evolve hydrogen than to reduce oxide.

It is convenient to work under **galvanostatic** conditions, obtainable by applying a high e.m.f. through a resistance high compared to that of the cell, so that the current remains constant. In that case, simple measurement of the time suffices to provide the number of coulombs needed for reduction – from which the film thickness can be calculated,* if the reaction is known; cuprous oxide may be reduced to metallic copper. Fe_2O_3 is reduced to Fe^{2+} ions (which enter the liquid) or to metallic iron – according to the solution chosen. The electrometric method is only suited for films which can be reduced without evolution of hydrogen.

Comparisons between the results of the different methods of measurement have been carried out at Cambridge. Wherever possible a single specimen has been used for all determinations, since two or more specimens, nominally identical, may not behave alike when treated in the same way. Bannister exposed weighed silver specimens in iodine solution, and then reweighed them; the mass increment enabled one value for the thickness of the iodide film to be calculated. He then determined the coulombs needed for the cathodic reduction of the film, obtaining a second measurement, and finally estimated the iodine which had passed into solution; this provided a third figure. Thus three measurements were obtained from each individual specimen, and, since the three were generally in good accord, confidence was felt in all three methods of estimation.

Similarly Price and Thomas[774] took weighed specimens of silver, and formed a film of sulphide on each, the gain in mass furnishing one measure of the thickness; the number of coulombs needed to reduce the film provided a second and the accompanying loss of mass a third. Here again, there were three measurements from a single specimen, and reasonable accord was obtained.

Eurof Davies in his work on the oxidation of iron used several methods – gravimetric, electrometric, chemical, X-ray and electron-diffraction – some of them for qualitative purposes only. His results are shown in Fig. 1.6 where it should be noted that the vertical scale rep-

*If reduction is complete in time t, using current I and a specimen of area A, the thickness of the film is ItJ/FAD, where J is the equivalent weight of the oxide (for the appropriate reaction), D its density and F Faraday's constant.

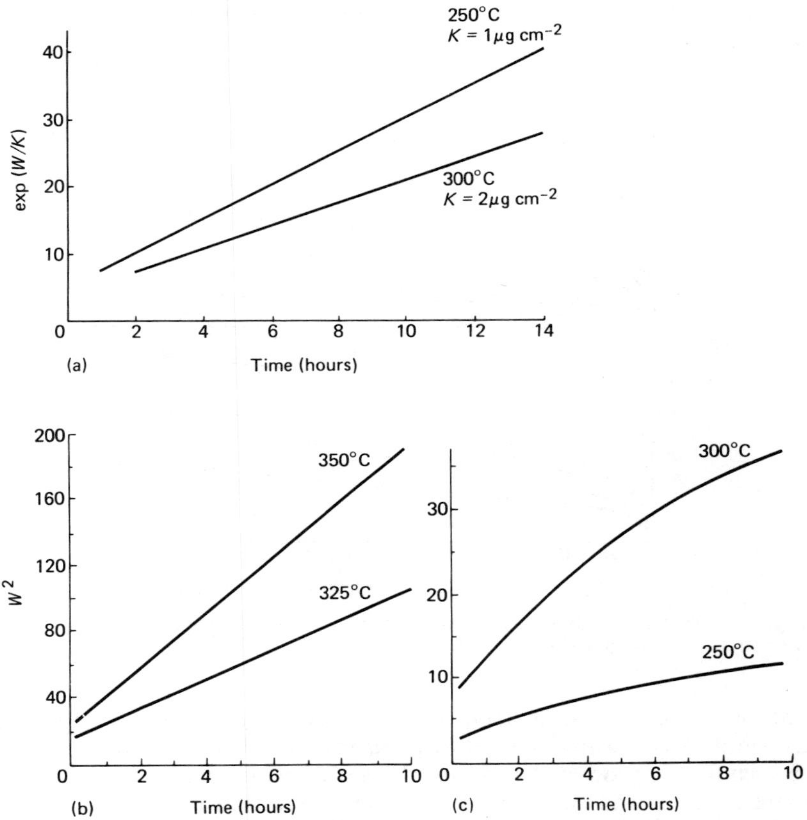

Fig. 1.6 (a) Logarithmic plots of oxidation at 250°C and 350°C. (b) and (c) Parabolic plots of oxidation of iron showing rectilinear relation between W^2 and t at 325°C and 350°C, but not at 250°C and 300°C. W represents the oxygen content of the α-Fe_2O_3 film in μg cm^{-2}. (D. Eurof Davies, U. R. Evans and J. N. Agar)

resents $e^{W/K}$ in Fig. 1.6(a) and W^2 in Fig. 1.6(b), W being the oxygen uptake and K a constant. It will be seen that whilst at 325 °C the parabolic law (p. 16) is obeyed, this fails at 300°C and 250°C, where thickening follows the logarithmic law (p. 20). The logarithmic equation $W = K' \ln (K''t + 1)$ can be written $e^{W/K} = K''t + 1$. Measurements on zinc, made by Vernon, Akeroyd and Stroud[835], are reproduced in Fig. 1.7.

The comparisons between the thicknesses of oxide films corresponding to a given colour, obtained by different experimenters using different methods, have not always shown good agreement, although the order of magnitude is not in doubt. There are several reasons for the discrepancies. Vernon,[57] studying abraded iron, obtained no tints, or

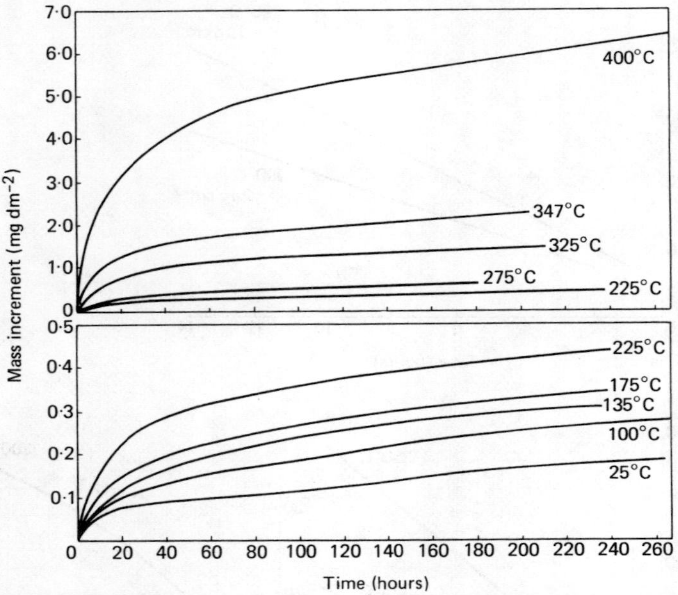

Fig. 1.7 Oxidation of abraded zinc exposed to air at different temperatures (W. H. J. Vernon, E. I. Akeroyd and E. G. Stroud).

only very early tints, at mass increments which, had the oxygen been distributed uniformly, should have produced advanced tints; much of the oxide was produced as a mixed zone of oxide and metal, below the main surface, and contributed nothing to the film responsible for the interference. In contrast, Eurof Davies,[56] using hydrogen-cleaned iron, obtained advanced tints at oxygen uptakes which, given uniform distribution, should have produced only early tints; here according to the mechanism for logarithmic growth presented on p. 239, the oxidation would cease over much of the surface, so that the film thickness needed for colour would be obtained over the remainder at an abnormally small oxygen uptake. Colour by itself is no criterion of thickness, and should only be used as a measuring method, if there has first been a careful calibration of the colour scale against one or more reliable methods; this was the principle adopted by Bannister. It should be added that even when a specimen appears to be of uniform colour, microscopic examination will reveal different colours on the different grains, showing that oxidation velocity can vary with crystal direction.

Mechanism of oxidation

Time-oxidation curves

All researches on freshly cleaned metals exposed to air or oxygen agree in showing that even at room temperatures the oxidation is very rapid in the opening stages, but soon slows down as the film begins to isolate metal and air from one another. The oxidation at room temperatures produces no change in the appearance of the metal (unless sulphur compounds are present in the air). At higher temperatures, the film quickly becomes sufficiently thick and sufficiently regular to give interference tints, but even in such cases the rate of thickening falls off as the film becomes thicker.

The two classes of metals

Pilling and Bedworth[39] were the first to divide metals into two groups according as the oxide occupies a volume smaller or larger than that of the metal destroyed in producing it. If the density of an oxide be D and its relative molecular mass M, the volume (in cm^3) occupied by one mole will be M/D; if m is the mass of metal in M of oxide, and d the density of the metal, then the volume of metal destroyed in producing M of oxide will be m/d. Clearly, if $M/D < m/d$, that is, if $Md/mD < 1$, the oxide is likely to contain voids and may fail to isolate effectually the metal from the air. Calculation indicates that for the very light metals Md/mD is less than unity; the values are Na 0·57, K 0·41, Mg 0·79, Ca 0·64. For the heavier metals and two fairly light metals (aluminium and beryllium) it exceeds unity, the values being Be 1·70, Al 1·24, Fe 2·16, Ni 1·60, Cu 1·71, Zn 1·58. Observation shows that the very light metals of the first class often build oxide films of limited protective value, so that these metals, if heated in air, catch fire and burn, giving out so much heat as to maintain their own temperature when the external source of heat is removed. The second class yield compact scales which clearly tend to shield the metal from the air; these metals do not burn in the massive state, although very thin wires (where the ratio of surface to volume is high) can be made to burn if strongly heated; the ignition temperature, as shown by Tammann and Boehme,[42] becomes lower as the diameter becomes smaller.

It was once believed that all metals of the first (light) class oxidize at a uniform rate (at constant temperature), owing to the non-protective character of the film, and that all those of the second (heavy) class suffer oxidation at a rate that falls off as the film thickens. Pilling and Bedworth's rectilinear curves for calcium (a light metal) and parabolic curves for copper (a heavy metal), reproduced in Fig. 1.8, seemed to support this view. However, Vernon showed that zinc (a heavy metal), exposed to polluted air at ordinary temperature, gives curves which remain straight for many weeks (Fig. 1.9). Moreover Gregg and Jepson, using pure calcium and dry oxygen, have obtained results different from those of Pilling and Bedworth; they found very little oxidation below

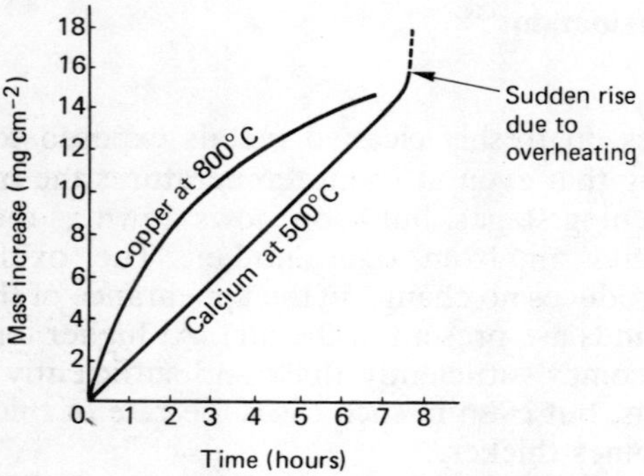

Fig. 1.8 Growth of oxide on calcium and copper (N. B. Pilling and R. E. Bedworth)

475°C, although it became measurable at 500°C; water vapour accelerated the reaction at lower temperatures – which suggests that the non-protective character of the Pilling–Bedworth films may have been due to moisture.

Some metals can be heated in air or oxygen without developing films. Molybdenum heated in the open remains bright, because the volatile oxide passes off as vapour; when heated in a closed tube the metal develops interference tints. On noble metals, oxidation may be absent altogether, if the temperature exceeds that at which the oxide would decompose. Silver, very gently heated, becomes slowly oxidized; but if the temperature is raised above about 180°C, the decomposition pressure exceeds that of the atmosphere, and the oxide decomposes. Mercury develops visible films when moderately heated; at still higher temperatures the oxide decomposes; this case is historically interesting in connection with the discovery of oxygen.

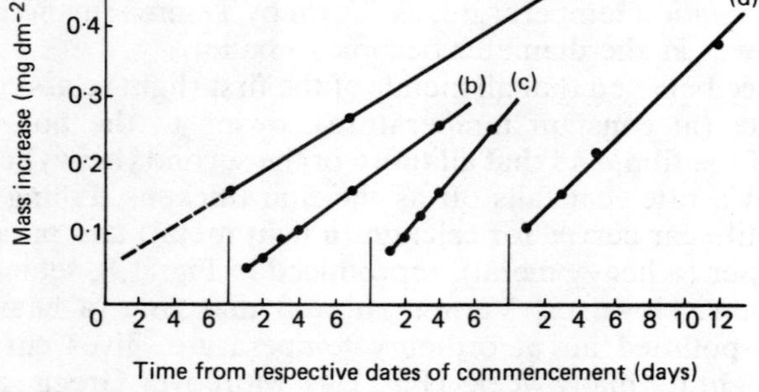

Fig. 1.9 Atmospheric corrosion of zinc. Specimens started respectively on (a) January 29, (b) April 24, (c) May 12, and (d) August 19, 1925 (W. H. J. Vernon)

Film growth in the brittle range

Where the rate of oxidation falls off with time, the phenomena may vary according to the temperature, since the film substance will usually be plastic at high temperatures, but brittle at low ones. In the latter case the film may crack when it exceeds a certain thickness, so that the rate of film growth, previously slow, may suddenly become fast again. This is illustrated by comparing the curves of Pilling and Bedworth for the oxidation of copper at 800°C (Fig. 1.8) with those at 500°C (Fig. 1.10). At 800°C a continuous curve is obtained, whereas at 500°C oxidation, after it has begun to slow down, suddenly starts up again; evidently the film has simultaneously failed over a large area. This is only likely to happen if a cavity has been formed between metal and film, leaving the latter unsupported; if the oxide is insufficiently plastic to subside into the cavity, cracking is bound to occur. The formation of cavities left when cations move out from the metal into the film have been demonstrated by several different investigators. The plasticity of films may be influenced by minor constituents. Tylecote[49] found pure cuprous oxide films to be remarkably ductile, but films produced on heating copper containing phosphorus are relatively brittle owing to the presence of cuprous phosphate; they exfoliate on cooling, being unable to withstand the strain imposed by differences between the thermal contractions of scale and metal.

Growth laws

Equations governing growth are discussed in detail in Chapter 8, where derivations are offered. In the present chapter, a brief statement of the chief equations is necessary for understanding, but only general explanation will be offered for obedience to different equations under different conditions. Certain other equations (e.g. the mixed parabolic,

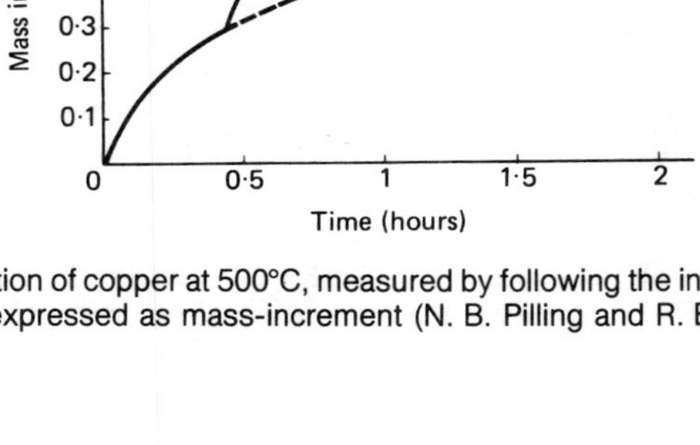

Fig. 1.10 Oxidation of copper at 500°C, measured by following the increase of electrical resistance; data expressed as mass-increment (N. B. Pilling and R. E. Bedworth)

the cubic and inverse logarithmic) will not be introduced into the present chapter; they are discussed in Chapter 8.

Rectilinear and parabolic growth

It has been explained that when film growth is controlled solely by passage across the film, the parabolic equation

$$dy/dt = k/y \text{ or } y^2 = Kt + K'$$

would be expected, and was indeed observed during the growth of silver iodide films – although not during the opening period, when the film is still very thin. The equation would predict an infinite value of dy/dt at the outset when a film is absent ($y = 0$). As long as the film is very thin, the growth rate is likely to be determined by factors unconnected with passage across the film, and thus be independent of film thickness, giving a rectilinear law,

$$dy/dt = k \text{ or } y = Kt + K'$$

Bénard[29] studying iron at 1000°C with a recording balance designed to provide a continuous trace for mass increase against time, obtained a straight line at the outset gradually passing into a curve of parabolic form. Probably most cases of oxidation commonly regarded as obeying the parabolic law would similarly show obedience to the rectilinear law at the outset.

In some cases, the rectilinear portion may continue almost indefinitely. This will occur if the film substance is porous, or at least non-protective. An example has been provided by Vernon's studies of the mass increase of zinc exposed indoors in London (Fig. 1.9). Another is provided by calcium under the particular experimental conditions observed by Pilling and Bedworth (Fig. 1.8). These investigators believed that all the lighter metals where the oxide, if unconstrained, would occupy a smaller volume than the metal destroyed in producing it, ought to oxidize according to the rectilinear law, since, they argued, the film must be porous in such cases.

It is now known that this is an over-simplification. Apart from the complications mentioned in connection with calcium, where the oxide can combine with water, the possibility of the metal vaporizing, with the vapour burning at a distance from the surface, has to be kept in mind. This seems to occur, under some circumstances, on calcium and also on magnesium. Pilling and Bedworth regarded the constant oxidation rate of their calcium as the velocity of a chemical combination between metal and oxygen. Dunn and Wilkins[828] however, pointed out that a chemical reaction rate would increase with temperature far more rapidly than is shown by the Pilling–Bedworth numbers. They suggested 'a thin film pseudomorphic with the underlying metal constantly breaking down to a granular film and constantly renewed'. This appears reasonable. If the oxide retains the structure of the metal, it must be in tension laterally, and cracks are likely to develop, extending down to a short distance of the metal–oxide interface – at which point the oxide, braced by contact with the metal, may resist further cracking. Thus we have a thin uncracked layer of constant thickness, covered with a cracked (non-protective) layer of a thickness increasing with time. That would predict an oxidation rate decreasing with time at first, until the thickness characteristic of the uncracked layer has been reached, and thereafter remaining constant. It is likely that the movement through the uncracked layer is a movement of oxygen anions inwards, not that of metallic cations outwards; the tension existing in the layer means that the interatomic spacing is abnormally wide, so that there will be less obstruction to the movement of the large oxygen ions than in unstressed oxide. Some of the objections raised against the Dunn–Wilkins picture disappear if inward movement is admitted.

Work at Oak Ridge[41] has shown that the oxidation of sodium in dry oxygen commences rapidly, but at low temperatures it ultimately becomes slow – suggesting that very thin films are uncracked and possess appreciable protective properties. The thickness at which growth practically ceases is only 0·005 μm at −79°C; at 48°C the thickness reaches 0·15 μm (much greater than the thickness reached by copper at similar temperatures) and even then reaction does not cease altogether; indeed there is an increase of oxidation rate after about 25 000 minutes. The results seem to fit in with the idea that the film – although distended – does not break under the internal stresses until a certain thickness has been reached. A somewhat similar interpretation has been adopted by McFarlane and Tompkins in explaining the formation of nitride films on solid lithium. (E. F. McFarlane and F. C. Tompkins, *Trans. Faraday Soc.,* 1962, **28**, 997.)

On heavy metals, we should normally expect rectilinear growth in the opening stages passing later into parabolic growth, but if the temperature is very high the rectilinear growth may well continue until the experiment comes to an end – perhaps because the specimen (which may consist of thin sheet) has been completely converted to oxide. Oxidation of furnace parts, which, although thick, will be exposed for long periods, may show apparent obedience to the rectilinear law for different reasons. This may arise because the operation of the furnace is continually cracking or scraping off the oxide scale, so that the curve relating wastage to time, although showing a number of little spurts if studied in detail, would, when considered over long periods, be represented roughly by a straight line. Another reason may be the slow supply of oxygen to the surface; if this arrives so slowly that passage of material through the scale can easily keep pace, oxygen replenishment will control the wastage rate – which remains independent of film thickness.

Passing to conditions where movement through the film controls growth (i.e. where the parabolic law is obtained), it must be emphasized that the composition of the film varies throughout its thickness, and rarely corresponds exactly to the textbook formula of the oxide. Thus iron is covered with three layers corresponding roughly to the three oxides FeO, Fe_3O_4 and Fe_2O_3. Within each of the three layers, the iron content varies, being highest nearest to the metal; ferric oxide contains more iron than corresponds to Fe_2O_3.

In contrast, the so-called cuprous oxide usually contains less copper than the formula Cu_2O would suggest. A crystal of 'perfect' cuprous oxide should consist of a lattice of oxygen anions interpenetrating a cation lattice consisting of a double number of cuprous ions. In 'imperfect' cuprous oxide, a certain number of points on the cation lattice are believed to be vacant, neutrality being preserved by the fact that at an equal number of points cupric ions exist where there would be cuprous ions in the perfect crystal. If this picture is correct, then clearly a mechanism is provided for the passage of metal outwards through the film; wherever there is a vacant space, a neighbouring cuprous ion can move into it, leaving another vacant space into which another cuprous ion can move, and so on. This movement, by itself, would involve an accumulation of electric charge, unless there is also a flow of electrons, which can come about in the following way. An oxygen atom attaching itself to the outer surface of a cuprous oxide film will absorb electrons from two cuprous ions, becoming an oxygen ion; the two cuprous ions will temporarily become cupric ions, but they can then absorb electrons from the cuprous ions farther in, and so on. Thus a regular passage of electrons, and of copper, occurs *outwards* through the film, and (provided that the supply of oxygen to the outer surface is maintained) the cuprous oxide layer steadily grows in thickness, the new oxide continually being deposited at the *outer* surface.

There is now much evidence that the oxidation of heavy metals depends on the movement of metal outwards through the film as well as of oxygen inwards. This was first shown by Pfeil,[34] using *markers* of chromic oxide; he applied an aqueous suspension of chromic oxide powder to an iron surface and found that, on heating, the particles of chromic oxide became embedded in the oxide which had grown outwards, surrounding the particles. Later, radioactive substances were used as markers in Mehl's laboratory at Pittsburgh[34] – a distinct improvement to the technique, since, as pointed out later by Sachs,[35] bulky markers may produce misleading results. On iron there seems little doubt that movement occurs in *both* directions. The iron cations are smaller than the oxygen anions, and outward movement predominates at first. This probably produces cavities at the base of the scale, reducing the area available to the outward movement and facilitating that of oxygen inwards.* On iron the situation is complicated by the presence of three oxides. On nickel, there is only one oxide formed, yet photographs of cross-sections show two layers, one being

* At the outer surface of a flat gap separating oxide from metal, the oxide will have a higher decomposition pressure than corresponds to equilibrium, so that oxygen will move across the gap and produce fresh oxidation on the opposite side.

probably formed by metal moving outwards and the other by oxygen moving inwards. The good resistance towards oxidation may perhaps thus be explained. A film formed solely by cations moving outwards is likely to be loose, owing to the gap left between metal and oxide; a film formed solely by anions moving inwards can develop explosive compressional stresses. Both gaps and stresses constitute sources of weakness, which can be avoided by a balance between the two types of movement.

On abraded nickel, however, the inward movement seems to be sufficient to produce serious compressional stresses. The author[43] heated specimens of sheet nickel to produce oxide films and then attached them to glass by means of petroleum jelly; the metal was then removed in a cell similar to that shown in Fig. 1.4, leaving the film on the glass. Warming to soften the jelly allowed the films to take up their unconstrained shapes. The thicker films developed wrinkles, showing that they had been compressed whilst attached to the metal; the thinner ones curled up into tight microscopic rolls; both changes provide evidence of film stress.

The parabolic equation is obeyed by most metals over an intermediate range of temperature. It was established on copper at about 800°C by Pilling and Bedworth in 1923, but has since been verified for lower temperatures ranges and for other metals. Some early estimations of K are reproduced in Table 1.3.

Evidence of ionic mechanism of film growth

There exists today good evidence – qualitative and quantitative – for the movement of ions through the film during growth. Several attempts have been made to ascertain whether film thickness can be influenced by the external application of an electric field. At least three sets of

Table 1.3 Parabolic oxidation rate constants ($K = m^2/t$) (where m is the mass increase in g cm^{-2} and t the time in hours) (J. S. Dunn and F. J. Wilkins)

Temperature	700° C	800° C	900° C	1000° C
Copper in O_2	$5{\cdot}86 \times 10^{-6}$	$3{\cdot}14 \times 10^{-5}$	$1{\cdot}27 \times 10^{-4}$	$6{\cdot}02 \times 10^{-4}$
Copper in air	$2{\cdot}86 \times 10^{-6}$	$2{\cdot}70 \times 10^{-5}$	$1{\cdot}10 \times 10^{-4}$	$4{\cdot}60 \times 10^{-4}$
Nickel (electrolytic) in O_2	—	$0{\cdot}093 \times 10^{-6}$	$0{\cdot}76 \times 10^{-6}$	$3{\cdot}4 \times 10^{-6}$
Nickel, grade A, in O_2	—	—	$1{\cdot}9 \times 10^{-6}$	$6{\cdot}8 \times 10^{-6}$
Cobalt in air	$5{\cdot}8 \times 10^{-7}$	$3{\cdot}3 \times 10^{-6}$	$2{\cdot}2 \times 10^{-5}$	$7{\cdot}4 \times 10^{-5}$
Tungsten in air	$1{\cdot}61 \times 10^{-5}$	$16{\cdot}6 \times 10^{-5}$	$17{\cdot}9 \times 10^{-5}$	—
Tungsten in air (second sample)	—		$13{\cdot}4 \times 10^{-5}$	461×10^{-5}
Brass (95% zinc) in air	—	$2{\cdot}8 \times 10^{-5}$	—	—
Brass (90% zinc) in air	—	$1{\cdot}69 \times 10^{-5}$	—	—

Table 1.4 Calculated and experimental values of the constant K' as defined by the equation $d\eta/dt = K'/y$, where η is the amount of oxide, sulphide or iodide in equivalents per cm^2. (Tabulated from papers by C. Wagner)

Reaction	Temperature	Calculated	Experimental
$2Cu + I_2 = 2CuI$	195° C	$3{\cdot}8 \times 10^{-10}$	$3{\cdot}4 \times 10^{-10}$
$4Cu + O_2 = 2Cu_2O$	1000° C	6×10^{-9}	7×10^{-9}
$2Ag + S = \alpha Ag_2S$	220° C	2 to 4×10^{-6}	$1{\cdot}6 \times 10^{-6}$

reliable experimenters (1976 vol., p. 24) seem to have produced the effect, and have shown that the application of a field can definitely increase or decrease the oxidation rate – according to the polarity.

If parabolic film growth is controlled by the movement of ions and electrons through the film substance, it should be possible to express K in terms of the electrical properties of the film substance (this is true whether there is migration under an electric potential-gradient or diffusion under a concentration-gradient, the two phenomena being closely related). An expression representing the dependence of K on electrical conductivity was obtained by Wagner in 1933; a simple proof, due to Hoar and Price, will be found on pp. 263–7.

In Table 1.4, Wagner's values of the velocity constant calculated from this expression are compared with the experimental determinations. The agreement is satisfactory, especially as the velocity constants are of different orders of magnitude in the different cases.

Non-parabolic growth

It is now possible to explain why the parabolic law will cease to be obeyed at low temperatures. At high temperatures, the ions within the film will be in motion, passing from one position of stability to another; but passage can only take place if the ion in question possesses sufficient energy to surmount the intervening energy barrier. Without further assumption, a general expression for the growth rate can be obtained, this being the difference between two exponentials representing the numbers of ions moving in the two opposite directions – helped or hindered respectively by the electric field due to the cell metal/oxygen; a simple proof is provided in Chapter 8 (p. 239), where it is shown that this general equation has two limiting cases. When the film thickness is sufficient, it reduces to the parabolic equation – explaining why the parabolic law is obeyed at reasonably high temperatures, once the appropriate thickness has been attained. At low temperatures where thickening is very slow, another limiting form may be expected – namely the inverse logarithmic equation, which can be written*

$$1/y - 1/y_0 = k \log [a(t - t_0) + 1]$$

where y_0 is the value of y when t is t_0.

* The reason for the apparently complicated form of the equation is due to the fact that obedience to the law will not start from the commencement of the experiment; it is necessary to measure time (t) from a moment t_0 when a film already exists.

At one time it was believed that another equation (the direct logarithmic equation) $y = k \log (at + 1)$ was obeyed over the low-temperature range. Now it is not easy to distinguish between the direct and inverse equations in short experiments for reasons explained elsewhere[840], but an important research by Mayne and Gilroy has shown that the exposure of iron to dry air at ambient temperature for one year gives results conforming to the inverse logarithmic law; they are definitely *not* consistent with the direct logarithmic law. (D. Gilroy and J. E. O. Mayne, *Corr. Sci.*, 1965, **5**, 55.)

Thus, the experimentally established facts seem to accord with the predictions of the simple theoretical argument – namely obedience to the parabolic equation at fairly high temperatures and to the inverse logarithmic equation at low temperatures. It may be noted that the idea that there should be a general equation of which these two other equations are limiting cases was put forward at about the same time by three different investigators – two working at Cambridge and the other (Hurlen) at Oslo; the Norwegian study carried the matter further than the others, and explained why, under intermediate conditions, a cubic equation is sometimes obeyed. Hurlen's paper, which deserves study, is reproduced in the 1976 vol., pp. 37–42.

The establishment of the inverse logarithmic equation at low temperatures does not disprove those experimenters whose measurements supported the direct logarithmic equation at temperatures just below the level where the parabolic law ceases to be obeyed. It is fairly certain that in this range the direct equation is indeed obeyed. Nor is this surprising. If film growth occurs mainly by cations moving outwards, vacancies must be left at the base of the film, and these may unite to produce definite gaps separating the film from the metal; over areas where gaps exist, little or no oxidation will take place, so that the oxidation of the whole area of a specimen will be less than what would be expected in the absence of gaps. At high temperatures, the film substance is plastic and the film will subside to fill the gaps, so that there is no serious departure from the parabolic law. If the temperature is too low the film will not be sufficiently plastic for subsidence, and the area available for oxidation will fall off. If we are concerned with small thicknesses where, in absence of gaps, a rectilinear law would be approximately obeyed, this reduction of available area will lead to the direct logarithmic law. If the thickness has reached the range where (given perfect plasticity) the parabolic law would be obeyed, a mixed logarithmic law is to be expected. Obedience to the direct logarithmic law has been established by several investigators; obedience to the mixed law was observed by Mills in an important research on copper; Mills' measurements were definitely not consistent with the direct logarithmic law, but were found to conform to the mixed law which had been arrived at theoretically before his experiments were carried out. All these matters are discussed elsewhere (1960 vol., pp. 829–37) where it is also pointed out that obedience to the direct logarithmic law can arise in other ways, even in cases where there is no formation of gaps between film and basis metal. However, there is little doubt that the formation of gaps is the cause of logarithmic growth in many cases; the gaps have been observed under the microscope by several investigators, whilst the sudden increase of oxidation rate which occurs when the film above a gap collapses has also been frequently recorded; Fig. 1.9 provides an example.

Although the slowing down of corrosion due to the formation of gaps between metal and film is often very marked, it is obvious that a gap will not provide reliable protection; the collapse of the unsupported film mentioned above is sufficient proof of that. If, instead of a gap, a thin layer of some impervious oxide could be formed at the base of the main film, that would be more reliable; and indeed such a layer provides a means of controlling oxidation by using alloys – a matter discussed on p. 24. Under certain conditions, mathematics would predict a logarithmic equation, but, as will be seen, there are many complicating factors, making simple theoretical treatment difficult. Nevertheless it is a fact that the form of the curves representing the oxidation of a resistant alloy often present the general shape associated with the logarithmic law.

Entry of oxygen into the metal

Metals like titanium and zirconium often oxidize mainly by oxygen passing inwards. This has been shown by marker experiments. On zirconium the marker is found after oxidation on the outside of the film, whereas on copper it is found within or below the film. On titanium, marker researches provide evidence of movement in both directions; further evidence for the inward movement of oxygen is given by work with radioactive tracers, but the most direct proof is the existence of a solid solution of oxygen in titanium below the lowest oxide layer. The scale is complicated, and three oxides may sometimes by found, nominally TiO, Ti_2O_3 and TiO_2, all of which have compositions varying over perceptible ranges; in 'TiO', which has the rock-salt structure, the composition can fall anywhere between $TiO_{0\cdot6}$ and $TiO_{1\cdot25}$. The growth-law varies with temperature and duration of oxidation, being in different cases (1) logarithmic (2) cubic (3) parabolic (4) linear and (5) broken, apparently due to cracking, with linear periods. Details will be found in the important paper by Kofstad and Hauffe.[44]

If oxygen enters the metallic basis, compressional stresses will arise, except in the ultra-light elements like sodium where the oxide occupies a smaller volume than the metal destroyed. The description of the film on titanium, provided by Jenkins[44] is highly instructive, 'Observations . . . indicate that the thin dense slate-grey scale formed at low temperature is replaced at high temperatures by a thick porous, yellow-brown scale. This scale is largely composed of layers of oxide which have been twisted and shattered like natural rock strata. The transformation . . . is believed to occur when the thin, dense scale grows beyond a certain maximum thickness.'

Pilling and Bedworth had thought that the non-porous film formed on those metals where the volume of the unstrained oxide would exceed that of the metal must always be protective, just as the porous film on the ultra-light metals must always be non-protective. Up to a point, they were right. Lateral compression in a film will tend to prevent the formation of pores, or close them up if they arise. But the film is like a compressed spring, and the internal stress per unit area will increase, perhaps proportionally, with increasing thickness. The energy needed to overcome adhesional forces if the film is to break away from the basis is roughly independent of thickness, so that at a certain thickness the strain energy will become sufficient for the breakaway to occur, and protection will suddenly fail. That is why, in wet corrosion, the protection afforded by films on titanium and zirconium is so insidious. Titanium generally resists red fuming nitric acid, but occasionally violent explosions have occurred. Again stores of zirconium scrap at factories after lying for long periods without incident, may suddenly take fire and explode when stirred – generally after slight wetting. The behaviour of zirconium to 'high-temperature water' (water heated under pressure far above 100°C) is important in nuclear energy plants. The film produced is protective until a certain thickness is needed, but then suddenly the compressional stresses – probably helped by hydrogen taken up – cause a breakaway. The 'zircaloy alloys'* have been introduced to avoid breakaway, but in general there is still a sudden increase in reaction rate at a certain thickness.

Magnesium also shows a breakaway when heated in flowing oxygen. At 550°C the oxidation rate falls off with time for several hours, and then suddenly the breakaway occurs with the formation of a white oxide scale which gradually extends over the surface; during this period the rate is increasing with time, but becomes constant when the whole surface is covered; the rectilinear oxidation may now be as rapid as 0·18 mg cm^{-2} h^{-1} at 550°C. Later a second breakaway to a rate ten times as great may occur, the oxide formed now showing a buff colour and differing in texture from the white oxide; it is probably formed by magnesium vapour breaking through the white oxide at weak points and burning outside, the buff colour being due to traces of metallic magnesium. Burning of vapour may be important in some circumstances in the oxidation of other metals, notably calcium. Castle, Gregg and Jepson (see p. 13) have shown that the breakaway does not occur if the oxygen is carefully purified; the impurities causing it appear to be hydrocarbons derived from lubricating oil; these deposit carbon which is built into the oxide and renders it non-protective. (J. E. Castle, S. J. Gregg and W. B. Jepson, *J. electrochem. Soc.*, 1962, **109**, 1018.)

In the case of the heavier metals, entry of a non-metal into the metallic phase is unusual, except where the atomic structure is loosened, as at grain boundaries. Nickel heated in a sulphur atmosphere develops a

* Zircaloy 2 contains 1·5% Sn, 0·12% Fe, 0·10% Cr, 0·05% Ni, whilst zircaloy 3 contains 0·25% Fe and 0·25% Sn.

network of sulphide along grain boundaries, and a rather similar oxide network in steel was noticed by Stead as early as 1921. Some of the nickel–chromium alloys used as furnace windings exhibit a 'pegging-in' of oxide into the metal by penetration along grain boundaries and other avenues of atomic disarray, which in moderation may be welcome since it prevents spalling on cooling or bending. Copper alloys containing silicon and manganese were found by Rhines[68] to develop on heating, below the true oxide scale, a sub-scale consisting of unchanged grains separated by a network of silica or manganese oxide.

Abraded metal shows penetration of oxygen between the blocks produced by the grinding. Stockdale[54] abraded specimens of copper sheet, heated them to give interference tints, and transferred the films to glass; he found, below the oxide film proper, a mixed zone of oxide and metal. Quantitative evidence was provided by Vernon who, after stripping his oxide films from iron, found that the amount of oxygen in them was insufficient to account for the mass increase measured during oxidation; this showed that parts of the oxygen must have penetrated into the interior. The penetration may prevent the proper development of interference tints. (W. H. J. Vernon, E. A. Calnan, C. J. B. Clews and T. J. Nurse, *Proc. roy. Soc. (A)*, 1953, **216**, 375–8.)

Factors influencing service life

Alloying of iron and copper to obtain protection

It might be thought that since, at high temperatures, the oxidation rate becomes increasingly slow as the oxide scale thickens, ordinary steel would become almost immune to oxidation when the scale had become sufficiently thick. In engineering practice, however, thick scale would break off when the part was rubbed or bent, or when the temperature was allowed to cool down; cooling produces stresses due to the unequal contraction of metal and oxide, and also renders the scale brittle, so that these stresses cannot be relieved by flow. For use at very high temperatures special steels are available containing alloying elements which make the scale protective whilst it is still very thin; thin films are unlikely to become detached, because the internal compressional energy per unit area available for detachment is much smaller than in thick films, whereas there is no diminution of the work needed to detach unit area.

Even on alloys which develop films possessing protective properties at very small thicknesses, the oxidation rate is much greater under conditions of fluctuating temperature than at constant temperature. The nickel–chromium wires used in electric heaters and furnace elements suffer far more oxidation if heated intermittently than continuously; the oxide film formed at high temperatures tends to crack off on cooling. In industry these wires are subjected to tests designed to gauge their power of resisting detachment on alternate heating and cooling, and sometimes on alternate coiling and straightening.

The elements commonly added to steel used in making furnace parts, engine components or heat-treatment boxes, in order to confer resistance against oxidation at high temperatures, are **chromium** and **aluminium**. Along with chromium, nickel is often introduced to give a one-phase, austenitic structure and, since the straight austenitic steels, such as the well-known 18/8 chromium–nickel stainless steel would be too weak at high temperatures for practical use, small quantities of other elements, such as silicon and tungsten, are added to improve the mechanical properties. Straight iron–aluminium alloys, although highly resistant to oxidation (provided the aluminium content is sufficient to give a white scale of alumina instead of the dark scale characteristic of unalloyed iron), are mechanically unsatisfactory. In practice, iron and steel furnace parts, annealing boxes and the like are often given a superficial layer of iron–aluminium alloy by heating them in a mixture of powdered aluminium, alumina and a little ammonium chloride; after this treatment, the oxidation-resistance is wonderfully improved, without serious deterioration of mechanical properties.

Special cast irons intended for parts which have to be heated and cooled alternately often contain nickel and chromium, in amounts chosen to stabilize the austenitic state. This avoids the susceptibility to the so-called 'growth' (volume increase) displayed by ordinary grey cast iron after being heated and cooled many times through the $\alpha \leftrightharpoons \gamma$ change-point. Growth is essentially an oxidation process, but the penetration of oxygen (especially along coarse graphite flakes) is facilitated by the volume changes associated with the $\alpha \leftrightharpoons \gamma$ transformation.

Of the protective elements mentioned above, chromium and aluminium produce oxides at the *base* of the film, in close contact with the metal, as shown by Portevin, Prétet and Jolivet.[63] These probably provide a barrier against the movement of cations from the metal into the iron oxide layer outside. The oxidation-resistance of iron–chromium alloys (including stainless steels) is too complicated a subject for detailed discussion. Oxide crystals containing both metals are met with, but generally the inner layers contain more chromium than the outer layers. The oxidation rate decreases with time more quickly than the parabolic law (roughly obeyed by unalloyed iron) would predict, but at irregular intervals sudden spurts occur. Both gradual decrease and sudden increase could be explained by the formation of gaps between scale and metal, left where cations have moved outwards. However, gaps are unlikely to provide reliable protection under service conditions, and the formation of an obstructive oxide layer is likely to be the important factor. On iron containing chromium the chromium oxide formed at the base will be largely Cr_2O_3, and this will decrease the area available for the movement of Fe^{2+} outwards.

Such explanations are not universally accepted, and indeed the published facts are in some cases contradictory (see 1976 vol., pp. 49–52). Some work at Leatherhead, however, deserves attention. It is found that 'chromic oxide' is a p-type conductor with less Cr than the formula

Cr_2O_3 would suggest; the vacant cation sites would clearly favour passage through the film. In contrast, 'ferric oxide' has interstitial iron and oxygen vacancies; it is an n-type conductor with more iron than the formula Fe_2O_3 would indicate. Evidently each pure oxide will permit passage of material, but at the composition where one type changes into the other, the film ought to be relatively protective; it is calculated that this should occur at about 20% Cr, and experiments show a minimum value of the growth constant at about 18% Cr – a reasonable agreement.

Aluminium greatly improves the resistance of copper and brass to oxidation at high temperatures and also to low-temperature corrosion by atmospheres polluted with sulphur compounds. The effect on brass is shown by Dunn's curves reproduced in Fig. 1.11.

The influence of different alloying elements in copper was studied by Fröhlich.[77] Qualitatively, his results seem to fulfil the predictions of theory; quantitatively there appeared, at first, to be discrepancies. Price and Thomas calculated that, if the corrosion resistance conferred by aluminium was due to an alumina film, the resistance should be raised at least 80 000 times; actually it was increased only 36 times. They suggested that this disappointing result might be due to the presence of copper oxide in the alumina film, which would lower the protective properties, and predicted that, if a film of copper-free alumina could be obtained on the alloy, the resistance to oxidation would be enhanced. Accordingly, they heated the alloy in hydrogen containing a trace of water vapour ($1{\cdot}3 \times 10^2\,Nm^{-2}$), which was expected to oxidize aluminium (an element with high affinity for oxygen) but not copper. This treatment produced no visible change in the alloy, but evidently an invisible protective film was really formed,

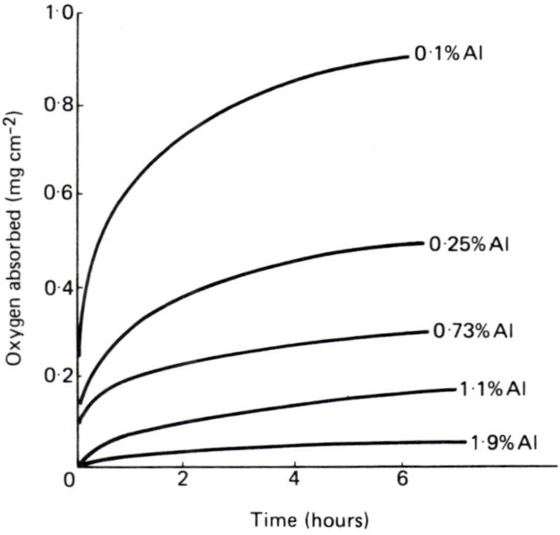

Fig. 1.11 Effect of aluminium on oxidation of brass at 775°C (J. S. Dunn)

since when subsequently the alloy was heated in oxygen under conditions which produced a dense black layer on the untreated alloy, the appearance remained unaltered. Quantitative experiments indicated that the resistance had improved about 200 000 times – the order of magnitude predicted by theory.

This success provided confirmation of the theoretical views which led up to the protective treatment now known as **selective oxidation**. Price and Thomas applied the same principle to the prevention of tarnishing on a silver alloy containing 6·5% copper and 1% of aluminium.

However the subject of alloys resistant to oxidation is complicated. As pointed out by Wood, no single or comprehensive theory can be provided; rather there is a series of special cases. Wood has classified alloy systems according to their behaviour on oxidation; in some cases both metals are oxidized, providing a mixture of oxides, a double oxide (e.g. a spinel) or a solid solution; in other cases only one metal is oxidized, sometimes leaving the metallic phase enriched with the other. His classification, summarized elsewhere (1976 vol., p. 44), deserves study – also his views on the causes of 'breakaway'.

Hauffe's principle

If the alloying element can dissolve in the oxide of the main metal, the oxidation rate may be increased or decreased – apparently owing to its effect on the number of vacancies. This effect should depend on the valency of the alloying element – as pointed out by Hauffe.[60] The oxide formed on nickel, nominally NiO, contains vacant cation sites. By introducing lithium (with a valency lower than nickel), the number of such vacancies should be diminished, since two Li^+ ions are needed to replace one Ni^{2+} if electrical neutrality is to be maintained. The presence of Li_2O in the gas phase is indeed found to reduce the rate of oxidation of nickel–presumably by reducing the number of vacant sites on which oxidation depends (it actually *increases* the *electronic* conductivity). Conversely, by introducing MoO_3 into the gas phase, oxidation is accelerated, the valency of molybdenum being higher than that of nickel. Oxidation is also accelerated by traces of chromium or manganese in the metallic nickel, since either metal will enter the oxide as trivalent ions. If we pass to zinc, a divalent metal with excess of metal in the oxide structure (as opposed to a deficiency), we find the oxidation accelerated by monovalent lithium and retarded by trivalent aluminium in the metallic phase. This principle, due to Hauffe, is useful in explaining the effect of minor impurities, but it does not invariably serve for prediction.

Wood recalls (1976 vol., p. 47) that the result of over-optimistic application of the Hauffe rule has been to cause disappointment, since the rule is not always obeyed, but he points out that universal obedience could not reasonably be expected. The rule as sometimes set out involves unrealistic assumptions; for instance, it is assumed that foreign cations will enter normal cation positions in the parent oxide at the anticipated valencies, and that parabolic growth of the Wagner type will occur rather than a mechanism involving, say, grain-boundary diffusion. Wood's logical discussion introduces several other causes for the failure of the rule to provide correct prediction; it may well leave the reader surprised that the rule is obeyed at all! But any disappointment felt today should not blind us to the great service rendered by Wagner and Hauffe in providing general understanding of the principles of oxidation.

Actually the case where the scale is a single phase (a solid solution of the oxides of the two metals), and scale growth occurs simply by movement outwards through the scale has been put on a satisfactory mathematical basis. (B. D. Bastow, D. P. Whittle and G. C. Wood, *Proc. roy. Soc. (A)*, 1977, **356**, 177. The comparison between calculated and observed values of the parabolic constants is shown in Fig. 13, p. 204.) By combining Wagner's oxidation equations with equations representing atomic diffusion in the metal, ionic movement in the scale, and mass-balance at the metal–scale interface, it is possible to calculate the parabolic growth constant. In the systems Co–Ni, Co–Fe and Fe–Mn (also Ni–Cr alloys dilute in Cr), experimental determinations are available. The pertinent diagram in which calculated and observed values are compared has a double logarithmic scale, and at first sight agreement may appear better than is really the case. Nevertheless, the discrepancy is generally well within one order of magnitude. The authors consider this acceptable in view of certain extrapolations necessary to obtain the diffusion data and the neglect of certain short-circuit diffusion effects. They conclude that 'the description provides a reliable interpretation of the growth'. That appears entirely justifiable.

Oxidation by products of combustion

The gas formed when coke is burnt in air consists mainly of nitrogen, carbon dioxide and a little water vapour; with coal or oil as the fuel, the products of combustion contain more water and less carbon dioxide, whilst there is still more water present in the mixture obtained on burning coal gas. If air has been admitted to a furnace in excess, oxygen will also be present, and this will enhance the oxidizing powers of the mixture; but even if free oxygen is absent, oxidation is still possible, as shown by the extensive studies of Cobb and his colleagues.[78] Carbon dioxide can oxidize steel, unless the ratio $CO:CO_2$ in the mixture exceeds a certain value which depends on the temperature. Equally dangerous is water vapour, and its presence in the combustion products of coal gas makes them very destructive, even if oxygen is absent. If the mixture is cooled so as to deposit most of its moisture, and if unburnt gas is then added, much of the scale-producing power disappears; if water is not eliminated, the amount of unburnt gas needed to render the mixture harmless is excessive. In all these gas mixtures, the presence of compounds of sulphur greatly increases the destructive power, as shown below.

In annealing non-ferrous metal parts, it is desirable to use an atmosphere which will cause neither oxidation nor tarnishing. The exclusion of sulphur compounds is particularly important, and in the days when coal gas was used in the 'bright annealing' of nickel, copper or their alloys, it was partially burnt, purified by passing through bog iron ore and peat and cooled to precipitate water vapour; a suitable analysis for the gas entering the furnace was 6% CO_2, 9% CO, 11% H_2, 1·5% H_2O (rest nitrogen). Butane, which contains no sulphur, is useful in bright annealing, or ammonia 'cracked' (decomposed) in presence of a catalyst may be used. Another method is to lead a mixture of carbon monoxide and dioxide over charcoal heated electrically at a temperature controlled to give the desired $CO:CO_2$ ratio. For copper, atmospheres very rich in hydrogen should be avoided, as hydrogen acts on intergranular oxide, producing steam internally, which may lead to brittleness.

For the heat-treatment of steel, the atmosphere – besides being non-oxidizing – should be 'neutral' in the sense that it will neither increase nor reduce the carbon content. It should also be one which does not introduce hydrogen. Burnt ammonia (i.e. ammonia burnt with a controlled quantity of air, so as to give a mixture of nitrogen with 5–15% hydrogen) has found employment for treating steel.

Sulphur compounds

The metal-deficit of a sulphide often exceeds that of an oxide, pointing to a larger number of holes in the lattice and diminished protective power. The mineral **pyrrhotite** is assigned the formula $Fe_{11}S_{12}$ in some old mineralogy books; it is now regarded as FeS with 1/12 of the cation sites vacant. Cuprous sulphide also contains less copper than the formula Cu_2S would indicate. It is not surprising to find that copper, exposed to air containing hydrogen sulphide, rapidly develops visible films at ordinary temperatures; the sequence of colours is the same as that produced in pure air at slightly elevated temperatures, but the films, containing copper sulphide, are much less protective, so that the interference colour range is reached and passed at ordinary temperatures.

The behaviour of iron in furnace gas varies greatly with the sulphur content, and wastage may be reduced by using fuel low in sulphur; coal, for instance, can have some of its sulphur-bearing minerals removed by washing. The presence of sulphur modifies the law governing oxidation. A relatively protective film would follow the parabolic law, so that the mass of oxide (m) formed in time t would be given by $m^2 = Kt$; a completely unprotective scale would permit an oxidation rate which did not diminish with time, $m = Kt$. For practical purposes, some authorities write $m^n = Kt$, where n varies between 1 and 2, diminishing with increase of sulphur compounds and with rise in temperature.

The presence of sulphur in the furnace gas may also cause the oxidation to penetrate down into the steel along the intergranular boundaries – apart from the formation of a scale; this is particularly important in steels containing nickel and copper.

Sulphur compounds may, however, affect oxidation in another way. The oxidation rate always suddenly increases when the scale melts; the pure oxide scale formed in the absence of sulphur melts about 1340°C, and above this temperature the oxidation becomes very rapid – being mainly determined by the rate at which gases containing oxygen are blown over the surface. If sulphur is present, a sulphide–oxide eutectic may be formed melting below 1000°C, so that in the range 1000–1340°C sulphur will greatly accelerate the attack. If nickel is present in the metallic phase, a molten eutectic may appear as low as 625°C; certain alloys rich in nickel are attacked much faster in presence of sulphur compounds than in their absence. Great care is also necessary in the fabrication of nickel equipment to ensure that sulphur compounds are absent from the atmosphere or compounds applied to the surface.

Catastrophic oxidation

Certain alloys containing molybdenum and vanadium suffer very rapid oxidation under conditions where even part of the corrosion product is a gas or a liquid. Thus whilst steels containing 0·5% molybdenum are in extensive use for steam-pipes and superheaters, molybdenum contents beyond 2–3% are considered dangerous at temperatures where MoO_3 would be volatile and disrupt the scale. Similarly V_2O_5 has a low melting point (about 685°C), whilst mixtures with Na_2SO_4 are liquid at even lower temperatures; vanadium is present in oil ashes, and many alloys including stainless steel suffer rapid deterioration in presence of such ashes. Nickel–chromium alloys resist better. Further information about catastrophic oxidation will be found in the 1968 vol., p. 40. This includes an account of catastrophic oxidation of tubes at an ammonia synthesis plant, and its prevention, provided by Keller. The tubes had been heated by flames provided by heavy fuel oils containing vanadium which produced compounds of low fusion-temperature (650–700°C); the normally protective scale layer was removed and rapid oxidation set in. The injection of very finely ground dolomite into the flame with the air stream proved effective, and this method, if skilfully managed, should, it is thought, provide protection for six years or more; it also converts SO_2 and SO_3 into relatively non-corrosive $MgSO_4$. It is possible to introduce the Mg as a magnesium oleate, stearate or palmitate; this procedure is convenient, since the soaps are soluble in oil, but it is more expensive than that based on ground dolomite.

Attack by ash containing sulphate and chloride

Shirley[84] points out that, although calcium and sodium sulphate are harmless in themselves, they can cause damage if a trace of chloride is present, sulphate being reduced to sulphide. Air-heater tubes and stator blades can suffer in this way. An alloy with 74% nickel, 20% chromium and 2% titanium has been found to develop small spheres of a eutectic containing Ni, NiS and NiO. Some of the failures commonly attributed to vanadium may really be due to sulphate–chloride mixtures.

Tarnishing

Whilst copper exposed indoors to town air below the critical humidity (p. 107) develops interference tints, its exposure indoors to the same air at high humidities generally produces an unattractive pale brown tint, due to a film containing both oxide and sulphide. If the ordinary air is filtered through silver powder, which removes the sulphur compounds, it loses its tarnishing power, as shown by Vernon.[76] He also demonstrated that a long exposure of copper to pure air greatly reduces its suceptibility to tarnishing when subsequently exposed to air containing sulphur compounds, since an invisible oxide film is formed in the first stage which is far more protective than the film containing sulphide.

Silver also develops an ugly tarnish colour when exposed to an atmosphere containing sulphur compounds. The reactions are complicated; Price and Thomas, studying the tarnish film produced on sterling silver (with 7·5% copper), detected four compounds, silver sulphide, cuprous sulphide, cuprous oxide and silver sulphate; the electrometric method is capable of identifying all these compounds, since they suffer reduction at different potentials.

2

Electrochemical Corrosion

Electrochemical action without applied e.m.f.

Corrosion at a break in oxide scale

Imagine a piece of iron which has been coated with a thick oxide scale by heating in air; suppose that the scale has cracked off over one small area, exposing the metal, and that the specimen is then immersed in sodium chloride solution containing oxygen (Fig. 2.1). An electric current will flow between the oxide scale as cathode and the bare metal as anode. The current will only remain strong if oxygen has access to the cathodic surface, where it will react according to some such equation as:

$$O + H_2O + 2e^- \rightarrow 2(OH)^-$$

At the anodic area, the iron will pass into solution, usually as ferrous ions, according to the equation

$$Fe \rightarrow Fe^{2+} + 2e^-$$

The electrons liberated in the anodic reaction are used up in the cathodic reaction. Since sodium ions and chlorine ions are already present in the solution, we can regard the cathodic product as being sodium hydroxide and the anodic product as ferrous chloride. Both of these are freely soluble bodies, and will not stifle attack. They will, however, yield solid substances where they meet: ferrous hydroxide $Fe(OH)_2$ may be momentarily precipitated, but if sufficient oxygen is present, it will be oxidized to ferric hydroxide; this substance, com-

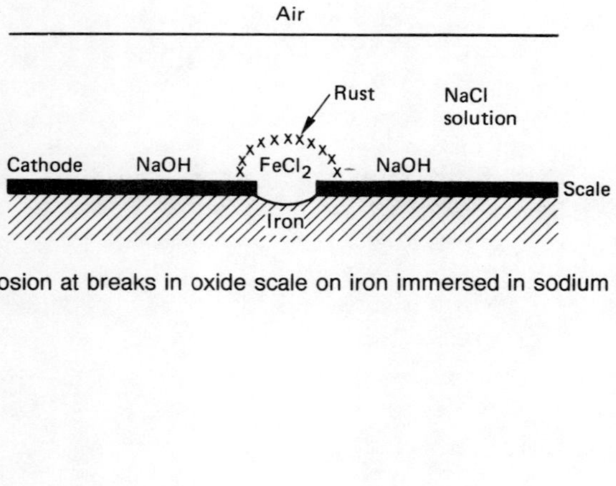

Fig. 2.1 Corrosion at breaks in oxide scale on iron immersed in sodium chloride solution

monly known as *yellow rust*, is FeO.OH or $Fe_2O_3.H_2O$, not $Fe(OH)_3$, but contains entrained water. Some authorities object to rust being described as 'ferric hydroxide', holding that this implies $Fe(OH)_3$. We need, however, a word to include *all* associations of Fe_2O_3 with water – no matter whether the water is held by capillarity or adsorption or combined as FeO.OH or $Fe(OH)_3$, and whether the rust is mainly amorphous or mainly crystalline. The word 'hydroxide' serves in this book to include all these things. In practice, the exact manner in which the water is held may not always be known. If the supply of oxygen is limited, the product may consist of green 'ferroso-ferric' hydroxides or black anhydrous magnetite, Fe_3O_4.

It will be noticed that this electrochemical corrosion ultimately brings about a combination of iron with oxygen (and usually with water) yielding an oxide or hydroxide. It differs from direct oxidation, however, in that the iron goes into solution at *one* place, the oxygen is taken up at a *second* place, and the oxide, or hydroxide, is formed at a *third* place; thus the solid corrosion product, formed at a distance from the point of attack, cannot stifle the action. It is on this account that electrochemical corrosion is often more dangerous than direct oxidation.

Another point deserves notice. The strength of the current flowing depends largely on the amount of oxygen reaching the cathode, and if the cathode, the scale-covered portion, is large, the current may be quite strong in relation to the size of the anodic area. The whole of the corrosive effect falls, however, on the anode, the bare portion of the metal. Thus the attack may become very intense when the exposed portion of the iron is small. *Intense attack* is frequently associated with this combination of *large cathode and small anode*.

Pitting is often met with in chemical and engineering practice at small gaps in the mill-scale covering mild steel plate, pipes or sheets. Removal of mill-scale from pipes or containers has been found to diminish the rate at which local corrosion bores down into the metal, leading ultimately to perforation. Since, however, scale protects areas where it is unbroken, the total production of rust (and possibly danger of internal blockage) may in some cases be increased by de-scaling.

It would, however, be wrong to imagine that pitting can be avoided merely by the removal of visible scale; it is indeed observed on materials like aluminium where there is no typical mill-scale. One feature of the situation is the appearance of acidity in pits, although opinions differ as to whether this should be regarded as the cause or the result of pitting. The acid results from the hydrolysis of a metallic salt (generally a chloride) produced by the anodic reaction in pits. Acidity was first observed in 1937 by Hoar in his study of the localized corrosion of tin. (Tin salts are particularly prone to hydrolysis.) Later Edeleanu attributed the pitting of aluminium to the acidity produced at a sensitive point where corrosion has started – so that it continues there rather than starting elsewhere (1960 vol., pp. 121–4). The liquid in pits on iron is also acid; Pourbaix found it contained 4·6M $FeCl_2$ with a pH value of 3·8. (M. Pourbaix, *Brit. corr. J.*, 1976, **11**.) Acidity is also found in other cases of localized corrosion. Acidity is produced in crevice attack (1960 vol., pp. 207–12) and in stress-corrosion cracking (1976 vol., p. 327); in the latter case an elegant method has been worked out by B. F. Brown for the estimation of the pH value prevailing at a crack tip; this depends on freezing in liquid nitrogen, so that the specimen can be broken open and the composition of the liquid in the crack established at leisure. The pH value noted in cracks on 0·45% carbon steel was 3·8, whilst 3·5 has been recorded on an aluminium alloy and 1·7 on a titanium alloy.

Other sources of corrosion currents

Clearly electric currents can be generated in other ways than by junction between oxide scale and metal. If the specimen in Fig. 2.1 had consisted of two different metals, such as copper and iron, a current would have been generated, the attack falling on the iron, the anode of the 'couple'. Intense attack is sometimes met with on iron carrying a copper coat which has been broken locally, perhaps at a bend, so that a small anode of bare iron is surrounded by a large cathode of copper. In a salt solution, the attack may then be locally more intense than if no attempt had been made to provide a protective coating.

Corrosion currents may also be set up if one part of a piece of iron has been bent, since deformation will damage any invisible oxide film present, and generally renders the metal itself less stable, by disturbing the crystal arrangement. But one of the most important ways in which corrosion currents arise is by differences of oxygen distribution in the liquid – as described below.

Whatever the origin of the current, the amount of corrosion can be obtained by means of Faraday's law (p. 277) if the current strength is known; thus a current of 1 ampere flowing for 1 hour will corrode 1·22 grams of zinc, 1·04 grams of iron (as a compound) or 3·86 grams of lead.

Differential aeration currents

Early experiments by the author[128] showed that if two electrodes of iron joined by wires to a milliammeter are placed in the two compartments of a divided cell, and if oxygen (or air) is bubbled through one compartment (Fig. 2.2), a 'differential aeration current' is registered, the aerated electrode being the cathode; when the current of oxygen is diverted to the other compartment, the direction of the current is reversed. If the experiment is repeated with zinc electrodes, the same result is obtained, the currents being stronger. With cadmium electrodes, the effect is much weaker, often no greater than the small current set up between two cadmium electrodes in the absence

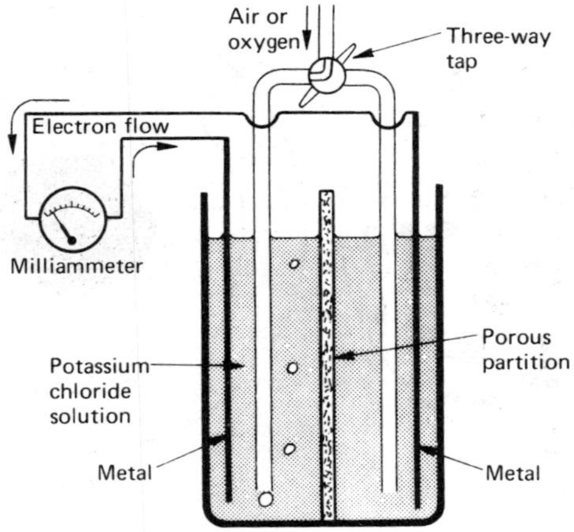

Fig. 2.2 Generation of differential aeration currents between electrodes of the same metal in the same liquid, oxygen being bubbled over one electrode only (schematic)

of bubbling (no two specimens of metal are exactly alike); but even in such cases a differential aeration effect can be observed, since the passage of oxygen over that electrode which, in the absence of bubbling, had been the cathode, increases the current, whilst the aeration of the electrode which had been the anode diminishes or even reverses the current. The case of copper appears anomalous, since the stirring set up by the bubbles removes the copper ions from the neighbourhood of the electrode over which bubbles are passing, thus shifting the potential in the negative direction (see p. 125); this renders the electrode concerned anodic towards the other. On copper, the differential aeration effect, although capable of demonstration, is feeble, and is usually outweighed by the stirring effect, whereas on zinc or iron the differential aeration effect far surpasses the stirring effect.

Corrosion by drops

The author's experiments[118] suggested that differential aeration currents are responsible for the distribution of corrosion produced by drops of sodium (or potassium) chloride (or sulphate) solution placed on horizontal sheets of iron or zinc. A drop of sodium chloride solution resting on iron sets up attack at the centre, where ferrous chloride is formed, whilst the peripheral zone remains immune from attack, sodium hydroxide being produced there; a ring of yellow rust appears where the alkali and iron salt meet and interact. This distribution is connected with electric currents flowing between a central anode and a cathode at the periphery, where oxygen, the cathodic reactant, can be replenished. In other experiments with large drops of

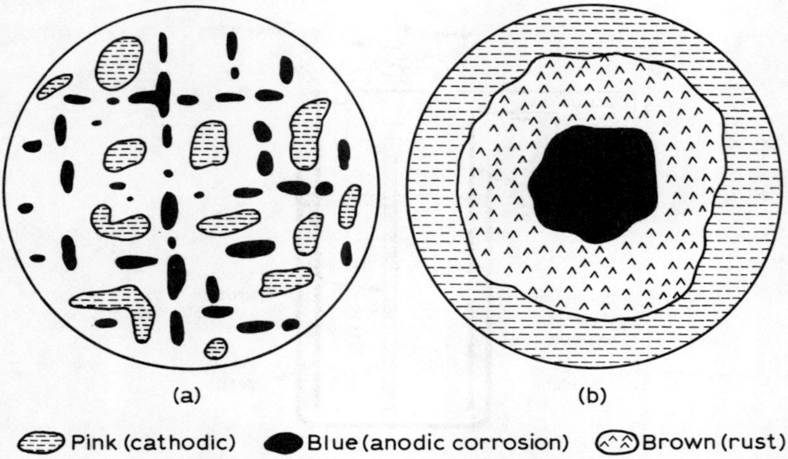

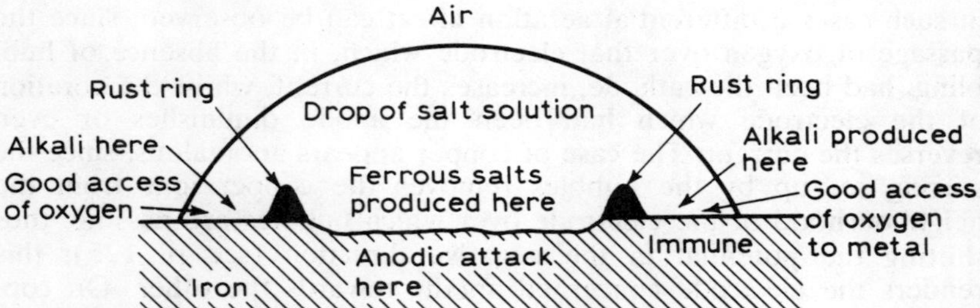

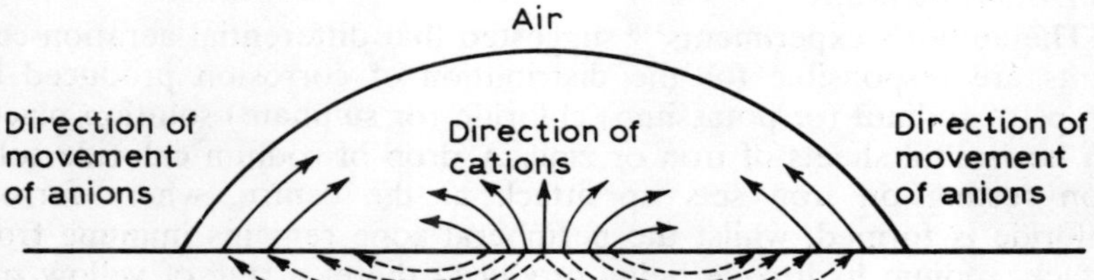

Fig. 2.3 Action of drops of potassium chloride on iron, with ferroxyl indicator to show up anodic and cathodic areas

sodium bicarbonate, it was found possible actually to detect the electric currents.

Blaha[128] demonstrated the current by placing an iron plate carrying a drop in a magnetic field. The drop was observed to rotate, just as a wheel carrying current along its spokes would rotate in a magnetic field; when the magnetic field was reversed, the drop rotated in the opposite direction.

Experiments were performed at Cambridge with drops of potassium chloride solution containing small amounts of **ferroxyl indicator** (phenolphthalein + potassium ferricyanide), which serves to show up the cathodic places (where alkali is produced) as pink patches and the anodic places (where iron salts appear) as blue points. If the solution contains oxygen at the outset, the primary distribution (Fig. 2.3(a)) is an irregular pattern of blue and red. When the original oxygen has been used up, the red colour becomes confined to the peripheral zone, since only here can oxygen be readily replenished; at the same time, the blue points disappear from the peripheral zone, since, once alkali (the cathodic product) has been formed in excess here, any iron salts formed momentarily will be precipitated in physical contact with the metal and will help to stifle attack. Thus the secondary distribution (Fig. 2.3(b)) will be a pink ring separated from a blue centre by a brown circle of rust. The time needed to pass from the primary to the secondary distribution naturally depends on the original oxygen concentration; if the liquid is supersaturated with oxygen, the primary distribution continues for some time, whereas, if dissolved oxygen is removed before the drop is placed on the metal, the secondary distribution arrives at once. Figs. 2.3(c) and 2.3(d) represent sections of the drop during the secondary stage, and show the chemical and electrical phenomena respectively.

Corrosion of vertical iron plates

Similar principles govern the distribution of corrosion on vertical iron plates partially immersed in a solution of sodium (or potassium) chloride (or sulphate).[117] During the first few days, the zone along the water-line, where oxygen can best be renewed, remains immune (being cathodic), whilst there is anodic corrosion a little lower down; along the line bounding the anodic region, there is a 'mantle' of membranous rust. With uniform, specially prepared, steel, the corroded (anodic) zone may (in M/10 solution) be confined to the cut edges, owing to breakage of the original invisible oxide film, or to the presence of stresses left by shearing; this pattern is called ideal distribution' (Fig. 2.4(a)). At lower concentrations, the attack starts also at points on the faces, and spreads *downwards* from each, owing to the sinking of the anodic product, which, as explained on p. 32, may be slightly acid, or otherwise capable of destroying the film. As the oxygen in the liquid becomes exhausted, breakdown occurs at still higher points, so that corrosion *as a whole* moves *upwards*, a se-

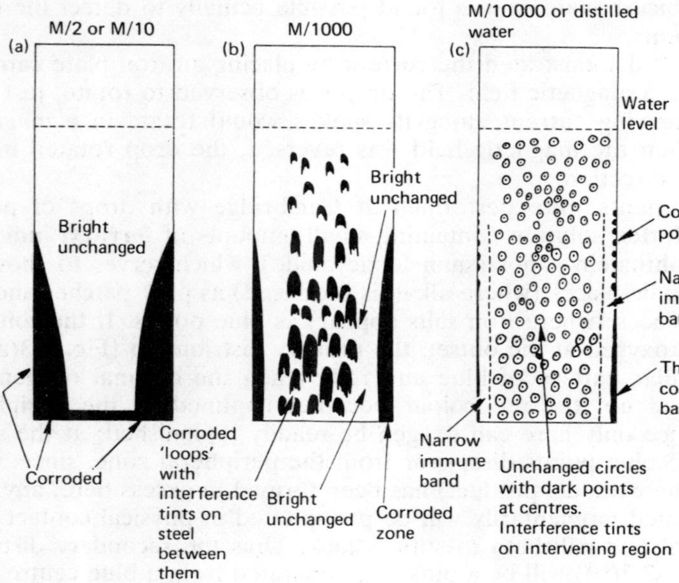

Fig. 2.4 Early stages of attack on an iron plate by sodium chloride solutions of different concentrations

quence of mantles being formed as suggested in Fig. 2.5(a). Under some conditions corrosion may reach the water-line, but, if thermal convection currents and access of carbon dioxide (which would dissipate or destroy the cathodically formed alkali) are avoided, a protected zone just below the water-line usually remains some weeks. Washing and drying reveals interference tints on this area (Fig.

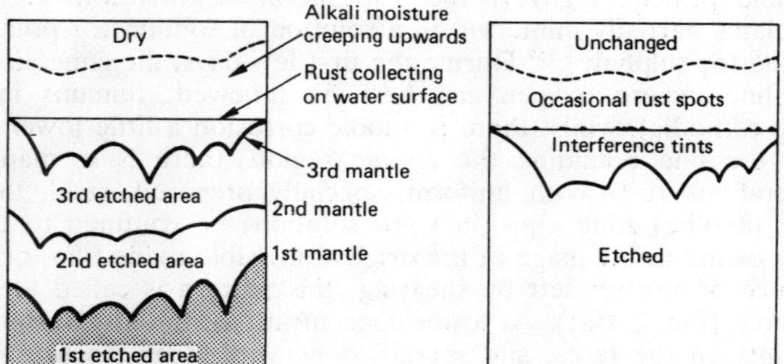

Fig. 2.5 (a) Discontinuous advance of corrosion upwards seen on some steels (b) Appearance after washing and drying

2.5(b)). At greater dilutions, distribution is irregular, with interference tints in the early stages on the main region between the numerous points of breakdown (Fig. 2.4(c)).

Corrosion of vertical zinc plates[116]

A rather similar distribution of attack is observed on zinc, although behaviour varies with surface condition. Corrosion starts at a larger number of points on a rough surface than on a smooth one, so that the mantle is more sinuous; it is less sinuous in sulphate than in chloride solution, and the white corrosion product (zinc hydroxide or basic salt) is precipitated closer to the metal when divalent $(SO_4)^{2-}$ ions are present, so that in sulphate solutions much white matter is found clinging to the zinc surface. As on iron, there is a tendency for the corroded portion to move upwards with time; sometimes on rolled zinc the upper boundary of the corroding area is perfectly horizontal (Fig. 2.6(a)) and moves upwards continuously with time (instead of intermittently with the formation of successive membranes). Often corrosion ultimately reaches the water-line, and is then usually most intense at that level; in the experiments of Bengough and Wormwell[96] perforation occurred along the water-line at a stage when there was no perforation anywhere else. In some cases, attack starts from susceptible points or from scratch-lines, if present (Fig. 2.6(b) and 2.6(c)). Often on rolled zinc, attack penetrates parallel to the surface along internal laminations, so that on bending a specimen which has been corroding for some weeks, the surface layers part company from the layers below them (Fig. 2.6(d)).

The experiments of Thornhill at Cambridge provide evidence of the electrochemical character of the corrosion on zinc. (R. S. Thornhill and U. R. Evans, *J. chem. Soc.*, 1938, **614**; 2109.) He used an

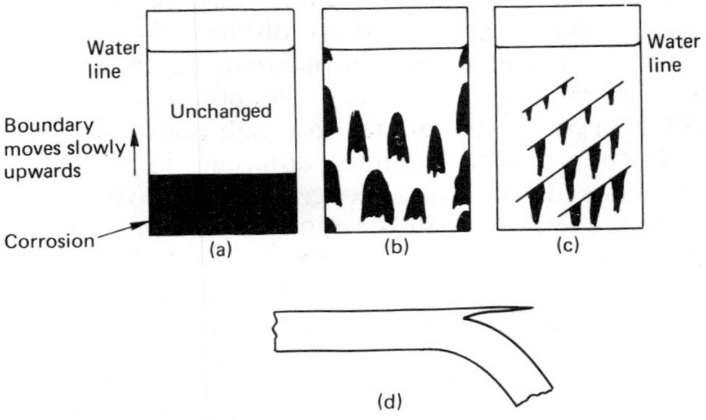

Fig. 2.6 Distribution of corrosion on zinc with rolled surface. (a) Regional attack (b) Point attack (c) Attack from scratch-lines (d) Result of bending corroded zinc

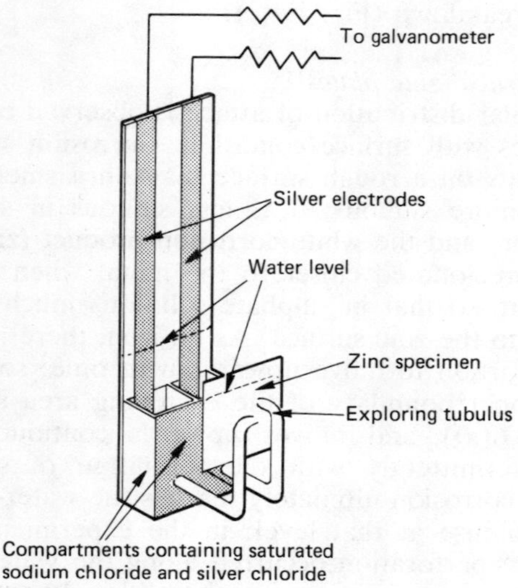

Fig. 2.7 Dielectrode for detecting anodic and cathodic points on zinc in sodium chloride solution

apparatus called a 'dielectrode' (Fig. 2.7) to distinguish anodes and cathodes. This consists of two Ag | AgCl electrodes provided with tubuli protruding to different distances and connected to a microammeter with central zero-point; one tubulus is pushed into contact with the zinc surface at the point where the polarity is to be tested. If current is flowing to or from the metal at this point, the potential at the tubulus mouth in contact with the surface must be different from that at the other, which remains a short distance from the surface; a deflection will be registered on the microammeter, and its sign will indicate whether the point is anodic or cathodic.

Thornhill worked with dilute salt solutions (M/1000 sodium chloride), which produced a rather different distribution of attack from that obtained at higher concentrations. After some initial irregularities, the electrical distribution became stabilized with the main cathodic zone extending along the top of the meniscus and the main anodic zone along the foot of the meniscus. Thornhill found that, even in the opening stages of irregular distribution, the principal cathodic zone was always above the anodic zone and that at any moment *attack was only observed at those points which the electrical apparatus showed to be anodic*; points which were *not being attacked* were found to be either *cathodic* or *neutral*. This seems definitely to establish the electrochemical mechanism.

Quantitative verification of the electrochemical mechanism on iron

The first accurate measurements of the currents flowing on a specimen consisting of a single metal were carried out at Cambridge by Hoar[863]. He used a good quality steel, which regularly furnished the ideal distribution (Fig. 2.4(a)) in potassium chloride solution. Hoar was thus able to cut a specimen along the lines which he knew would come to represent the boundaries of the anodic area, when later the specimen was immersed vertically in the solution; after carefully covering with wax the new edges produced by the cutting, he mounted the two parts in their proper relative positions in a containing vessel, but connected to the two terminals of a milliammeter; the arrangement was that shown in Fig. 2.8, with P and Q joined. The liquid was then poured in and the current passing was measured; this was found to correspond – in the sense of Faraday's law – to the corrosion velocity as obtained by loss of mass.

Hoar performed further experiments in which currents of known strength were applied from an external source joined up between P and Q. For each value of the applied current, he measured, by means

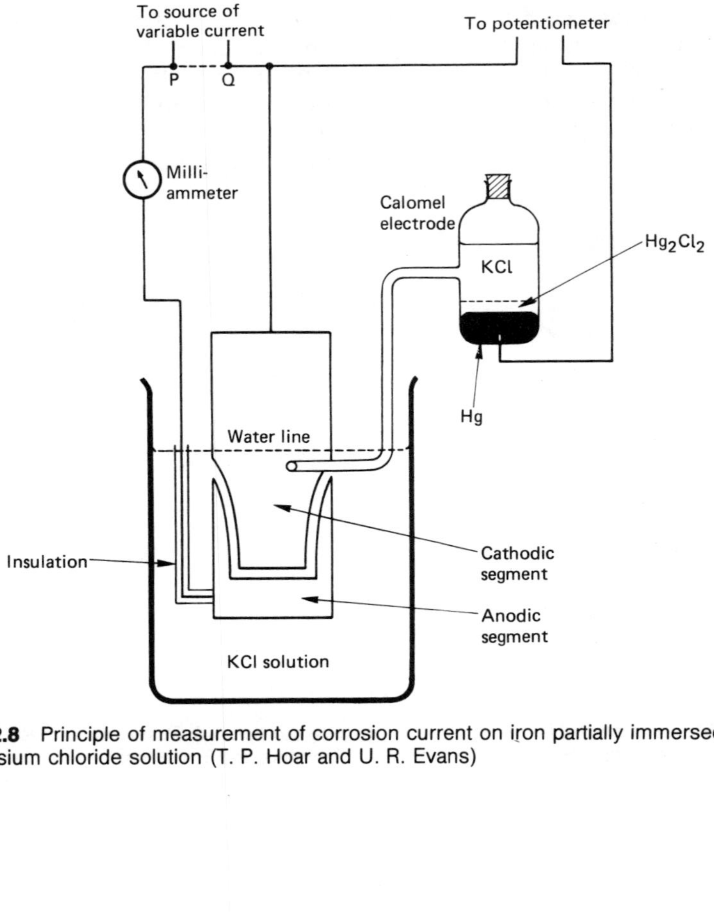

Fig. 2.8 Principle of measurement of corrosion current on iron partially immersed in potassium chloride solution (T. P. Hoar and U. R. Evans)

of a calomel electrode and potentiometer (p. 286), the value of the potential difference between metal and liquid at the cathodic zone situated near the water-line. Owing to polarization, this p.d. dropped as the applied current was increased, and curves could be plotted showing the relation between p.d. and current. It thus became possible to measure the current flowing on *uncut* specimens, because the p.d. could be measured on the water-line zone and the corresponding current read off from the curves. Division by Faraday's number gave the corrosion velocity, and the values thus obtained at various concentrations (Fig 2.9) showed remarkable agreement with the experimental values obtained by loss of mass. No arbitrary constant was involved in these calculations; if the corrosion had not been electrochemical in character, there would have been no reason to expect that the two sets of numbers would even be of the same *order* of magnitude.

Hoar's experiments – just quoted – refer to corrosion set up by differential aeration on partly immersed specimens. Another comparison has been carried out on corrosion produced by sodium bicarbonate solution at a scratch-line. Sodium bicarbonate is peculiar in that its dilute solutions are highly corrosive to steel, although concentrated solutions cause no attack; this is probably due to the reduction in the solubility of ferrous bicarbonate by the presence of excess $(HCO_3)-$ ions, which permits the formation of protective films in concentrated solutions. At intermediate concentrations, behaviour depends on the previous history of the metal. An abraded surface, exposed to air for some time before the bicarbonate solution is applied, is not attacked;

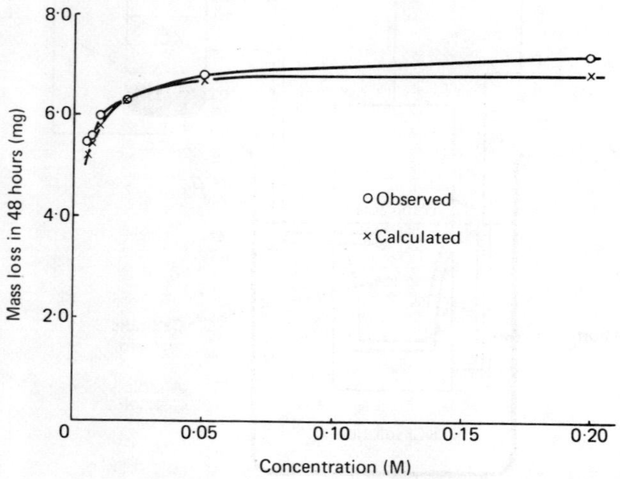

Fig. 29 Corrosion of iron in potassium chloride solution. (a) Observed by mass loss (b) Calculated from purely electrical data (U. R. Evans and T. P. Hoar)

but if a scratch-line is traced on the surface just before the liquid is applied, rusting sets in along that line. Mears,[937] in his statistical studies, found that the probability of corrosion by 0·07M sodium bicarbonate (measured by the proportion of the scratch-lines producing rust) fell off steadily as he increased the period of exposure to air before application of the liquid.

If a single fresh scratch-line is made on a surface which has long been exposed to air, and a piece of filter paper soaked in dilute sodium bicarbonate is applied, a rust stain is produced along the length of the scratch-line. The author,[866] by means of a special dielectrode, showed that an electric current was really passing between the scratch-line as anode and the surrounding region as cathode. Later, Thornhill[867] succeeded in calibrating the apparatus and thus obtained an accurate measurement of the current. At the outset, the current corresponded to the corrosion velocity on the assumption that the iron was passing into solution in the ferrous condition; later it increased in relation to the corrosion – which was explained on the supposition that, when once a quantity of ferrous bicarbonate had accumulated, it underwent anodic oxidation to the ferric condition, thus accounting for the surplus current.

Quantitative verification of the electrochemical mechanism on zinc

In the case of zinc partly immersed in chloride or sulphate solution, Agar[861] obtained a measure of current by studying the distribution of

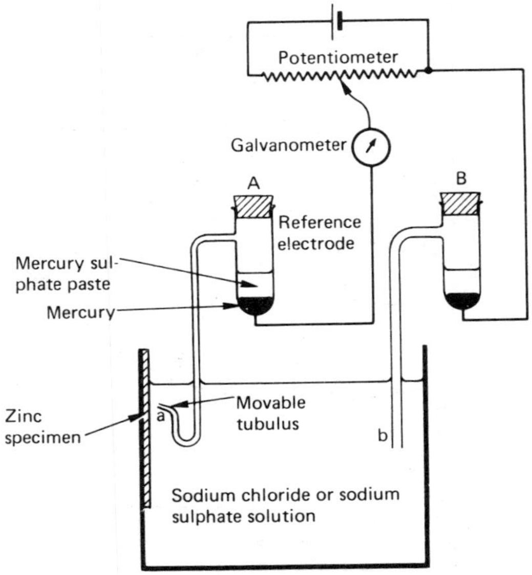

Fig. 2.10 Apparatus for exploring the distribution of potential around corroding zinc (J. N. Agar and U. R. Evans)

potential in the liquid. He used an electrode supported in a stand, provided with screw-gear which allowed motion in three directions at right angles; the apparatus carried scales which indicated the exact position assumed by the tubulus tip (Fig. 2.10). The electrode was joined through a potentiometer to another electrode fixed at a distance from the metal. Agar traced the shape of surfaces upon which all points were at the same potential; the 'equipotential' curves of Fig. 2.11 which refer to M sodium chloride solution, represent the intersection of such surfaces with the plane of the paper. Since the equipotentials must be crowded together if the current is strong, but must lie far apart if it is weak, they can be used for the calculation of current strength. If two equipotentials differing by ΔV are separated by distance s, the current flowing from one to the other is approximately $\kappa A \Delta V / s$ where κ is the specific conductivity of the liquid, and A the mean area of the two surfaces. Considering equipotential surfaces extending for 1 cm at right angles to the paper, the area of any such surface is l cm², where l is the length of the curved lines, such as those shown in Fig. 2.11. If two such lines, separated by a mean distance s, have lengths l_1 and l_2, the current per centimetre length of water-line will be approximately

$$\kappa \frac{(l_1 + l_2)}{2} \frac{\Delta V}{s}$$

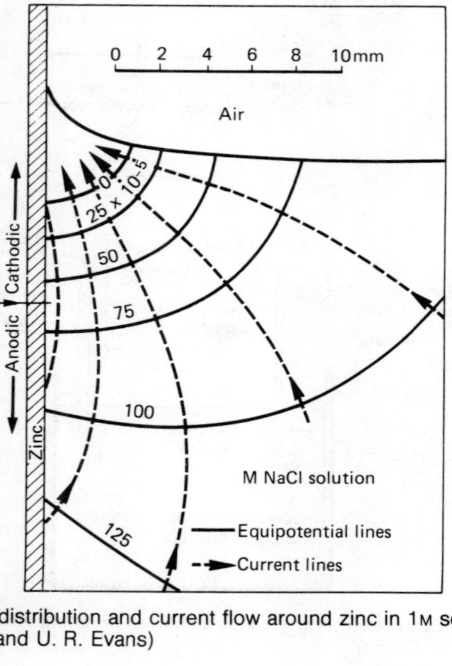

Fig. 2.11 Potential distribution and current flow around zinc in 1 M sodium chloride solution (J. N. Agar and U. R. Evans)

The currents obtained by measuring l_1, l_2, ΔV and s were found to be sufficient to account for the whole of the attack leading to a loose corrosion product. A small amount of adherent product formed above the water-line may be connected with upward creepage of alkali from the cathodic zone. Agar's work furnished quantitative evidence for the electrochemical mechanism of the reactions studied.

The manner in which the corrosion current depends on the polarization characteristics of the metal and oxygen and on the resistance of the circuit is discussed on p. 79 in connection with acid attack.

Causes of breakdown

The position of the susceptible spots where corrosion starts received close study in early Cambridge researches[108] – mainly on iron and zinc, but also on aluminium and copper. Corrosion almost always starts at places where internal stresses are present. On carefully prepared specimens of high-quality material, it may be confined to the neighbourhood of the cut edges where stresses have been produced by shearing; in liquids of high conductivity, the attack here developing provides cathodic protection to the face which, in the early stages, will remain immune (Fig. 2.2). On surfaces with no cut edges, such as round bars, or wide flat surfaces protection in unstored near-neutral solutions is confined to a band at the water level of uniform width. In more dilute solutions, protection will not extend so far. On material which has been badly rolled, attack often starts at points situated on straight lines parallel to the rolling direction; sometimes these represent visible surface defects. On abraded specimens the points of attack may be arranged along the deeper grinding grooves. On bent specimens, attack often starts on the bending axis, either on the convex or concave side. On zinc, it is often possible, by tracing a scratch-line on a smooth surface, to produce attack along this line, although in general it will start at certain isolated points – not along the entire length (Fig. 2.6(c)). Later Cambridge studies[111] of corrosion at a scratch-line on nickel anodically treated with an external e.m.f. (leading ultimately to perforation along the scratch-lines without attack anywhere else), showed that often the starting points were not situated in the trench left by the scratching, but on the two 'banks', where the metal would be left in tension; at such places any oxide film produced would come to be in tension, and persistent rupture is easily understood.

For iron placed in acidified potassium chloride or sulphate, soluble ferrous chloride or sulphate is the expected anodic product from the first (according to the principle set forth on p. 50), but a neutral solution, acting on iron or zinc, should at the outset produce oxide; this will, however, render the liquid locally acid, and soluble salt formation should later become possible. That is observed to occur. Experiments on zinc, interrupted in the early stages, show no cloudiness or loose precipitate, but only clinging solid, producing a white appearance or interference tints according to the angle of viewing; later, loose matter (evidently formed by precipitation of zinc chloride with alkali) appears – more quickly in chloride than in sulphate solution, more quickly on coarsely than finely abraded zinc (probably owing to the greater internal stresses). The sequence (film-thickening followed by film-breakdown) is, however, best shown by tin – as was brought out in Hoar's study of the phenomenon known as 'black spots'. Tin salts are only stable in definitely acid solution; in absence of acid they hydrolyse. Tin immersed in M/10 potassium chloride solution developed circles of colours around the susceptible spots; after 72 hours, there were high-order colours at the centre of each circle, whilst first-order colours had spread over the whole specimen. Then, when sufficient acidity had developed, the formation of soluble stannous salts became possible, and the black spots appeared at the centres of the circles, showing that pitting had started. The blackness was due to absence of reflection and is comparable to the blackness of the mouth of a rock cavern; there is no need to postulate a black tin compound.

Pitting

It is obvious that corrosion concentrated at a few points, causing pits, is something more dangerous than attack well spread out – even though the latter may cause a greater total destruction of metal. Much enquiry has been pursued regarding the factors determining the sites of pits. Where the material is one which carries a highly protective oxide film, the pits will generally appear at flaws in that film. That is the conclusion reached by Wood as a result of several extensive researches carried out by Richardson and other colleagues on aluminium (1976 vol., pp. 77–9). The evidence for such a view rests on work with a electron microscope capable of detecting changes – such as cracking, peeling and undermining – on a scale so small that the optical microscope fails to reval them. The pits are found to arise in an otherwise inert oxide film, and when once started, proceed to undermine that film. A distinction is drawn between different types of flaws. 'Residual flaws' – such as copper-rich segregates – provide points on which the cathodic reaction can take place (alumina is not an efficient cathode), whilst mechanical flaws, produced during the relief of internal stress, provide gaps in the film, so that anodic attack on the exposed metal is possible. When once pitting attack has been established at a point, other factors may become important; acidity produced by hydrolysis of aluminium salt (the anodic product) will itself cause attack on the metal.

Where, as on iron, the film has little protective power, corrosion, although undoubtedly starting at isolated points, will generally spread out, and pitting is not usually a menace; the most serious feature of the situation is the large amount of voluminous rust or other product, which may obstruct flow in a water-system. If, in order to avoid that trouble, a soluble inhibitor is added to the water, and the amount needed is underestimated, the small amount of attack which occurs may be localized, and something in the nature of pitting can be set up. Here the starting points are frequently connected with inclusions (1976 vol., pp. 66–8, 73–5). Early work by Homer on steel placed in a solution of NaCl with sufficient Na_2CO_3 to localize, but not entirely to prevent corrosion, showed that the attack generally started at inclusions of iron or manganese sulphide. (Not all sulphide inclusions present acted as starting points whilst most other inclusions, such as silicates and alumina produced no pits at all). On the upper surface of an iron specimen, the attack often spread out, becoming less intense, since the rust produced screened a constantly increasing area from replenishment with the inhibitor (Na_2CO_3) on the lower surface, this did not occur to the same extent because the rust precipitated fell away; even on an upper surface, however, Homer obtained localized attack if the ratio of Na_2CO_3 to NaCl was carefully adjusted.

The importance of sulphide inclusions in connection with the inception of pitting has been brought out by more recent researches in many countries – especially France, Poland and Sweden, where Wranglen has pointed out that in soil and water the sulphur content of iron has little effect on general corrosion, but that it increases localized attack especially at welds and in crevices, as well as ordinary pitting.

Sulphide inclusions are especially objectionable in stainless steel which may suffer serious pitting. The pits on stainless steel are sometimes found to be covered with a thin film; this appears to be the original passive film which has been undermined.

Research work in Bianchi's laboratory (1976 vol., p. 74) shows that the pitting of austenitic stainless steel depends greatly on the state of the film before contact with the pitting liquid. If a specimen had been placed for 140 hours at 150°C in dry air, the pitting susceptibility was small, whereas a specimen pre-heated at 300°C showed a high susceptibility. The difference was not due to the greater film thickness in the latter case since a specimen heated at 300°C and then at 150°C assumed the same pitting characteristics as one heated simply at 150°C. It is believed that the difference of behaviour is connected with the fact that the oxide found at 150° is an n-type conductor whereas that found at 300°C is a p-type conductor.

Chlorides as a factor in pitting

It has sometimes been stated that the pitting of aluminium only occurs if copper and Cl^- are present; that is probably an over-simplification (cases are known of pitting in their absence); but in practice pitting is usually associated with this combination of factors. The copper may arise if water has run through copper pipes, acquiring a trace of Cu^{2+}, and then into an aluminium vessel on which metallic copper is deposited by simple exchange. Or the copper may be in solid solution on the aluminium, and be dissolved and then redeposited. In either case, there is no mystery about the cause; the bimetallic cell Cu/Al will generally cause anodic attack on the Al.

Regarding the part played by chlorides in starting pits, there is less general agreement. The matter is really part of a wider problem, since Cl^- can set up attack on various materials normally resistant to corrosion, such as stainless steel, and can also prevent protection by recognized inhibitors; if Cl^- is present in a quantity sufficient to set up attack at a few points only, the combination of large cathodic and small anodic areas will produce intense attack. It is evidently important to agree upon the cause of the specific action of Cl^-. Various explanations have been suggested – some of them based on the idea of competitive adsorption; if Cl^- is adsorbed more readily than ions or molecules having a protective character, that might explain its effect. Without quantitative information about the adsorption characteristics of all the ions concerned, it is difficult either to accept or reject such an explanation.

An interesting explanation proposed by Pryor is based on the difference between the charges on Cl^- and O^{2-} respectively (1976 vol., p. 76). It has had an unfavourable reception in some quarters, but expressed in the manner stated below, it would seem to explain the facts. No doubt the local attack starts owing to some malformation in the film, as suggested by the studies of Wood's research group. If, for instance, there is tensile stress in film and metal at some point, so that the interatomic distance (measured parallel to the surface) is slightly greater than that in unstressed material, conditions will be locally favourable for movement through the film, and also there will be a few metallic atoms below the film which, owing to the abnormally great distance from their various neighbours, will require less energy to move out through the film ions; at such points, a certain small amount of corrosive action (cations moving outwards, anions inwards) may be expected. However, if all the anions possess the same charge (as would be the case if the liquid were a sulphate solution), the action would probably cease as soon as these 'loose' atoms had passed into the film. In a chloride solution, it will not cease when ionic movement occurs through a film, only part of the current is moving outwards; the rest is carried by trains of O^{2-} moving inwards, followed by any anions from the solution small enough to enter the film. If then the electrical situation is changed, owing to the replacement of O^{2-} by Cl^-, an electrical charge will accumulate, unless Al^{3+} moves out into the liquid more quickly than would otherwise be the case. If the cation expulsion is accelerated, the setting up of a charge will be avoided by the formation of an appropriate number of vacant cation sites. This, however, provides a situation much more favourable for the outward movement of cations than anything which existed at the outset; wherever a vacant cation site occurs, a cation can move into it, leaving a new vacant site into which another cation can move – and so on. In effect, there is now, instead of a train of Al^{3+} moving outwards, a train of vacancies moving inwards. Therefore, in calculating the rate of movement through the film, we are no longer concerned merely with the mobilities of Al^{3+} and Cl^-; the important factor is now the mobility of a vacancy, which is likely to be greater than that of either ion. The conclusion is reached that at a point where, owing to some slight defect structure, movement starts to occur, conditions will become steadily more favourable to the movement of cations outward as Cl^- ions move inwards; and this will continue even after the few original 'loose' atoms have been used up. It would seem that the difference between the charges on Cl^- and O^{2-} does provide an explanation of the power of Cl^- to develop local deterioration in the character of a normally protective oxide film.

Pitting on Copper

Certain waters have given trouble by the production of pits on copper tubes, especially tubes annealed in a reducing atmosphere which leaves a deposit of carbon on the surface derived from the cracking of oils; today precautions to avoid or remove such carbon deposits are being taken. In France, where hard (unannealed) tubes have generally been used, this type of trouble has not been widely experienced.

The explanation at first favoured was that the carbon film acted as a large cathodic area, so that the anodic attack, concentrated at small gaps in the film, was very intense. This, however, is not the whole story. The matter has received authoritative discussion from Pourbaix who points out that the anodic product, cuprous chloride, will hydrolyse, producing HCl, and that unless the acidity is washed away or otherwise destroyed, attack will continue. (M. Pourbaix, *J. electrochem. Soc.*, 1976, 123, 25C; *British non-ferrous Metals Res. Assoc. Misc. Publication*, 1971, 538; 1972, 568.) If, however, the potential is sufficiently low, or can be kept low by appropriate means, pitting should be avoided. One method, used in certain districts of the U.K. where the pitting of copper hot-water tanks placed in the roofs of houses had been a matter of concern, has been cathodic protection by means of sacrificial rods of aluminium, which depress the potential below +100 mV on the standard hydrogen scale. This appears to have been very successful.

Electrochemical action with applied e.m.f.

General

In the cases of electrochemical action hitherto considered, the metal has provided its own e.m.f. The rate of attack is often limited by the rate at which oxygen, the cathodic reactant, can reach the cathodic area. Cases of attack by acid and alkali which proceed rapidly in absence of oxygen are discussed in Chapter 3. Where an e.m.f. is applied from an external source, much greater rates of anodic attack become possible.

Consider, for instance, two iron plates dipping into sodium chloride solution and connected to a dynamo or external battery. The rate of corrosion of the anode is, in the absence of passivity (see below), proportional to the current flowing, and may be very rapid. On the other hand, the cathode will suffer less corrosion than in the absence of external e.m.f., and, if the cathodic current density is sufficient, may be protected completely (see p. 117).

Factors deciding between the formation of salts or oxides on an anode

Any aqueous salt solution contains, in addition to anions such as SO_4^{2-} or Cl^- (according to the salt chosen), a small concentration of hydroxyl ions (OH^-) due to the dissociation of the water. It is important to know whether the anodic product will, in effect, be a soluble salt (sulphate or chloride), or whether it will be a sparingly soluble oxide (or hydroxide), since in the latter case even electrochemical attack may be stifled.

If a zinc or iron electrode is subjected to anodic attack in a weakly acidified solution of sodium chloride at a very low current density, there will be active corrosion, producing zinc or ferrous chloride in solution. The formation of oxide would here be improbable on grounds of energy, since the system 'metallic oxide + acid' represents a state of higher energy (free energy would be liberated when the acid dissolved the oxide); if the potential has been raised just sufficiently for the reaction $Zn \rightarrow Zn^{2+} + 2e^-$ to be possible, it will not be high enough to supply energy for the formation of oxide and acid. In contrast, if the solution were weakly alkaline, oxide formation is the expected reaction; the co-existence of Zn^{2+} (or Fe^{2+}) with OH^- represents an energy state which, at low current density, could not be produced; hence a soluble zinc or iron salt will not be formed.

The formation of *oxide* by anodic action in weakly alkaline solution (which seems to have puzzled some people, who maintain that *hydroxide* would be expected) is simple. The hydroxyl ions must be oriented somewhat as in Fig. 2.12(a) by the electric potential gradient connected with the current flow, and probably also by surface forces. Thus a cation which formerly was part of the solid metal takes its appropriate place among the oxygen atoms (giving the start of oxide-formation), and two H^+ ions (assuming the metal to be divalent) are enabled to join other OH^- ions to form water molecules. This produces precisely the same electrical transfer as would have occurred if the cation had moved right out into the solution, as would occur in acid solution. When once a film has started forming, it can continue to thicken (Fig. 2.12(b)), cations entering at the inner surface, and others leaving to take their places in the O zone at the outer surface.

Now, however, consider a case where a constant current is supplied

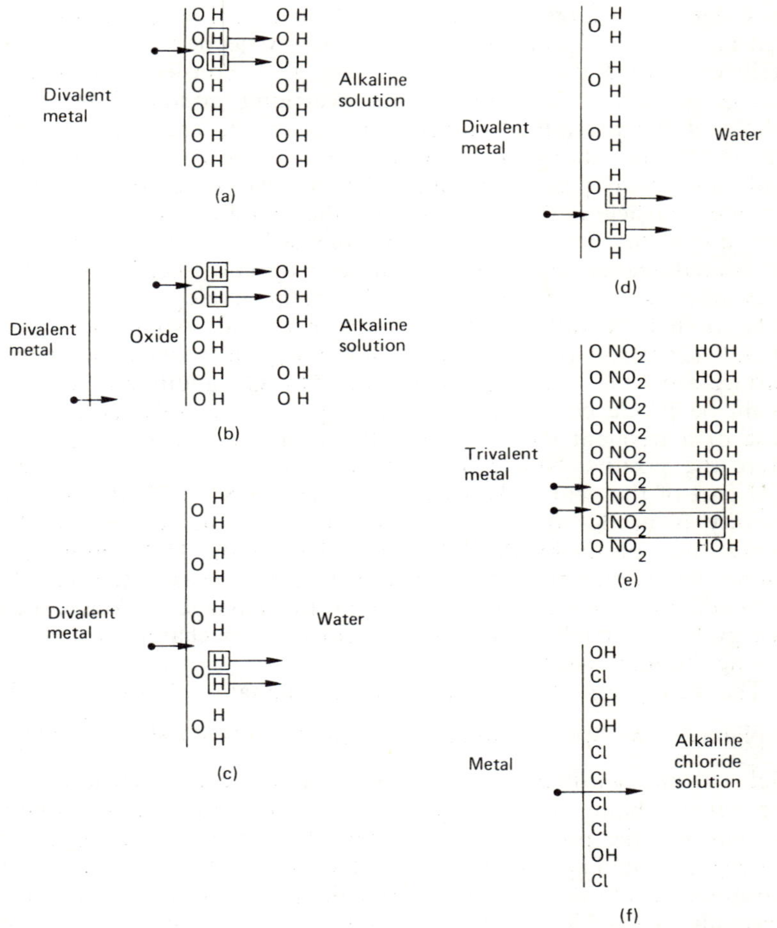

Fig. 2.12 Anodic change. (a) in alkali (first stage) (b) in alkali (later stage) (c) in water, giving oxide (d) in water, giving hydroxide (e) in nitric acid (f) in alkaline chloride solution

from a galvanostatic circuit as explained on p. 176 designed to keep the current strength constant, and assume the liquid to be water or very dilute acid solution. If the current density is fairly low, the cations will pass into the body of the liquid; to do this they must pass first through the negative zone where there are O atoms, and then through the positive zone of H atoms. But if the current density is fixed at a high value, there may be a deficiency of cations possessing sufficient energy to pass through the positive zone of the oriented water molecules (or oriented oxygen-containing anions, if present), and an alternative reaction *must* be called into play, even though it may demand a larger consumption of energy. The galvanostatic circuit has been designed to supply any energy needed to force the prescribed current through the cell, and thus considerations of free energy no longer concern us; the overriding condition is that the whole of the prescribed current must be used. The only way to do this is for the cations to remain in the negative zone, forming oxide, and thus avoid passage through the positive zone; the rest of the oriented particle is then sent on into the liquid. If the oriented particles are water molecules, two H^+ ions will be released, producing additional acidity, and providing the appropriate electrical transfer; if they both come from the same OH group, oxide is produced (Fig. 2.12(c); if from different OH groups, hydroxide would be expected (Fig. 2.12(d)); if (as in a nitrate solution), the particles next to the surface are NO_3^- probably oriented as O–NO_2, additional acidity will again be produced, as suggested in Fig. 2.12(e); this represents the case of a trivalent metal, so that the sesquioxide (M_2O_3) is formed. Here the positive NO_2 group will leave its O partner and join the OH part of the nearest HOH molecule giving NO_2OH, that is HNO_3. Acidity is produced in two ways, partly by the ionization of the HNO_3, and partly by the H^+ displaced from the HOH molecule. It cannot be claimed that the details of the reaction as stated have been verified, but it is difficult to see how the passage of ions over the energy barrier could be avoided except by a mechanism roughly following the lines suggested.

The behaviour of an iron anode in sulphate solution is similarly explained. The SO_4^{2-} is probably oriented as O–SO_3H or $\genfrac{}{}{0pt}{}{O}{O}\!\!>SO_2$. At high current densities, the dissolution of the iron practically ceases, the current being used in the formation of oxide, the evolution of oxygen and the production of acidity. The metal is then said to be *passive*. It would appear, therefore, that, whilst at low current density, the metal would be expected to remain *active* and pass into solution smoothly as soluble salt, this will not be the case on imposition of a high current density, if the metal is one which provides ions sluggishly – as is the case of the 'abnormal' metals such as iron and nickel (see p. 290). Such a metal will become passive and covered with oxide, and the current thereafter will be devoted to the evolution of oxygen; that is just what occurs. With zinc, a 'normal'* metal, a more plentiful supply of cations possessing the necessary activation energy will be available, and the anode will not easily become passive. Even on 'normal' metals, however, passivity may occur if the current density is very high. Aluminium, 'anodized' in sulphuric acid at an appropriate current, develops an oxide film (porous in its outer portions), much thicker than the film formed on iron. The difference arises from

* A normal metal is one in which the electrons of the outer orbital readily detach themselves leaving positive ions.

the fact that alumina is a non-conductor of electrons, and the current is carried by cations moving outwards, so that thickening proceeds; part of the rest of the current is carried by anions moving inwards. On iron electrons can pass through the oxide film; the movement of cations outwards – when once a steady state has been reached – is only that needed to compensate for the slow (chemical) dissolution of oxide by acid.

There are, however, conditions where a high current density can be maintained without passivity setting in. If the liquid contains chloride, Cl^- ions will be present near the metallic surface, and may interrupt the phalanx otherwise interposed by the oriented molecules; there is good reason to believe that halogen ions are attached to metal by strong adsorptive forces, and in any case charged ions should be pushed towards the metal by the electric potential gradient in preference to water molecules, which (as a whole) carry no charge. If we picture the phalanx interrupted by Cl^- ions, as in Fig. 2.12(f) we can visualize metallic cations passing out into the liquid without much activation energy being needed. If some film is formed, it will be less protective than in the absence of chloride. As explained on p. 48, Cl^- ions moving inwards through the film will provide conditions specially favourable for the movement of cations in the opposite direction.

Passivity on a horizontal anode

In an acid liquid, an anode generally dissolves freely at low current densities, giving a soluble salt, but at high current densities passivity supervenes, as explained above; when once an oxide has been formed, most of the current which continues to pass is devoted to the production of oxygen. Under very stagnant conditions, even a low current density can ultimately bring about passivity; but the time needed for active dissolution to give place to passivity increases as the current density diminishes. This was shown by W. J. Müller (1976 vol., pp. 166–70), who used a horizontal anodic surface protected from chance stirring by a hood (Fig. 2.13(a)). With a vertical anode (Fig. 2.13(b)), particularly if there is rapid movement of the liquid over it (Fig. 2.13(c)), conditions are less favourable to passivity, which will only set in if the current density is high.

Müller's extensive experiments with 'protected' anodes (Fig. 2.13(a)) showed that the first stage in 'passivation' is often the production of crystals consisting of a simple salt of the metal. Consider an iron anode in dilute sulphuric acid. At first ferrous sulphate solution is formed over the anode surface and accumulates, since the stagnant conditions preclude its dispersal. Sooner or later the solution must here become so supersaturated with ferrous sulphate that the salt begins to crystallize out; Müller detected the tiny crystals (in this and similar cases) by means of a polarizing microscope. As the crystallization extends, an increasing fraction of the anode surface is covered up, and (for a fixed applied

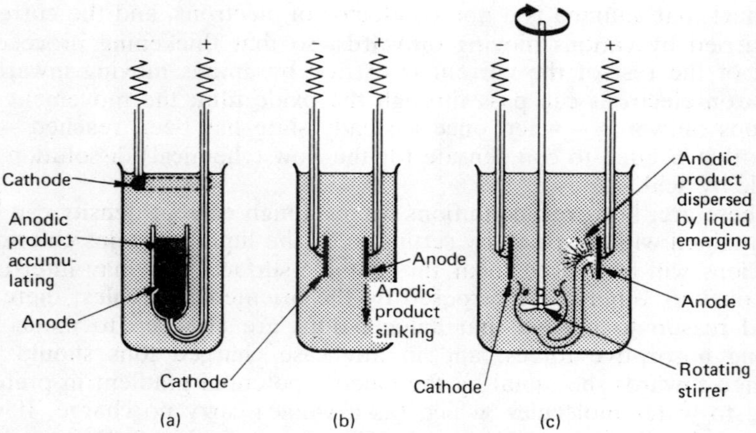

Fig. 2.13 Cells for studying anodic behaviour. (a) With horizontal anode protected from chance stirring (W. J. Müller's Method) – an arrangement favouring passivity (b) With vertical anode, allowing the heavy anodic products to sink – an arrangement less favourable to passivity (c) With a rapid movement of the liquid over the anode (Shutt and Walton's method) – an arrangement preventing passivity except at high current densities

e.m.f.) the current flowing decreases, although the current density on the portion remaining uncovered may actually increase. Since hydrogen ions are migrating away from the anode all the time, the solution in the interstices of the crystalline deposit becomes less acid. As soon as an oxide film appears over the surface, the dissolution of metal is obstructed, and the crystals of sulphate dissolve.

Passivity on a vertical anode

At low current densities, a vertical anode never becomes passive at all. The heavy anodic product sinks down (Fig. 2.13(b)) and there is no supersaturation. Above a certain limiting current density, however, passivity becomes possible. If the current density employed is well above this limit, the passivity, when once it has set in, persists indefinitely, and the passive anode behaves as an unattackable electrode, approaching platinum in behaviour and potential, although a very small amount of corrosion can be detected. If the current density has been only just sufficient to bring about passivity, we obtain fluctuations to and fro between the active and passive states. This **periodicity** has been studied by Hedges[232] and is further considered on p. 178. In general it is found that

(1) *low* current densities produce *active corrosion*;

(2) *intermediate* current densities produce *periodicity*; and

(3) *high* current densities produce *passivity*. The formation of oxide is only possible at a high potential, as shown in the Pourbaix diagram (p. 247); the high current density allows this potential to be obtained.

Influence of ions

The presence of chloride ions favours active corrosion; this is true not only of iron, but of most other metals. Generally, a metal tends to become passive if subjected to anodic action in the presence of ions which produce sparingly soluble salts, and to remain or become active in the presence of those which produce highly soluble salts. On iron, and all metals whose oxides possess a basic character and are soluble in acids, an acid reaction favours the active state, whilst an alkaline reaction will tend to maintain passivity. For metals like tungsten and molybdenum, where the main oxides have an acidic character (forming tungstates or molybdates), the reverse is true; such metals can suffer anodic corrosion in alkaline solutions sometimes at a rate corresponding approximately to Faraday's law, whereas in acid solution passivity commonly sets in.

Behaviour of ferric oxide films in acid

At one time it used to be argued that the passivity produced on an iron anode in dilute sulphuric acid could not be due to an oxide film, since – it was urged – any oxide would quickly be dissolved by the acid. It is now known that ferric oxide is not *quickly* dissolved by dilute acid.

Pryor showed that carefully prepared 'ferric oxide' was only very slowly dissolved, and that the slight dissolution observed took place at lattice defects where the Fe^{2+} ions were present, since much ferrous iron appeared in the liquid; as stated on p. 18, ferric oxide generally contains less oxygen than the formula Fe_2O_3 would imply. But, whilst the *direct dissolution* is *slow*, the *reductive dissolution* at any point where the oxide is made a cathode in acid proceeds *rapidly* and, in suitable circumstances, quantitatively, according to the equation

$$Fe_2O_3 + 6H^+ + 2e^- = 2Fe^{2+} + 3H_2O$$

Thus on iron carrying a ferric oxide film which at any point is 'leaky' owing either to actual cracks or to lattice defects penetrating the thickness, the cell

$$\text{Fe (anode)} \quad | \quad \text{acid} \quad | \quad Fe_2O_3 \text{ (cathode)}$$

is established, and the film will be destroyed around the leak-point, passing into the liquid as ferrous ions*; the gap in the film, initially small, will become extensive, and the iron, left unprotected, will quickly be attacked.

The author's[224] early experiments on iron coated with *visible* oxide may serve to clarify the situation. Freshly abraded iron was heated to produce first-order tints and tested in dilute acid; the colours quickly disappeared. This disappearance was ascribed to the cells (Fe | Acid | Fe_2O_3) set up at cracks in the film; if the explanation suggested was correct, it was felt

* Probably ferrous *ions* are formed, and pass directly into solution without intermediate production of *solid* ferrous oxide. Films on nickel (already divalent as NiO) and on aluminium (invariably trivalent as Al_2O_3) are only very slowly attacked by dilute acids.

that the destruction of the film should be prevented by the presence of an oxidizing agent. Experiments confirmed that expectation. It was found that M/100 sulphuric acid containing chromic acid (M/10) produced practically no alteration to tinted iron, although M/100 sulphuric acid, in the absence of chromic acid, caused appreciable change in five seconds, and complete destruction of the earlier colours in one minute.

Another rational method of preventing destruction was to submit the tinted specimen as a whole to anodic treatment, thus preventing local cathodic action. Iron tinted to a first-order colour and subjected to anodic treatment from the moment of its introduction into the acid, suffered no change of colour and no corrosion; the specimen behaved as a passive anode, oxygen gas being evolved. An interruption of current for a fraction of a second was possible without dispelling passivity or destroying the colour; the evolution of oxygen recommenced when the current was turned on again. With unheated specimens, an interruption of some seconds can be tolerated without loss of passivity. During such an interruption, bubbles continue to appear on the iron, even though no current is flowing; apparently there is a super-charge of high-energy oxygen, and this seems essential to the maintenance of passivity, since if the interruption continues until the formation of bubbles has ceased, the iron is found to be active; probably the oxygen-charge acts in a manner analagous to chromic acid (see above). The author has discussed the matter elsewhere. Nagayama and Cohen suggest that the supercharged layer is Fe_2O_3 with vacant cation sites and hexavalent iron ions. (*J. electrochem. Soc*., 1962, **109**, 781; see also U. R. Evans, *Chem. Ind*., 1962, 1779.) Whatever the nature of the high-energy oxygen, the destruction of the film by reductive dissolution will not occur until this oxygen-charge is used up. Long interruptions of current destroyed the colour, revealing bright iron below, and ordinary anodic dissolution began when current was again applied. An anode heated to show second-order colours behaved similarly, but interruptions of current for fifteen seconds did not destroy the film completely, although they allowed attack on the iron to begin at certain points. A series of such interruptions, with intervening periods of current flow, produced complete undermining of the film, which peeled off in curling flakes; when once out of contact with the metal, the oxide film remained unchanged in the acid for hours.

Isolation of films

The experiments just described suggested a method of isolating the invisible film which was believed to be present on untinted iron after it had been rendered passive by anodic treatment in dilute sulphuric acid at high current density. A specimen of clean (unheated) iron was subjected to anodic action in M/2 sulphuric acid at 6 volts; at first it suffered active corrosion, but soon became passive, oxygen gas appearing. If the circuit was now broken for a short time and restored, the anode was found once more to be active, and iron passed into solution; in due course passivity returned, and thus the iron could be made active and passive alternately.

The experiments were then repeated in a cell built on the stage of a microscope and furnished with a sloping anode. During the passive periods, the anodic surface remained bright; but when, after an interruption just sufficient to destroy the film at the edges and weak spots, but not elsewhere, the current was renewed, a series of horizontal shadow fringes passed upwards over the surface, due to the wrinkling of a thin surface film, which had been present in optical contact with the metal during the passive stage, but only became visible when the metal immediately below it was dissolved away. By timing the interruption at the suitable moment, it was possible to undermine the whole film, which frequently, as it peeled off from the anode, rolled up like a carpet into long tight microscopic rolls. Although mechanically flimsy, the films were chemically stable, and could survive an hour in M/12 sulphuric acid, except where particles of metallic iron adhered to them. These experiments showed that anodic passivity is due to a thin protective film, invisible on the metal, but perfectly visible when stripped from it. The rapid destruction of oxide films, and the accompanying loss of passivity, produced when metal is dipped into acid without applied current, is due *not* to the *direct dissolution* of the ferric oxide film (yielding *ferric* ions) but to its *reductive dissolution* (yielding *ferrous* ions).

Corrosion at contacts between dissimilar metals

General

When two dissimilar metals in electrical contact are introduced into a corrosive salt solution, the anodic metal will, in general, suffer more corrosion, and the cathodic less corrosion, than if they were placed unconnected in the same solution. This principle is brought out in Table 2.1 for couples of iron and a second metal, studied in early work by Bauer and Vogel.[188]

Table 2.1 Corrosion of plates of iron and a second metal in electric contact totally immersed in 1% sodium chloride solution (O. Bauer and O. Vogel)

Second metal	Corrosion (iron) mg	Corrosion (second metal) mg
Magnesium	0·0	3104·3
Zinc	0·4	688·0
Cadmium	0·4	307·9
Aluminium	9·8	105·9
Antimony	153·1	13·8
Tungsten	176·0	5·2
Lead	183·2	3·6
Tin	171·1	2·5
Nickel	181·1	0·2
Copper	183·1	0·0

Unalloyed aluminium is anodic towards most aluminium alloys which contain copper. At a junction with unalloyed aluminium, the alloy is often protected at the expense of the aluminium, although when both are disconnected, the alloy is the more rapidly corroded of the two materials. Likewise, on specimens of steel bolted to wrought iron and exposed five years in the sea at Plymouth by Friend, the steel was protected at the expense of the wrought iron, although disconnected specimens of wrought iron were corroded less than disconnected specimens of steel.

Copper as a stimulator of corrosion

Kaesche has shown that contact with copper stimulates attack on aluminium by chloride solution more than contact with platinum, which stands well above copper in the potential series (see endpaper table). (H. Kaesche, Williamsburg Conf. 1971, p. 516, esp. Fig. 2.) The idea, once held, that the potential series represents the 'order of dangerousness' of the various metals is clearly wrong; the normal electrode potential is the potential of a **film-free** metal in a solution containing *normal* concentration (or, more strictly normal activity) of its cations. Under practical conditions, metals are not film-free and cations are not usually present in normal concentration. But it is reasonable to enquire why copper stimulates corrosion more strongly than most other metals.

Doubtless oxygen is adsorbed to some extent on the surface of a relatively noble metal, although it is unlikely that a complete monolayer will be maintained. However, the adsorbed oxygen along with the outermost layer of metallic atoms constitute the start of a thin film of oxide. Even though this film is incomplete, the positions of the oxygen atoms (or ions) will presumably be such as would represent the start of a three-dimensional film, which would appear if the temperature was higher or the time available longer; the placing of the oxygen atoms (or ions) will be such as to conform with the structure of a three-dimensional oxide. If so, we should expect that on copper there will be the beginnings of a Cu_2O film with one oxygen atom for every *two* atoms of metal, whilst on nickel there will be the beginnings of a NiO film, with one oxygen atom to every *one* metal atom. The cathodic reduction of oxygen requires in effect free oxygen atoms:

$$O + H_2O + 2e^- \rightarrow 2OH^-$$

If the oxygen atoms are already far apart, the activation energy needed for the dissociation of O_2 into 2 O will be less when they are far apart in the adsorbed state than when they are close together. That is probably the reason why the cathodic reduction of O_2 to OH^- proceeds more easily on copper than on nickel or platinum. This crude statement is doubtless an over-simplification, but may serve to indicate a reason why, for reasons quite independent of the electrode potential, some metals form a more effective substrate of oxygen-reduction than others.

Stimulation by direct contact

If iron is joined to a cathodic metal like copper, the attack on the iron is usually stimulated, especially if the area of the second metal (copper) is large compared to that of the iron; for the amount of oxygen reaching a large cathode will exceed that collected by a small cathode. Serious intensification of corrosion caused by the combination of small anode and large cathode will, however, arise only when the control is 'cathodic', the current being determined mainly by events at the cathodic metal (in this case, by the rate of arrival of oxygen). If the current were limited by the existence of a nearly continuous insulating film on the anodic metal, which might occur if the anodic metal were stainless steel ('anodic control'), the cathode/anode ratio would become less important.

In cases where the control is purely cathodic, a curious result is obtained. Whitman and Russell[193] compared the corrosion of bare plates of steel with similar steel plates in which three-quarters of the surface was coated with copper. The total attack was found to be the same in both cases, but on the coated plate it was concentrated upon the bare quarter, where the intensity of attack was consequently four times as great. The reason is clear. The copper received the oxygen which (in the absence of the coat) would have reached the underlying iron; being a cathode, it was not itself attacked, but passed on the corrosion to the anode, the bare iron; in other words, the copper acted as a **catchment area** for the oxygen. The catchment area principle must, however, not be pushed too far; had 99·9% of the iron been coated with copper, the attack on the tiny bare area probably would not have been 999 times as intense as that of an uncoated plate, since in this extreme case, the anodic polarization, or the resistance of the bottleneck approach to the anode, would almost certainly have restricted the current.

Many cases of serious attack through contact between dissimilar metals have been reported from industry and structural engineering. Aluminium alloys – resistant to many liquids owing to the presence of an alumina film which is itself too bad a conductor to act efficiently as a cathode towards the metal exposed at discontinuities in it – may become rather rapidly attacked by water at contact with copper. The attack produced by a globule of mercury resting on wet aluminium is due to a different cause, the film formed on amalgamated aluminium being non-protective; the breaking of a mercury thermometer in an aluminium vessel at a brewery has led to rapid perforation.

Indirect stimulation

Serious consequences are sometimes met with where copper pipes and galvanized tanks or cisterns have been used together in a domestic water system. Rapid failure of the galvanized vessels has often been reported, and was at one time attributed to a cell with the copper as cathode and with the zinc or iron as anode. Kenworthy[205] has shown that the trouble is due *not* to this *macroscopic* cell, but to numerous *microscopic* cells. Waters containing free carbonic acid act on copper, at a rate insufficient to damage the copper pipes, but sufficient to take up traces of copper bicarbonate, and thus deposit metallic copper on the galvanized iron (metallic zinc and copper salt give metallic copper and zinc salt). Little corrosion cells of the type zinc | copper (and later iron | copper) are set up, with disastrous results. Comparing two housing estates supplied with the same water, Kenworthy states that on one estate, where copper pipes were used with galvanized tanks, 50% of the installations failed in four years, although on the other estate, where both pipes and tanks were galvanized, there were no failures. He mentions two other neighbouring estates *both* fitted with copper pipes and galvanized tanks; on one, employing a water containing 4·1 ppm free carbonic acid (which took up 0·32 ppm of copper), *every* installation failed in four years; on the other estate, supplied with a less cuprisolvent water (1·1 ppm free carbonic acid, taking up only 0·03 ppm of copper), no failures were reported. He suggests as a remedy reduction of the cuprisolvency by lime addition or by aeration, which would remove much of the free carbonic acid.

A more recent method of preventing attack on copper by a water which is normally cuprisolvent, depending on the installation of sacrificial aluminium protectors was mentioned on p. 49, in connection with the pitting of copper. However, the type of trouble discussed by Kenworthy is not often met with today. The choice of materials has changed; in cases where the hot water cylinder is of copper, the pipes are generally copper also.

Variable Polarity

Despite the warning already given that the potential series does not represent an estimate of dangerous character, it will in most cases serve to indicate the relative polarity of two metals. Noble metals like silver and copper are cathodic to iron in nearly all liquids; base metals like zinc and aluminium are anodic to iron, at least if freshly abraded. Metals like tin and lead, which stand close to iron in the potential series (endpaper table), display a variable polarity; the presence of an oxide film on the one metal may render it cathodic, whereas the presence of a substance in the water forming *complex* ions with one metal may make it anodic; when a small current is passing in one direction, vigorous aeration of the metal functioning as anode often reverses the flow.

Tin behaves as cathode to iron in hot distilled water, but as anode in solutions of many organic acids, such as occur in fruit juice. Thus fruit preserved in canisters of tin plate (steel coated with a thin tin layer usually interrupted by pinholes) has little action on the steel exposed. The tin gives protection in two ways, (1) by direct cathodic protection at the expense of the tin, which suffers enhanced attack, (2) by the tin salts thus formed retarding the attack on the steel, possibly by fixing in stable form the sulphur which would stimulate corrosion. The situation is, however, complicated; it is discussed elsewhere. (1960 vol., p. 648; 1976 vol., p. 308.)

Although freshly abraded zinc is at ordinary temperatures strongly anodic to iron in all ordinary waters or solutions, the polarity may be reversed at high temperatures. Kenworthy and Smith found that, whereas a plate of galvanized (zinc-coated) iron, placed in cold water, suffers no corrosion of the iron exposed at the cut edges, a similar protection of exposed iron does not occur in hot water; this should be remembered in planning hot-water systems, where it is clearly important that the steel basis should not be exposed at cut edges or elsewhere. Actually, zinc often becomes cathodic to steel at high temperatures – a reversal attributed by some (but not all) authorities to the loss of water from zinc hydroxide, giving zinc oxide – a fairly good electronic conductor, especially at elevated temperatures.

Crevice corrosion

General

Although no surprise has ever been occasioned by corrosion occurring at contacts between dissimilar metals, considerable mystification has at times been felt at the intense attack sometimes noticed where non-conducting substances pressed against metal. The explanation is to be found in the fact that where a 'foreign body' (whether a glass rod, a stone, or a piece of fabric) rests against metal, an inaccessible cranny is formed in which replenishment of certain substances can take place only slowly.

Lenticular glass on lead

One case studied by the author[208] is that of a lenticular glass resting on a horizontal lead plate below the surface of potassium chloride solution (Fig. 2.14); alternatively a lenticular piece of cast lead can be placed in a porcelain dish containing the same solution. Over an

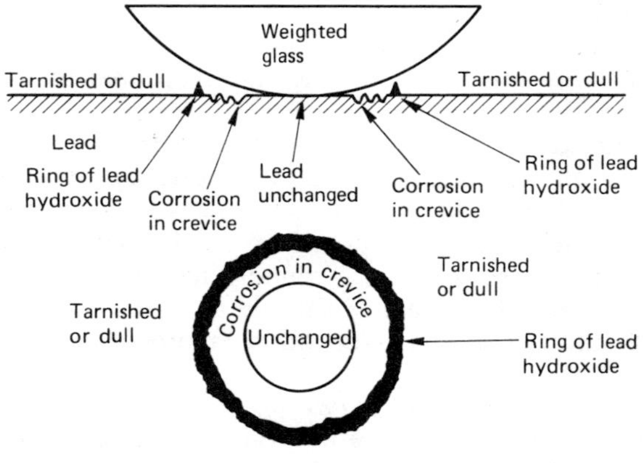

Fig. 2.14 Corrosion produced on lead in a chloride solution where a lenticular glass rests upon it

annular area around the centre the lead suffers marked attack; beyond this comes a ring of lead hydroxide, outside which the lead, although dulled or tarnished, remains unattacked. This is explained by the fact that oxygen is less easily replenished in the annular crevice than elsewhere, and consequently anodic attack occurs here, yielding lead chloride, which, interacting with the sodium hydroxide formed at the better aerated cathodic regions, produces the ring of lead hydroxide. Other metals show similar phenomena. A brass plate leaning against a glass rod in sodium chloride solution often suffers a line of corrosion along the cranny between glass and metal.

Crevice corrosion in liquid containing an inhibitor

In the case of iron, crevice corrosion is most easily produced when enough inhibitor has been added to the liquid to bring conditions near to the boundary dividing passivity from corrosion. For instance, sodium carbonate solution, if very dilute, produces corrosion, whilst higher concentrations lead to passivity (cf, p. 187). On a certain steel used in the author's laboratory by Mears,[209] the borderline came about 0·025M; when, in liquid of that concentration, a series of horizontal glass rods was laid resting on a series of horizontal steel rods (the axes of the two series being at right angles to each other), it was found that corrosion was generally produced at the crevice between steel and glass, but not elsewhere. The reason is that inhibition involves a consumption of inhibitor. Since at the cranny the inhibitor will be replenished less quickly than elsewhere, the probability of corrosion developing in the cranny area is greater than that of development at any area *of the same size* outside the crevice. Nevertheless, the corrosion does not invariably appear in the cranny. It will not occur if there chances to be no susceptible spot within the cranny area; nor will attack develop within the cranny if there happens to exist, just outside the cranny, an *exceptionally* susceptible spot which develops anodic attack (despite the good supply of inhibitor), and then confers cathodic protection on the cranny area itself by altering the potential distribution in a manner unfavourable to the incipience of attack.

Mears found attack at points where steel rods rested on other steel rods (arranged at right angles), and the frequency of attack was then actually greater than in the case of glass rods laid on steel. This was an important result, since it disposed of an alternative explanation of corrosion at steel–glass contacts – namely that films cling to glass in preference to steel, thus leaving the metal bare; there is no reason why a film should desert one steel rod in favour of another.

Attack at crannies in industrial plants

Corrosion is sometimes noticed in badly designed chemical plant, particularly at places where heating coils or the like come close together. Austenitic stainless steel, a substance wonderfully resistant when a trace of oxygen is present, is liable to break down under anaerobic conditions, such as may exist at inaccessible corners. In designs likely to produce attack at anaerobic points, it may sometimes be better to choose a material like Monel metal, the corrosion of which diminishes with diminishing oxygen concentration, rather than stainless steel, where the passivity may break down when the last trace of oxygen is removed. Clearly for highly aerated conditions, the opposite choice may be indicated.

Corrosion at inaccessible anaerobic spots may in some cases be due to sulphate-reducing bacteria (p. 114), which flourish at such places.

Cranny corrosion in marine environments

Aluminium alloys are liable to suffer crevice attack in marine environments, although resistant to attack on the free surface; this is attributed by most authorities to differential aeration; the voluminous character of the corrosion product (if precipitated within the cranny) may cause serious bulging.

Cranny corrosion in structural steelwork exposed to the atmosphere

Probably the main reason why atmospheric corrosion occurs rapidly in crevices left between steel plates is that moisture is retained there long after it has dried up on the external surfaces. The dangerous feature of such crevice corrosion is that the rust occupies a much larger volume than the steel destroyed in producing it; the volume increase is partly due to enmeshed water present even in rust which looks 'dry'. The effect is similar to that of a wedge driven in between the plates, forcing them apart. A case is known of the failure of a line of rivets holding two plates together on a bridge; these had been spaced too widely apart, so that they snapped one after the other by the lever action of the rust formed in the crevices. It is most important to fill up all such crevices; red lead paste is suitable for steel, and certain mixtures containing zinc chromate (or other chromate) for aircraft alloys.

3

Corrosion by Acids, Alkalis and Pure Water

Action of non-oxidizing acids

Mutual replacement of metals

The attack upon a metal by a non-oxidizing acid is essentially a replacement of hydrogen by the metal in question. It may be better understood if some other cases of mutual replacement are first considered.

When a base metal is placed in the solution of a salt of a more noble metal, i.e. one standing above it in the potential series (endpaper table) mutual replacement usually occurs. Thus metallic iron placed in a copper sulphate solution becomes coated with metallic copper, iron sulphate appearing in solution; zinc placed in lead acetate produces a tree of metallic lead; copper precipitates mercury from a salt, and so on. Exceptions are met with where a resistant oxide film is present. If the normal potentials of the two metals stand close together, the replacement may be incomplete and reversible. Tin throws out metallic lead if placed in a lead salt solution, but lead deposits tin from a tin salt solution; in both cases, replacement should cease when the liquid contains tin and lead in the ratio 2·98:1, since the potential of both metals is the same, and the equilibrium

$$Sn^{2+} + Pb \rightleftharpoons Pb^{2+} + Sn$$

is established.*

Hydrogen evolution

If a metal is placed in acid, contained in a closed vessel, replacement of hydrogen ions by metal ions may usually be expected to proceed until equilibrium is established, which in the case of a divalent metal can be written

$$M + 2H^+ \rightleftharpoons H_2 + M^{2+}$$

This will occur when the concentration of M^{2+} ions and of H_2 molecules has become such as to make the potential $M \mid M^{2+}$ equal

*It is generally believed that some of these cations are in close combination with adjacent water molecules, and some authors write for example Sn^{2+}. Aq. For simplicity this aquation sign is omitted throughout this book.

to the potential $H_2 \mid H^+$ at the metallic surface. The necessary concentration of hydrogen in the liquid involves a certain pressure of hydrogen in the gas space above the liquid. In the case of the nobler metals the pressure involved is low, and equilibrium being soon established, all reaction ceases. But for the baser metals, equilibrium will only be reached when the pressure has become high (sometimes 'astronomically' high). If the walls cannot stand this pressure, or if the vessel is open to the air, equilibrium will be impossible, and evolution of hydrogen will continue (rapidly in bubbles from the baser metals, or perhaps by slow diffusion through the liquid to the gas space from the intermediate metals).

In this corrosion process, the reactions are

$$M \rightarrow M^{2+} + 2e^-$$

at the anode, and

$$2H^+ + 2e^- \rightarrow H_2$$

In general, the cathodic reaction will take place at any points of especially low overpotential, where molecular hydrogen can be formed most readily (see p. 291). The cathodic and anodic reactions may sometimes occur at contiguous points (or even at the same point); this presents a contrast to corrosion of the oxygen-reduction type, where the anodes and cathodes are relatively large areas (the former bare and the latter usually oxide-covered).

Noble metals

In the case of the metals at the top of the potential series, the formation of a very small concentration of cations will serve to bring about the equilibrium

$$M + 2H^+ \rightleftharpoons H_2 + M^{2+}$$

and the passage of metal into liquid will cease long before the hydrogen concentration has approached the saturation value corresponding to 1 atmosphere. Thus metals such as gold, silver or copper do not commonly liberate hydrogen from acids. Apparent exceptions, however, occur in solutions which produce complex ions, so that the equilibrium is not reached. If copper is placed in boiling hydrochloric acid, the vapour contains a little hydrogen, chlorocuprous acid $HCuCl_2$ being formed, with copper locked up mainly as the complex anion $[CuCl_2]^-$. Potassium cyanide solution, despite its alkaline reaction, attacks copper, forming a complex cyanide with the anions $[Cu(CN)_2]^-$. In dilute hydrochloric or sulphuric acid, copper is only attacked if air or an oxidizing agent is present*; sulphuric acid con-

*A specimen of copper, partly plated with platinum, was sealed in a tube containing 0·25M sulphuric acid under anaerobic conditions, by the late Professor O. P. Watts of Wisconsin University. This he kindly presented to the Author. After 25 years, the liquid remained colourless, and there was no sign of attack on the metal.

taining potassium chlorate or permanganate produces relatively rapid attack.

Intermediate metals

Tin and lead, which stand below hydrogen in the potential series, might be expected to react freely with dilute hydrochloric acid in the absence of oxygen, liberating hydrogen. Actually, however, the hydrogen overpotential of these metals is high (Table A.3, p. 291), and a very small cathodic current density would suffice to depress the potential of hydrogen evolution to the potential provided by tin or lead. Thus, in the absence of air, corrosion is slow in dilute hydrochloric acid; hydrogen evolution can be increased by contact with platinum black, which provides a cathodic surface of low overpotential. Towards sulphuric acid, lead is immune, owing to the sparing solubility of lead sulphate (in hot concentrated acid, the film sometimes breaks down). In hot concentrated hydrochloric acid, lead suffers appreciable corrosion; still quicker is the attack by hydrobromic acid, which can be obtained at high concentrations (14M), and builds complex anions containing lead.

Iron

The first metal in the potential series to expel hydrogen rapidly is iron. Its normal potential stands well to the negative side of that of hydrogen, whilst the overpotential of hydrogen evolution on an iron cathode is fairly low; thus, although the overpotential attending the anodic dissolution is appreciable, the working e.m.f. of the cell

Iron | Acid | Hydrogen

is sufficient to cause a lively effervescence. The attack is accelerated if a trace of platinum salt is added to the acid, which deposits platinum on the metal and thus provides a cathode of still lower overpotential. Foreign particles in the iron can act in the same way, e.g. graphite; grey cast iron is attacked by dilute sulphuric or hydrochloric acid much more quickly than relatively pure iron, and here the addition of platinum salt causes practically no further acceleration. Dead mild steels containing most of the cementite in *massive* form were found by Hoar and Havenhand[316] to be attacked somewhat more quickly (*ceteris paribus*) than those where the cementite was finely dispersed. The massive cementite probably provides cathodic particles from which hydrogen can readily be evolved; dispersed cementite has no such effect, possibly because the very small particles are quickly undermined and carried away by the hydrogen bubbles. There are, however, more important causes for the variation in the behaviour of different steels. One factor favouring rapid corrosion is the presence of sulphur, which in certain forms catalyses the (otherwise sluggish) anodic reaction. Sulphur present as iron or manganese sulphide is dissolved by the acid, yielding hydrogen sulphide, the real catalyst. If sufficient copper is present in the steel, this appears to 'fix' the sulphur, so that the attack on the metal is retarded. Opinion seems divided as to whether the sulphide is present in the steel as stable copper sulphide, or whether the hydrogen sulphide formed by attack on iron or manganese sulphide is precipitated by the copper salts which will have appeared in the liquid. Hoar and Havenhand found that if the ratio of copper to sulphur exceeded 1·38, the attack by citric acid became slower. Tin salts added to the solution produced the same effect, possibly by precipitating the hydrogen sulphide as stable tin sulphide.

The presence of copper (up to about 0·3%) in structural steel increases its resistance towards the acid rain water and condensed moisture typical of industrial districts (see p. 104). More than one explanation has been suggested, but the fixation of sulphur seems a probable one.

Zinc

Zinc stands far below hydrogen in the potential series, and if brought into contact with platinum below dilute acid will pass readily into solution, hydrogen being eliminated from the platinum. In absence of such a contact the high hydrogen overpotential of zinc makes the attack on *pure* zinc slow; commercial zinc is also attacked slowly at first, but, as impurities (mainly lead and iron) pass into solution and are reprecipitated as a black metallic sponge, the corrosion and evolution of hydrogen are speeded up. This black sponge presents a large surface area and acts as an effective cathode, even though its main constituent is lead, which has a high overpotential in the massive state; possibly the lead sponge provides a substrate for a small amount of iron, on which hydrogen can be liberated without much overpotential. If the black sponge is brushed away, the evolution of gas slows down. The author[318] has detected currents flowing between the zinc as anode and the sponge as cathode. The period of slow dissolution, which elapses before the attack becomes rapid, is called the **induction period**.

Naturally, those minor constituents are most effective which have low overpotentials. The curves of Fig. 3.1 due to Vondráček and Izák-Križko[310] show that copper, iron and antimony cause most acceleration, the reaction rate increasing with time, whereas on relatively pure zinc the rate remains unchanged. Only reprecipitated impurities work in this way; tin in solid solution at first retards the attack, but, after it has begun to dissolve in the acid and has been redeposited in metallic form, the reaction rate rises above that of unalloyed zinc. Lead also retards attack at first, but later hardly influences the rate, its overpotential being higher than that of zinc. Mercury greatly retards attack, raising the overpotential.

The attack of acids on zinc and on iron is greatly accelerated by the presence of oxidizing agents such as potassium nitrate. Unlike iron, zinc evolves hydrogen from sodium hydroxide solution, forming 'zin-

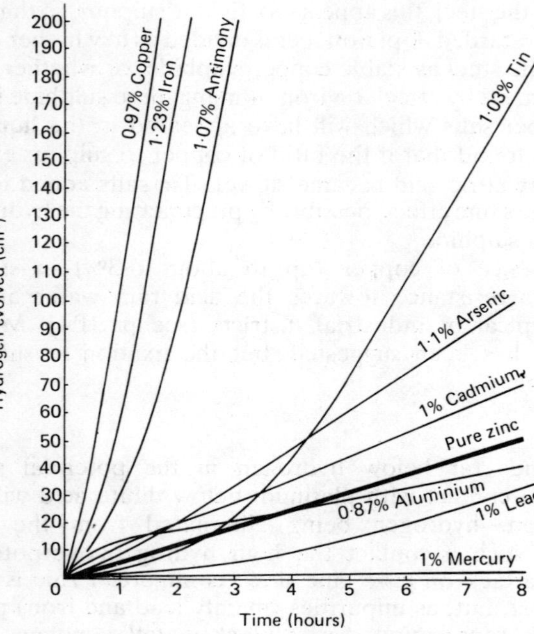

Fig. 3.1 Effect of minor constituents on the corrosion of zinc by 0·25M sulphuric acid (R. Vondráček and J. Izák-Križko)

cates' such as Na_2ZnO_2; despite the low hydrogen ion concentration (which will depress the cathode potential), the fact that zinc enters solution as complex anions $[ZnO_2]^{2-}$ or $[HZnO_2]^-$, depresses the anode potential, and consequently the anode potential, and consequently a considerable e.m.f. remains to drive the reaction.

Although in the author's work on the corrosion of solid zinc by acids there is definite evidence that the anodic and cathodic reactions are proceeding at different places, the situation might be different on a surface which is chemically and physically uniform. Wagner and Traud[881], working on zinc amalgam, reached the conclusion that the cathodic and anodic reactions can take place anywhere on the whole surface; each point can be anode or cathode at different moments – perhaps even at the same moment. This conclusion is easily accepted. Even in such a case, however, it is still legitimate to divide the gross reaction

$$Zn + 2HCl \rightarrow ZnCl_2 + H_2$$

into anodic and cathodic components:

$$Zn \rightarrow Zn^{2+} + 2e^- \text{ (anodic)}$$
$$2H^+ + 2e^- \rightarrow H_2 \text{ (cathodic)},$$

the potential measured being the **mixed potential** – namely, the level at which the anodic and cathodic reactions can proceed at equal velocities.

The question arises as to whether the concept of mixed potential (with the whole surface acting as anode and cathode) is valid for solid metals. Possibly in the case of very pure metals presenting a surface which is physically as well as chemically uniform, the Wagner–Traud picture is still appropriate. It has been applied in Hackerman's laboratory[881] to the corrosion of iron in acid in the presence or absence of oxidizing agents, and especially to the action of amine restrainers. The conclusion reached was that the picture is valid for fairly pure iron, but that it might not be appropriate for steel 'where secondary phases may have considerable influence'.

Aluminium

As explained on p. 55, sesquioxides (M_2O_3) are only very slowly dissolved through direct attack by acids. The 'reductive dissolution' which causes such a rapid destruction of ferric oxide films cannot affect alumina films, since aluminium exists in one oxidation state only. Consequently aluminium, introduced into acid after air-exposure, exhibits a period of induction, which (in contrast to that met with on zinc) is due to the slow dissolution of alumina. On very pure aluminium the induction period may be measured in days; on the impure metal, it is much shorter, possibly because the sites of impurities constitute defects in the film. Straumanis[320] has pictured both the anodic and cathodic actions taking place at pores in the skin; pores filled with acid provide anodes, whilst those containing residual metallic impurities furnish cathodes. If pure aluminium is etched in concentrated hydrochloric acid and then placed in more dilute acid, there is no period of induction, the film having been destroyed by the stronger acid.

Aluminium is attacked by caustic alkali without the induction period found in acids, although the final attack is slower; alkalis seem to destroy the protective alumina film very quickly, possibly because OH^- ions are positively adsorbed.

Magnesium and alkali metals

Metals at the 'base' end of the potential series evolve hydrogen in absence of acid or alkali. The violent reaction of potassium or sodium with water is well known. Magnesium is remarkably inert towards pure water, doubtless because its oxide and hydroxide are sparingly soluble, so that film formation is possible. Attack is greatly stimulated by chloride, but little increased by sulphate or nitrate; possibly anions containing oxygen produce fresh oxide by anodic action at places where the film is faulty, for reasons analogous to those operating in the anodizing of aluminium. The effect of impurities in increasing attack by chlorides becomes marked if the impurity content exceeds a certain 'tolerance limit' (0·017% for iron, but only 0·0005% for nickel); below this the hydroxide formed by interaction between the anodically produced magnesium chloride and the sodium hydroxide formed on the cathodic specks of impurities seems to isolate the specks, so that attack ceases; above the tolerance limit, the cathodic specks are closer together and fresh ones can be laid bare by anodic attack on the metal before the original ones are covered up.[323]

Attack by nitric acid

Base metals

Although free hydrogen appears when *very dilute* nitric acid acts on magnesium, the majority of base metals, on dissolving as nitrates, yield (in lieu of free hydrogen) hydrogen-rich compounds formed by interaction between cathodically discharged hydrogen ions and the nitric acid. These include ammonia (NH_3) and hydroxylamine (NH_2OH), both present as nitrates, along with nitrogen (possibly produced by decomposition of ammonium nitrite) and nitrous oxide (N_2O, possibly produced from ammonium nitrate).

Noble metals

In the corrosion of noble metals, the reduction of nitric acid does not proceed so far. The anodic reaction (for a divalent metal) may as usual be written

$$M \rightarrow M^{2+} + 2e^-$$

but there is less agreement about the cathodic reaction. The following line of argument has been put forward by the author.[326]

The complicated equations published in some textbooks must really take place in steps, and it seems logical that the cathodic reduction of nitric acid should come about in simple stages thus:

$$(1)\ HNO_3 + H^+ + e^- \rightarrow NO_2 + H_2O$$
$$(2)\ NO_2 + H^+ + e^- \rightarrow HNO_2$$
$$(3)\ HNO_2 + H^+ + e^- \rightarrow NO + H_2O$$

each of which might also occur in the ionic form:

$$(1a)\ NO_3^- + 2H^+ + e^- \rightarrow NO_2 + H_2O$$
$$(2a)\ NO_2 + e^- \rightarrow NO_2^-$$
$$(3a)\ NO_2^- + 2H^+ + e^- \rightarrow NO + H_2O$$

Of these six equations, (2a) involves only two particles (including the electron), and may be regarded as taking place more easily and frequently than the others which involve the meeting of three or four particles. So soon as a trace of NO_2 has been formed by (1) or (1a), it will quickly be reduced to NO_2^- by (2a), which will combine with hydrogen ions to give nitrous acid (HNO_2). This can readily react with nitric acid to regenerate *twice the original quantity* of NO_2 by the reaction

$$(4)\ HNO_2 + HNO_3 \rightarrow 2NO_2 + H_2O$$

which also involves only two particles and is therefore likely to occur easily. The $2NO_2$ will then rapidly be reduced by (2a) and rapidly re-oxidized by (4), yielding $4NO_2$, and so on. *At each cycle*, the quantity of the NO_2 (and HNO_2) involved will be *doubled*, and the reaction, which at the start involves only a few molecules, will soon become violent, involving large numbers of molecules, provided that the NO_2 (and HNO_2) remain on the spot. The increasing violence of this 'autocatalytic' reaction will be obtained at crevices where one piece of metal rests on another, or at the cranny formed where the metal rests against glass. At such points the products will not easily be dispersed, and the action will increase in violence (and with it the anodic attack on the metal which must keep pace with the cathodic reactions), until the loss of HNO_2 through other reactions exactly balances the increase due to the cycle constituted by (2a) and (4).

One way in which HNO_2 can be destroyed is through reduction to nitric oxide (NO), by (3) or (3a). In the steady state when the total HNO_2 ceases to increase, the destruction of one molecule by (3)

$$HNO_2 + H^+ + e^- \rightarrow NO + H_2O$$

must balance the *net* production of one molecule by (4) and (2)

$$HNO_2 + HNO_3 \rightarrow 2NO_2 + H_2O$$
$$2NO_2 + 2H^+ + 2e^- \rightarrow 2HNO_2$$

In the steady state, the total change is obtained by adding these three equations, giving

$$HNO_3 + 3H^+ + 3e^- \rightarrow NO + 2H_2O$$

Alternatively, especially if the volume of liquid is small, so that the reaction raises the temperature, nitrous acid may be destroyed by decomposition to NO and NO_2, which will then regenerate *half* the original quantity of HNO_2. Thus the *net* loss of one molecule by the cycle

$$2HNO_2 \rightarrow NO + NO_2 + H_2O$$
$$NO_2 + H^+ + e^- \rightarrow HNO_2$$

must balance the *net* gain of one molecule by the cycle

$$HNO_2 + HNO_3 \rightarrow 2NO_2 + H_2O$$
$$2NO_2 + 2H^+ + 2e^- \rightarrow 2HNO_2$$

Adding these four equations, we *again* arrive at

$$HNO_3 + 3H^+ + 3e^- \rightarrow NO + 2H_2O$$

Thus the same result is obtained whether HNO_2 is lost by reduction, decomposition or both. This, the final sum of the reactions at the cathode in the steady state, must balance the electrical transfer of the anodic change

$$M \rightarrow M^{2+} + 2e^-$$

To obtain electrical balance, it is necessary to double the cathodic equation, treble the anodic equation and add them together, so that the $6e^-$ transfer cancels out, giving

$$2HNO_3 + 6H^+ + 3M \rightarrow 3M^{2+} + 2NO + 4H_2O$$

To preserve electric neutrality in the liquid, $6(NO_3^-)$ may be introduced on each side, so that the reaction can be written

$$2HNO_3 + 6H^+ + 6NO_3^- + 3M \rightarrow 3M^{2+} + 6NO_3^- + 2NO + 4H_2O$$

or in non-ionic language

$$3M + 8HNO_3 \rightarrow 3M(NO_3)_2 + 2NO + 4H_2O$$

– the equation usually printed in textbooks for the action of nitric acid on copper (where M = Cu).

We now understand why the action of nitric acid on such metals proceeds most quickly at crevices. If the products of reaction are dispersed, the autocatalytic cycle will fail to develop the same violence, and accordingly rapid stirring of the liquid is found to slow down the attack on copper or silver. A substance which breaks the autocatalytic cycle should have a similar effect; urea reacts with and destroys nitrous acid, a key substance in the cycle, and it is found that the addition of urea to nitric acid greatly retards the corrosion of metals such as copper. These autocatalytic characteristics are confined to the noble metals; Hedges states that stirring, which greatly retards the action of nitric acid on silver or copper, stimulates that on tin, zinc and magnesium. (E. S. Hedges, *J. chem. Soc.*, 1928, 969; 1930, 561.)

Iron

Particularly interesting is the action of nitric acid on iron. At low concentrations, the attack is most violent, and the behaviour is intermediate between that of the two classes of metals mentioned above. The metal is attacked with greater violence in crevices than elsewhere, but the products, besides nitrogen peroxide (NO_2) and nitric oxide (NO), include nitrous oxide (N_2O), nitrogen and ammonia. The behaviour is altered by cold-deformation which, by distorting the lattice and producing an unstable condition, makes the iron behave more like zinc. The proportion of $N_2 + NH_3$ to $NO + NO_2$ increases if the metal contains internal stresses.

The powerful oxidizing action of nitric acid will cause violent attack only if the anodic reaction can keep pace with the cathodic reduction of the nitric acid. Passivity will set in where that condition is not fulfilled; if the supply of iron atoms at the surface of the metal possessing sufficient energy to cross the positive part of the phalanx formed by oriented $O\text{–}NO_2$ ions is insufficient to provide the current fixed by the cathodic reduction, they are likely to get no further than the negative O^- zone, so that the product is iron oxide with a slight increase of acidity. This is shown in Fig. 2.12(e). It occurs in concentrated acid, which produces a slight attack on the iron just after introduction, but soon renders it passive and inert. In dilute acid, there is violent attack, unless chromium has been introduced into the iron, as in stainless steel.

At intermediate concentrations, behaviour depends on past history. Iron which has first been dipped in concentrated acid can resist weaker acid – such as would attack iron which had not been pretreated in concentrated acid. In acid of borderline concentration, contact with zinc (i.e. cathodic treatment) can cause passive iron to become active, whereas contact with platinum (anodic treatment) may render active iron passive.

The passivity of iron in nitric acid is very liable to break down, especially at a water-line. The author[328] found that pure iron foil, dipped in concentrated nitric acid, withdrawn and allowed to drain suffered no attack at first; then suddenly a violent reaction would set in, usually at the top of the wetted area, and travel rapidly downwards, until the whole surface was evolving bubbles. Apart from the usual factors which cause protective films to fail at a water-line (p. 197), there is here a special mechanism involved. When once attack has set in at a point from which oxides of nitrogen can escape, the ratio N_2O_5:H_2O falls, which must favour corrosion as opposed to passivity. Now an escape of gas is easiest at a water-line, and it is here that violent reaction so often sets in.

In another respect, the attack of nitric acid on iron differs from that on other metals. Any nitric oxide (NO) formed by reduction of nitrous acid, instead of escaping from the system, can be retained by combination with ferrous salts present, forming brown nitroso-compounds, such as $Fe(NO_3)_2.NO$, and remains available to enter an autocatalytic cycle. The brown coloration often seen on iron during experiments involving nitric acid is due to such compounds.

Choice of materials for chemical works

Resistance to acids

In choosing materials to withstand acids or alkalis, a metal can sometimes be selected which forms a sparingly soluble compound with the principal anion. Thus lead is often used to withstand sulphuric acid, silver to resist hydrochloric acid, whilst magnesium (which reacts violently with most acids) has been recommended for handling hydrofluoric acid (a reagent which attacks many otherwise resistant materials); the behaviour of the three metals is connected with the low solubilities of lead sulphate, silver chloride and magnesium fluoride.

In cases where a metal forms oxides which possess acidic but no marked basic properties, the action of acids is likely to produce an oxide film rather than a soluble salt, and good resistance may be expected. Thus tantalum, molybdenum and tungsten form well-defined tantalates, molybdates and tungstates, but the sulphates of these metals are unknown or unstable.

Accordingly, all three metals offer considerable resistance to non-oxidizing acids; tantalum, for instance, has proved useful for handling hot hydrochloric acid.

Non-metals, the oxides of which possess no basic properties, naturally suggest themselves as acid-resistant materials, but their physical properties are generally unsuitable. However, by introducing silicon into cast iron, alloys are obtained which, although brittle, can be made into acid-proof castings suitable for use at chemical works. For handling sulphuric acid, iron containing about 14·2% silicon is generally suitable.

It might be expected that for handling concentrated nitric acid, unalloyed iron or steel, which usually becomes passive in that liquid, might be used. However, although small pieces of iron can usually be kept in concentrated nitric acid without attack, there is always a small chance of breakdown, and, with large areas, the probability of attack – especially at surface imperfections or in the event of scraping – becomes serious. This may be one of those cases where – for statistical reasons discussed on p. 261 – laboratory experiments provide results in apparent conflict with large scale behaviour. For quite different reasons, iron or steel cannot generally be relied upon as a safe material for the concentrated acid; if, for instance, accidental dilution occurs locally (e.g. from condensed moisture flowing down into the acid), attack may start and spread. If, however, sulphuric acid (which depresses the activity of any water) is present mixed with the nitric acid, ordinary steel can generally be used in contact with the mixture, at least at high concentrations; it is advisable to keep the sulphur content of the steel low and the carbon fairly high.

Sulphuric acid, without nitric, renders iron passive if sufficiently concentrated, being itself an oxidizing agent. Iron and steel vessels are frequently used for handling the acid above 70%; there is a danger, however, of absorption of moisture from the air, and if the concentration falls below 60% attack may become serious.

In the case of normal metals (using the classification of p. 52) the oxidizing character of concentrated sulphuric acid leads to attack rather than to passivity. Copper, which the dilute acid attacks only in presence of oxygen, reacts readily with the hot concentrated acid. The reaction clearly takes place in steps, but the final result is usually written

$$Cu + 2H_2SO_4 = CuSO_4 + SO_2 + 2H_2O$$

Alloys based on nickel

Among materials which resist acid, alloys with nickel as main component are prominent. This is probably due to the fact that such alloys become covered with NiO, which, unlike Fe_2O_3, cannot suffer reductive dissolution. There are, of course, other metals, like aluminium, forming oxides unlikely to suffer reductive dissolution; but these are mostly essentially reactive elements which would be quickly destroyed if the film were to fail in some other way; nickel, in contrast, stands fairly high in the order of electrode potentials.

An early example of a resistant nickel alloy was **Monel metal**, made by direct reduction of a nickel–copper alloy without separation of the two main metals; the composition is about 67% Ni, 30% Cu, with small (but carefully controlled) quantities of other elements, such as Mn, Fe, C and Si. Synthetic alloys, made to the same composition, sometimes behave differently, apparently because the carbon is in a different condition; graphite flakes promote attack.

Naturally, experimental work has long been carried out to introduce new alloys, and the series known in the U.S.A. as Hastelloys and in the U.K. as Langalloys have long been used where resistance to HCl is needed; they contain Mo as well as Ni. Hastelloy C, which resists HCl even in presence of oxygen and chlorine, contains 58% Ni, 17% Mo, 14% Cr, 6% Fe and 5% W. Not all Hastelloys contain molybdenum; Hastelloy D is a nickel-silicon alloy, analagous to silicon iron.

For resisting H_2SO_4, smaller contents of Mo suffice. Interest attaches to an alloy with 9·5% Si, 2·8% Ti, 2·5% Cu and 3·0% Mo. The resistance depends on the structure. Two phases are generally present, and in boiling 28% H_2SO_4, serious attack occurs if the γ-phase provides continuous paths into the interior; in the corrosion cell formed by the α-combination, the γ-phase is the anode in 28% acid, but the α-phase is preferentially corroded by boiling 80% and 94% acid. (W. Barker, T. E. Evans and K. J. Williams, *Brit. corr. J.*, 1970, **5**, 76.) Another alloy containing 25% Cr and 10% Mo was originally introduced for high-temperature situations, such as the flame-tubes of gas-turbine engines; later it had other applications, since it was found to resist acids. (A. C. Hart, *Brit, corr. J.*, 1972, **7**, 105.)

Alloys based on cobalt

Recently there has been much development of cobalt alloys. They have proved useful for surgical implants and other medical purposes. (T. H. Devine and J. Wulff, *J. electrochem. Soc.*, 1976, **123**, 1433.) Resistance to oxidation is also a valuable feature. (C. M. Chen, A. Barthoff and E. D. Verink, *J. electrochem. Soc.*, 1976, **123**, 245C; M. E. Dahshan, D. P. Whittle and J. Stringer, *Corr. Sci.*, 1976, **16**, 77, 83.)

Stainless steel

The ferric oxide film present on unalloyed iron provides no protection against dilute sulphuric acid, being quickly destroyed by reductive dissolution, as explained on p. 55. If sufficient chromic acid is present, this is reduced in preference to ferric oxide at any imperfection in the film, and at the elevated potential maintained the anodic process will repair the gap with fresh oxide, preventing any attack on the metal. If the film contains chromium (an element in which the divalent state is unstable) oxygen can play the same part as chromic acid; a reduction of Cr_2O_3 to give Cr^{2+} ions would demand a potential more depressed than the reduction of oxygen. An alloy rich in chromium is found to resist dilute sulphuric acid, provided that oxygen is present. The best results are obtained in a one-phase austenitic alloy, since phase-boundaries are liable to be a source of weakness. The alloy commonly used contains 18% of chromium along with 8% nickel, added to stabilize the austenitic condition. An important use of this well-known 18/8 alloy is for resisting atmospheric corrosion (p. 108), but it shows satisfactory stability towards oxidizing acid, including dilute nitric acid which would violently attack ordinary iron; it even resists dilute sulphuric acid, provided that oxygen is present. In order to obtain the desired one-phase condition, the material is heated at 1100°C and then quenched. During welding, the strips heated to about 700°C on each side of the weld-line may become sensitive to attack; the causes of this weld-decay, and methods of preventing it, are discussed on p. 108.

The behaviour of 18/8 stainless steel in dilute sulphuric acid containing oxygen was studied in the author's laboratory by Berwick[334] who found that the time (t_p) needed for an initially active alloy to become passive decreased with the oxygen concentration C; under Berwick's conditions of working, $1/t_p$ was a rectilinear function of C, and when C was reduced to a certain value (C_0), the value of $1/t_p$ became zero, which meant that below C_0 the alloy would never become passive at all. (I. D. G. Berwick and U. R. Evans, *J. appl. Chem*., 1952, **2**, 576.)

The concentration of oxygen in the acid needed to maintain the passive state can be lowered if the liquid is stirred, as is easily understood, since it is the rate of replenishment which will determine behaviour. Under stagnant conditions, and particularly at crevices, where there will be practically no replenishment, passivity is likely to break down. That explains why stainless steel is suceptible to crevice attack, and also why pitting, once started, will generally continue. As in the case of aluminium (p. 78), protection may fail in the presence of chlorides.

Industrial experience shows that accidental removal of oxygen from dilute acid can set up attack on stainless steel; if once this starts at some relatively inaccessible corner, provision of oxygen to the bulk liquid, far from stopping attack, will increase it; the same is true of attempts to halt corrosion by violent stirring. Two methods are available to render the use of stainless steel safer for dilute acids.

One is to apply **anodic protection** – raising the potential by means of an applied e.m.f. and keeping it in the range where reductive dissolution of the film is impossible and where any gap in the film would automatically be healed by the anodic formation of fresh oxide. The practical possibility of such a method of protection was first demonstrated by Edeleanu,[336] and important industrial applications of the principle in the United States are due to Sudbury, Riggs and Shock. (D. A. Shock, J. D. Sudbury and O. L. Riggs, *First International Congress on Metallic Corrosion*, London 1961; report (Butterworth), 1962, 363). Success depends on exact control of the potential by reliable potentiostatic equipment, since treatment at the wrong level can be disastrous; for instance, a potential that is too high may cause dissolution of chromium in the soluble, hexavalent condition. The safe range of potential depends on the composition of the alloy.

The other method is to use an alloy containing molybdenum, a metal which possesses practically no salt-forming properties, as mentioned on p. 73. The 18/8 alloys containing 3–4% molybdenum are essentially designed to withstand oxygen-free or reducing acid conditions; they are not well suited for strongly oxidizing acid liquids, where the molybdenum may pass into the liquid in the hexavalent condition. Molybdenum stainless steels are specially useful for resisting sulphurous acid – a reagent important in the paper and textile industry and many other situations where a higher corrosion resistance than that of the 18/8 austenitic alloy is required. The possibility of using silicon, a non-metal devoid of salt-forming properties, for the same purpose has aroused much interest.

Not all the stainless steels are austenitic. The ferritic 13% chromium steel, used in cutlery, was the first stainless steel to be developed. Other ferritic steels include the low-carbon 16–17% chromium alloy used extensively for car trim, architectural applications and domestic ware, as well as for some applications in the chemical industry, e.g. nitric acid manufacture. Up to 1% molybdenum may be added for increased resistance to chloride or reducing acids.

A ferritic steel containing 20% Cr, 2% Mo and 1% Nb has been developed to give general corrosion resistance similar to the austenitic 18/8 Cr–Ni alloy, but with a very much greater resistance to stress corrosion under conditions where failure from that type of attack is probable.

The resistance of stainless steels is diminished by surface defects and particularly by foreign matter rolled into the material; the atmosphere of the rolling mill must be kept free from all suspended particles. Similarly mill-scale rolled into the surface may cause a breakdown of passivity.

Aluminium

Whilst in chromium the divalent condition is unstable, in aluminium it is absent altogether, which would suggest this trivalent metal as a material suitable to resist oxidizing acids. Unfortunately aluminium occupies a low place in the potential series, and the essentially reactive character of the metal itself must be set against the protective character of the alumina film. For handling nitric acid of concentration above about 96%, aluminium is used with satisfaction. Below that concentration, the attack becomes rather too fast to make the employment of aluminium a practical proposition, although even at the concentration where the attack is most rapid (about 30% nitric acid), the action is sluggish compared to that on iron. The attack on aluminium at any given concentration is greatly stimulated by the presence of traces of chlorides.

Aluminium is also much used for resisting organic acids. An early report by Seligman and Williams, on the effect of hot organic acids, contains some curious features. Dilute (1%) acetic acid attacks aluminium somewhat readily, but the corrosion rate diminishes as concentration increases; 99% acid is almost without action, but the removal of the last 0·05% of water increases the attack one hundred times. Rather similar results have been obtained with propanoic and butanoic acids, with phenol and with ethanol, butanol and pentanol. The absolutely anhydrous substances attack the metal, but a mere trace of water serves to protect, presumably because aluminium can displace hydrogen from water, forming a film of alumina.

The film-forming power of small amounts of water is of practical importance. For making ethanoic, propanoic or butanoic acids of high concentration, aluminium plant should enjoy a reasonable life, provided that there is no risk of complete dehydration – or of undue dilution.

The resistance of aluminium to acids depends largely on the purity of the metal and the exclusion of chlorides and metallic salts from the acid. Great care must be taken not to introduce impurities into originally pure aluminium either from tools, or by the rolling in of metallic particles which have settled on the surface. Even emery particles embedded in an abraded surface can increase the rate of attack, whilst inclusions near the surface – due to inverse segregation – can be responsible for pitting; sometimes the casting skin, which contains such impurities, is removed before rolling. Porosity, due to evolution of hydrogen in the molten aluminium before casting, may favour attack; hydrogen can originate from corroded scrap aluminium included in the charge, being formed when the moisture in the corrosion product acts on the hot metal; any hydrogen present should be removed by washing the molten metal with chlorine and/or nitrogen before casting.

Welding may also be the cause of corrosion trouble on aluminium. When making welds in aluminium of commercial purity, there is some benefit in using filler rods of still higher purity. In making storage tanks for nitric acid, the welds should be hammered, annealed and thoroughly washed before use, especially if chlorides have been present in the flux used. Some fluxes contain alkali fluorides, often with chlorides to reduce the melting point, but even with potassium bisulphate flux, washing is important. It should be remembered that particles of flux entrapped in the molten metal may remain after washing, and constitute sources of weakness, besides setting up corrosion. Closed blow-holes within welds on aluminium – whether due to hydrogen or another cause – are rapidly picked out by nitric acid.

Titanium

A material which is being increasingly used to avoid corrosion in industrial processes is titanium; its extensive occurrence in the earth's crust places it among the more common elements, but at one time the cost of the reduction process was such that its price in the metallic state was high. The price has now been reduced by technological improvements and today titanium may be cheaper than many corrosion-resistant alloys. The resistance is due to an oxide film, which generally requires a trace of water for maintenance. Thus certain alcohols which have no action on titanium in the condition supplied by the manufacturer may attack it if completely dehydrated. Breakdown of resistance may also occur in completely dry chlorine gas or in dry red fuming nitric acid. Titanium has a high affinity for hydrogen and may absorb hydrogen produced in corrosion reactions. The result is embrittlement due to the presence of titanium hydride. An outstanding use for titanium is with sea-water, particularly on coolers; it does not pit under shellfish, sufficient oxygen getting through to keep the film in repair. There is also a temperature limit to the use of titanium in chlorides, which can be raised by the employment of palladium-bearing material.

Titanium is largely used in the oil and petrochemical industries, where its resistance to chlorides and sulphides is welcome. An important application is in organic syntheses involving chloride catalysts, where stainless steel is too sensitive to Cl, whilst the high temperature and pressure rule out most non-metallic material; it is used in the manufacture of ethanal (acetaldehyde) from ethene, with copper chloride as catalyst.

Titanium anodes, carrying a thin surface layer of platinum or ruthenium oxide to maintain the potential within the passivity range, are used for cathodic protection processes (p. 118). They are also used in the electrochemical manufacture of chlorine or chlorate, in the place of the traditional graphite anodes; this change has introduced various advantages, since power consumption is lower, maintenance easier and the products purer.

Metals to withstand alkali

The solubility of metallic oxides in alkali is much more widespread than is generally supposed. Zinc, aluminium, lead, tin and even copper are all attacked by sodium hydroxide solution, at least if oxygen is present. The metals most resistant to alkali are nickel, silver and magnesium. Nickel is largely employed in reactions involving hot alkali, whilst silver is sometimes used in contact with caustic soda up to 80%; the resistance of silver deteriorates if nitrates are present.

Iron becomes passive in dilute sodium hydroxide, but in concentrated alkali, at least at high temperatures, it can pass into solution as sodium *ferroate*, Na_2FeO_2, – a compound of divalent iron. This change is important in connection with 'caustic cracking' (p. 96).

Graphical construction for corrosion velocity

Non-oxidizing acids

The reason why iron placed in dilute sulphuric acid suffers steady attack, with expulsion of hydrogen, is simply because, under the concentration conditions set up at the metallic surface, the p.d. of $H \mid H^+$ lies on the positive side of $Fe \mid Fe^{2+}$, so that there is an e.m.f. (represented by the distance between these two p.d.s) to maintain current flow, and thus corrosion. But fortunately the current flowing is far less than the value obtained on dividing the unpolarized value of the e.m.f. by the resistance of the circuit. When current flows between the anode and cathode, both potentials (*A* and *C*) are shifted so as to approach one another, as shown in Fig. 3.2. However low the resistance, the current flowing could never exceed the value (representing the abscissa of the intersection point, P) which would render the acting e.m.f. vanishingly small. In acid corrosion

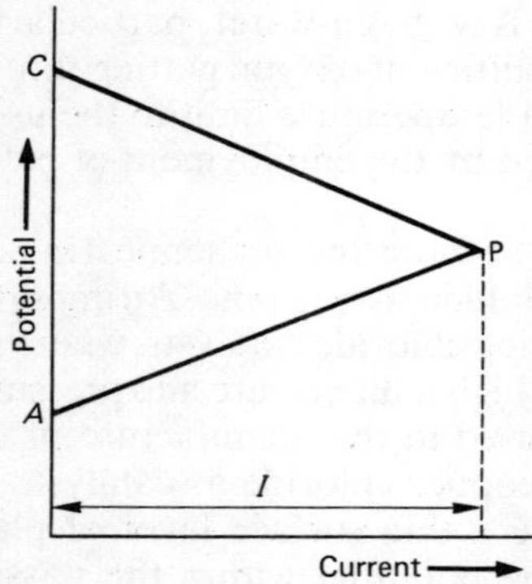

Fig. 3.2 Maximum current obtainable from a low-resistance corrosion cell

(where the solutions are usually good conductors whilst the anodes and cathodes are generally close together), no great error is involved in the assumption that the intersection point does represent the current flowing; if the current is I amperes, the corrosion rate will be I/F gram-equivalents per second, where F is Faraday's constant (96 484).

The slope of the hydrogen curve will depend on overpotential, and will be steeper for zinc, which has a high overpotential, than for iron (Fig. 3.2). Thus, notwithstanding the fact that zinc (the more reactive metal) will have an anodic curve lower than iron, it will corrode (in the absence of contact material) more slowly than iron, the rates being indicated by the abscissæ of P and Q respectively. If a little platinum salt is added, the black platinum deposited will have an overpotential lower than that of either metal, and the corrosion rate of iron will be increased to a value represented by the abscissa of R.

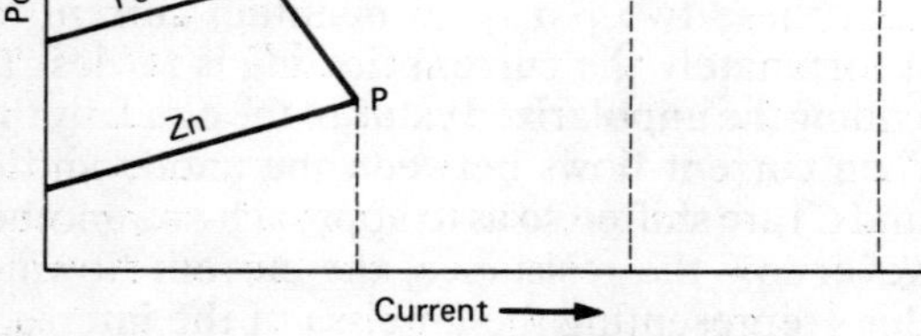

Fig. 3.3 The action of non-oxidizing acids on zinc and iron

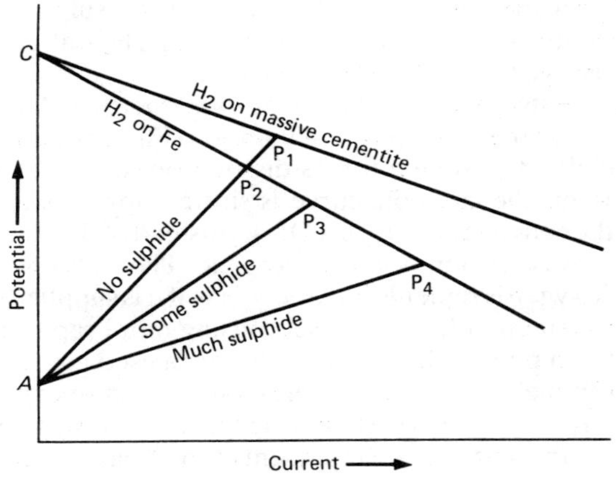

Fig. 3.4 The action of non-oxidizing acids on steel

In steel, as explained on p. 66, the presence of massive cementite facilitates the cathodic reaction, probably by providing points of low overpotential; thus (Fig. 3.4) the cathodic curve in presence of massive cementite is less steep than in its absence (CP_1 instead of CP_2) and the corrosion rate becomes faster (P_1 instead of P_2). In the absence of massive cementite, hydrogen sulphide (produced by the attack on sulphide inclusions) will reduce the 'stickiness' of the anodic reaction, rendering the anodic curve less steep and increasing the corrosion rate (P_4 instead of P_2). In a number of steels all free from massive cementite but differing in the

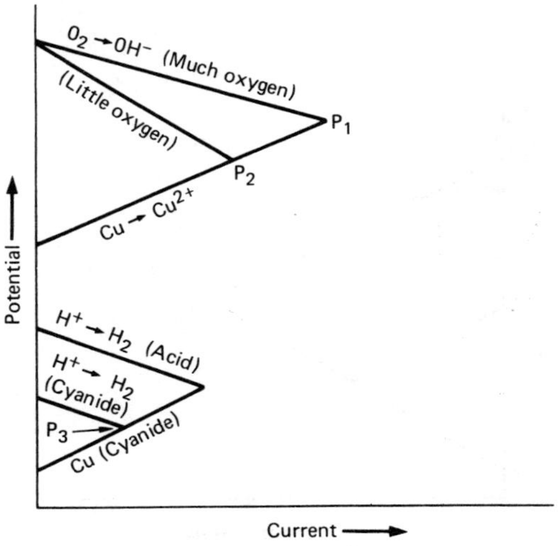

Fig. 3.5 The action of non-oxidizing acids and cyanide solutions on copper

content of sulphur and copper (which serves to render sulphur unavailable, fixing it in stable form), the cathodic curves will be CP_4 in all cases, although the anodic curves (AP_2, AP_3, AP_4) will differ.

Copper, in the absence of complex-forming bodies, exhibits a positive potential, far above that of hydrogen-evolution (Fig. 3.5), and there will be no e.m.f. available to provide for its dissolution in oxygen-free acid. If oxygen is present, the cathodic curve is shifted upwards to a new level representing the conversion of O_2 to $(OH)^-$, instead of the conversion of H^+ to H_2. In this way corrosion becomes possible. The cathodic reaction will polarize less if oxygen is replenished freely than if it is supplied scantily, and the curve is less steep in the first case, causing more rapid corrosion (P_1 instead of P_2). If a potassium cyanide solution is used instead of acid, the formation of complex ions causes a depression of the copper potential, and, although the hydrogen curve is also depressed owing to the alkaline reaction, slow corrosion at a rate represented by P_3 can occur in absence of oxygen.

Oxidizing acids

If, in considering the corrosion of iron, nitric acid is substituted for sulphuric, the cathodic reaction, instead of being the evolution of hydrogen (curve CP_1, Fig. 3.6), will be the reduction of nitric acid to various products, which will take place at much higher potentials (curve $C'P_2$). Thus the corrosion will be much faster (P_2 instead of P_1), and will tend to become greater as the acid concentration increases (since this will render the cathodic curve less steep). If, however, passivity sets in, the anodic potential will jump from the negative value shown by AP_2 to a positive (noble) value indicated by $A'P_3$; thus the rate of corrosion will become very slow (represented by P_3).

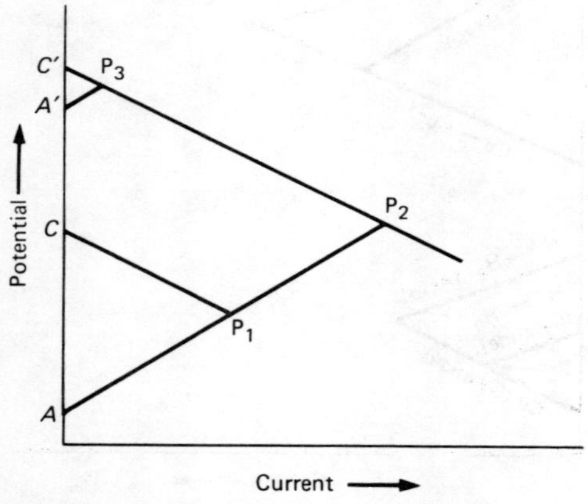

Fig. 3.6 The action of oxidizing acids on iron

Salt solutions in presence of oxygen

The graphical method can be applied to corrosion by neutral salt solutions, but, except at high concentrations, the resistance must now be taken into account. In general, the current will be such a value (I) as will produce an intercept between the cathodic and anodic polarization curves equal to IR, where R is the resistance of the cell (Fig. 3.7).

If we consider a series of vertical iron specimens of the same shape, partially immersed in salt solution of different concentrations the cathodic curve will of course represent the reduction of O_2 to OH^- and the anodic curve the change from Fe to Fe^{2+}. At very low concentrations R will be high, and a large IR intercept being necessary, the corrosion rate will be small; but as the concentration increases smaller intercepts will be needed, and the corrosion rate will increase towards that represented by the intersection point. When the intersection point is closely approached, further decreases of resistance will clearly have little effect, since the rate of attack can never exceed that represented by the intersection point – however low the resistance may become. At first sight it would seem that, with increasing concentration, the corrosion rate would asymptotically approach the value given by the intersection point. However, at high concentrations, another factor comes in: the oxygen-solubility of salt solutions falls off with the salt concentration, so that the cathodic curve (which depends on oxygen supply) steepens, and the position of the intersection point itself shifts towards the left. Thus the corrosion velocity, after first increasing with concentration, will later start to diminish again. In the researches of Hoar and the author[870] on partly immersed specimens the maximum corrosion velocity was reached at concentrations between 0·3 and 0·6M according to the variety of iron and salt employed. In the experiments of Bengough, Lee and Wormwell[796] on fully immersed specimens, where the velocity was controlled essentially by oxygen-supply, resistance being unimportant, the maximum was placed at a very low concentration, and

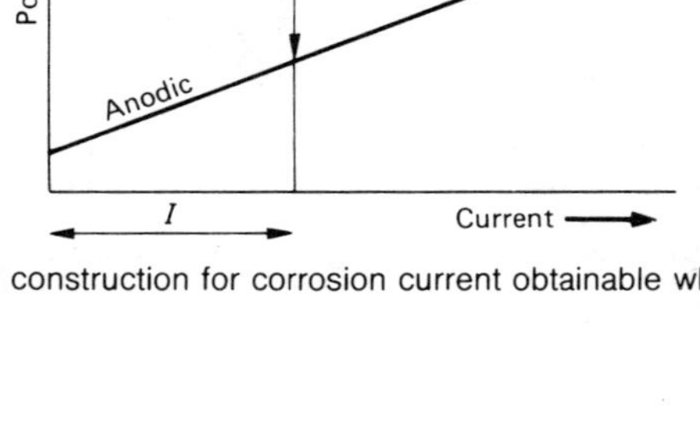

Fig. 3.7 Graphical construction for corrosion current obtainable where resistance is appreciable

certainly from 0·01M upwards, the corrosion rate *declined* with increasing salt concentration.

Cathodic and anodic control

Let us consider, for simplicity, a case where the resistance is low, so that the intersection point defines the corrosion velocity. If the cathodic curve is much steeper than the anodic (Fig. 3.8(a)), it is clear that an influence which stimulates the anodic reaction (say by reducing the steepness of the anodic curve, giving *A′* instead of *A*) will hardly move the intersection point, and thus will not appreciably alter the corrosion rate. But an influence stimulating the cathodic reaction (i.e. reducing the steepness of the cathodic curve) will greatly shift the intersection point and thus increase the corrosion velocity (e.g. from P to P′). This state of affairs, where the corrosion velocity is affected by happenings at the cathode but not appreciably by occurrences at the anode, is called **cathodic control**. An example is provided by the corrosion of iron in concentrated potassium chloride or sulphate, where, as just explained, the corrosion velocity is dependent mainly on the rate of supply of oxygen, the cathodic stimulant, whereas the nature of the anion hardly affects the rate directly, and any influence is indirect – due to changes in the oxygen-solubility. It is, however, wrong to suppose that the nature of the anodic reaction has no effect, even when its curve is not steep. Change of metal shifts the anodic curve from *A′* to *A″* and alters the corrosion velocity from P′ to P″.

If the anodic curve is much steeper than the cathodic, then factors stimulating the anodic reaction will greatly increase the corrosion rate, but those stimulating the cathodic reaction will hardly affect the corrosion velocity. This is called **anodic control** (Fig. 3·8(b)), and examples are provided by the corrosion of aluminium by sodium chloride solution containing chromate in quantity just insufficient to prevent attack. The anodic areas are here small defects in a protective film, and the introduction of small quantities of stimulating anions (such as chlorine ions) will

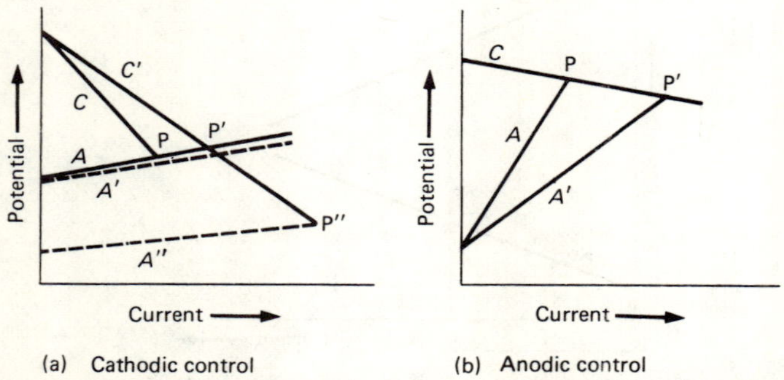

Fig. 3.8 Cathodic and anodic control

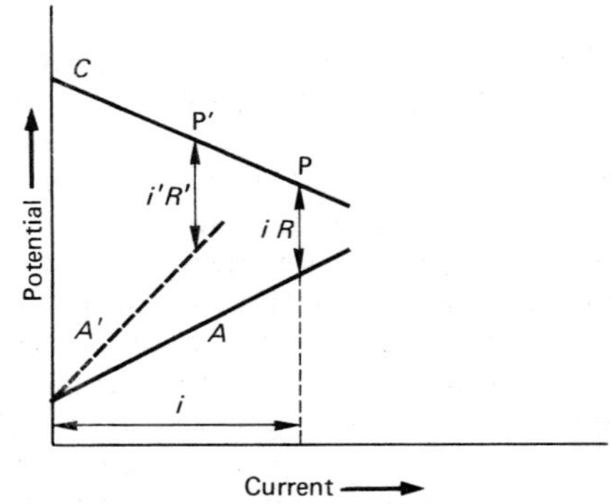

Fig. 3.9 Potential of metal covered with a porous film

greatly accelerate the velocity of corrosion, whereas cathodic stimulators (oxygen or oxidizing agents) will not increase the attack; sometimes by repairing the film, they may retard it.

In practice, cases are common where the gradients of both curves are of the same order of magnitude. This is known as **mixed control**.

Potential movement with time on an oxide-covered metal

Consider a specimen of metal covered with a nearly complete conducting oxide film (as a result of long exposure to air) plunged into a salt solution. Let the potential be measured just after immersion, and then again at suitable intervals. At first the current will be passing between the film as cathode and the metal exposed at occasional cracks in it as anodes. If the film is nearly complete, the resistance (R) will probably be somewhat high, and the current will have the value i just sufficient to produce an intercept iR (Fig. 3.9). The potential measured by a tubulus pressed against the surface will be represented by P.

Now if the solution is one which tends to repair the defects in the skin (i.e. to render the metal passive), the anodic area will become smaller and the anodic curve will steepen from A to A′; the resistance will rise (to R'), so that the intercept needed ($i'R'$) will probably be as long or longer than iR. Thus the measured potential must tend to rise also (e.g. to P′). But if the solution is an activating one, causing the damage to extend, the anodic curve will become less steep, the resistance will fall, the intercept will become smaller; thus the potential will fall. Consequently, the direction of movement of the potential will indicate whether the liquid is tending to make the metal passive or active, whether it is stifling corrosion or promoting it. This is valuable information, and accordingly potential–time curves have been measured for many metals and alloys in many liquids. They do not, in general, provide any substitute for measurements of corrosion velocity.

Velocity of corrosion

In Figs. 3.2 to 3.7 the cathodic and anodic polarization curves have, for simplicity, been drawn straight, although this will rarely represent the truth. If we know the equations governing the curves, it should, in principle, be possible to calculate the corrosion velocity, assuming that the liquid resistance can be neglected, so that the intersection point represents the current flowing. The equations governing movement of current with potential are discussed on pp. 289–291. If the cathodic reaction is evolution of hydrogen, then, in general, both anodic and cathodic reactions should obey Tafel's logarithmic equation; but if the cathodic reaction is the reduction of oxygen, concentration polarization is likely to control the cathodic current, if oxygen is used up as quickly as it is supplied. The calculation of corrosion velocity from basic electrochemical principles in the general case is too complicated a subject for this book, but in suitable instances excellent agreement has been obtained between observed and calculated values; apart from the examples from early Cambridge researches mentioned on p. 41, work from the laboratories of Stern[879] and Hackerman[882] deserves study.*

Protective films of salt-like oxides (1976 vol., p. 199).

Hoar has pointed out that many of the binary alloys found to resist corrosion contain a metal possessing a basic oxide combined with another possessing an acidic oxide; he suggests that resistance is attributable to a salt-like compound formed by union of the two oxides. His list of examples is reproduced in Table 3.1. The materials included are used for a number of purposes, involving many different environments, and doubtless in most cases there is some specific explanation of the good behaviour observed under the conditions which the material is required to withstand. A salt-like character is clearly not the sole criterion for resistant behaviour; it would not guarantee protection if the film were discontinuous, or if it carried internal stress liable to cause cracking. But if physical stability has been provided, it would seem that the chemical stability would be expected from a salt-like oxide; for if the union of the basic and acidic oxides is accompanied by a decrease of free energy, this energy would have to be provided from an external source if the oxide film were to be dissolved away.

*R. F. Steigerwald and N. D. Green (*J. electrochem. Soc.*, 1962, **109**, 1026) have worked out equations for the general case where each phase can take part in both anodic and cathodic reaction, using Tafel's equation to calculate the contribution of each phase, adjusting this for the fraction of area occupied by that phase, and adding the two contributions. They point out that if the active phase is dispersed as small particles in a more inert matrix, the preferential anodic attack on the active particles exposes an increasing area of the inert matrix, and diminishes the area of the active phase. They consider the attack on a one-phase alloy as a special case where the dispersed particles are of atomic size. They have studied the anodic polarization curves of the two components of alloys, calculated that of the alloy and obtained fairly good agreement with experimental values in suitable cases.

Table 3.1 The use of 'salt-like oxides' in resistant films (T. P. Hoar)

Component with 'basic' oxide	Component with 'acidic' oxide	Alloy(s) more resistant than 'basic' component
Cu	Zn	Brasses
	Al	Aluminium bronzes
	Sn	Tin bronzes
	Si	Silicon bronzes
Fe	Cr	Chromium irons; stainless steels
	Sn	$FeSn_2$
	Si	Silicon irons
Ni	Cr	Nichrome, Nimonics
	Sn	NiSn
	Mo	Hastelloys etc.
Co	Sn	CoSn
	Cr	Vitallium
Ti	Mo	16% (mass) Mo alloy
	Ta	5% (mass) Ta alloy
Zr	Al	35–50% (mass) alloys

An interesting case is the compound NiSn, which can be obtained by electrodeposition at almost stoichiometric composition, and exhibits passivity over a wide range – becoming passive at a lower potential than either Ni or Sn. There appear to be three layers in the film, and these, Hoar suggests, are $NiSnO_2$, $NiSnO_3$, and $NiSnO_4$; all three compounds contain Ni and Sn in the same ratio as the alloy.

Corrosion by pure water

General behaviour of heavy metals in water containing oxygen

Parson's principle (p. 1) suggests that salt-free water containing oxygen in contact with a heavy metal should build up a protective film and develop little attack, since the oxides and hydroxides are (except in the case of thallium) sparingly soluble. Yet the purest water obtainable is found to produce attack in presence of oxygen; there has at times been a tendency to attribute this to traces of salt, but, although absolutely salt-free water cannot be obtained, it is difficult to think that the very small content of salt present in the best distilled or demineralized water is responsible.

If account is taken of nucleation, the apparent conflict with Parson's principle becomes understandable. The concentrations of metallic cations and OH^- anions in a saturated solution of a metallic hydroxide is very small. The coming together of the number of ions needed to produce a stable crystal of hydroxide or oxide is an infrequent event (very small crystals are unstable and would quickly redissolve). Thus when once a stable crystal-nucleus has appeared, fresh hydroxide or oxide is more likely to be deposited on it than to start fresh crystals. The reaction, therefore, may lead to a limited number of comparatively large, isolated crystals, instead of a continuous protective film, so that slow attack continues. That is what is found to occur.

It is likely that attack by pure water is electrochemical, metallic cations entering the liquid by the anodic reaction, whilst the cathodic reaction is the reduction of oxygen, first to hydrogen peroxide (which is often detected) and then to water. The anodic and cathodic reactions may occur on distinct areas, at contiguous points, or even at the same point at different moments. Evidence for the electrochemical mechanism has been provided in the case of zinc; it is probably operative on other metals.

Lead in water containing oxygen

Mayne* obtained a bright lead surface by casting it against a polished stainless steel plate, and placed it in distilled water. After some hours the metal became etched, and subsequently oxide crystals were observed under the microscope; after some days, the surface became covered with an open network of beautiful little crystals, but no continuous film was formed, so that attack continued. The low nucleation rate was indicated by the comparatively large size of the crystals; the fact that sometimes the red form of litharge appeared, and sometimes the yellow form, indicates that events depend on chance formation of suitable atomic groupings.

Water of low calcium-salt content, if conveyed in lead pipes, takes up sufficient lead to be dangerous for drinking; the danger is especially serious in the case of soft waters drawn from moorland districts; these usually contain organic acids derived from moorland plants which form soluble lead salts. Hard waters containing calcium bicarbonate do not as a rule take up dangerous quantities of lead, since the relatively large concentration of $(HCO_3)^-$ ions allow a continuous, protective film of basic lead carbonate to be produced. The situation is, however, complicated, and probably minor organic constituents of natural water affect behaviour. Often, soft acid waters can be rendered suitable for carriage in lead pipes by percolation through limestone or treatment with lime water; this is to be preferred to neutralization with soda, since the presence of $Ca(HCO_3)_2$ is desirable. Sometimes sodium silicate has been effective in rendering **plumbo-solvent** waters safe for conveyance in lead pipes.

Zinc in water containing oxygen

Bengough, Stuart and Lee[109] studied zinc fully immersed in very pure water. The attack (measured by the disappearance of oxygen from the gas-space above the liquid) started fast but soon slowed off, finally coming almost to an end; small amounts of potassium chloride postponed the slowing down, whilst in M/10 KCl the corrosion rate was found to be constant with time. In pure water the attack slowed off on annealed zinc specimens more quickly than on surfaces prepared by turning or abrasion. With both those two surfaces, attack, after it had become slow, kept

*J. E. O. Mayne. Experiment described by U. R. Evans, *Metallic Corrosion, Passivity and Protection*, Edward Arnold, London, 1946, p. 353.

starting up again. The results illustrate the effect of internal stresses in cracking the film – probably at places where it is separated by a gap from the zinc basis – and causing attack suddenly to recommence.

Despite the fact that the total attack on zinc is small, it is localized at points, and the pitting burrows quickly into the metal; zinc sheet, exposed to distilled water, may suffer perforation at certain points with the intervening area remaining practically unattacked. It was noticed by Bengough and O. F. Hudson[101] that the pits were arranged on vertical lines. This was confirmed by Eurof Davies,[99] who found that the vertical arrangement had no relation to rolling direction or abrasion direction, but was determined by gravity. Specks of solid corrosion product, loosening themselves from a pit which had developed at one point, would slide down the surface and lodge at some point immediately below it, shielding the zinc from oxygen, and setting up new anodic attack. Special experiments with a model representing the pit (Fig. 3.10(b)), but with the zinc representing the pit bottom insulated from that representing the face, showed electric currents passing between the pit bottom as anode and the face, where oxygen could be replenished, as cathode. Evidently, differential aeration currents can be important even in pure water, and the combination of large cathode and small anode leads to great intensity of attack at the latter, explaining the rapid boring into the metal. Davies produced pitting by stretching a nylon thread over a horizontal zinc surface (Fig. 3.10(a)) – which gave two crevices where oxygen will not easily be renewed; on immersion in pure water, two lines of pits were produced.

When a zinc disc was whirled in water containing oxygen, so that there was no area shielded from oxygen, it remained practically unchanged in appearance; there was no pitting and the water remained clear. Under stagnant conditions, however, much zinc entered the water, which was soon found to contain colloidal particles, presumably zinc hydroxide, and

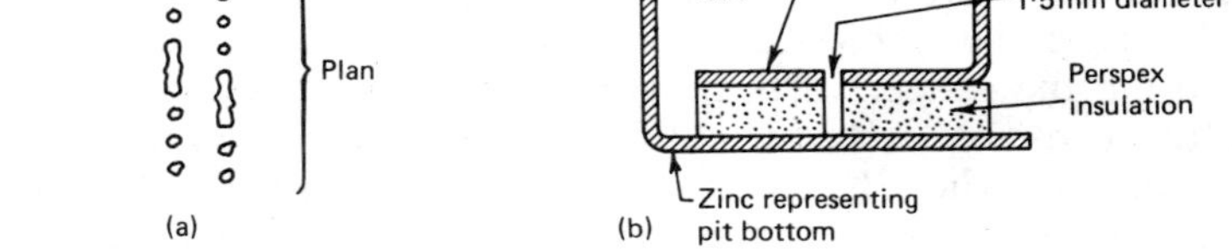

Fig. 3.10 (a) Two lines of pits set up on zinc in distilled water by contact with polythene thread (b) Model to demonstrate differential aeration current at a pit (D. Eurof Davies and U. R. Evans)

gave a strong Tyndall cone; pitting then occurred; the white product scraped from around the pits, examined by X-rays was found to be ZnO with some β-$Zn(OH)_2$.

Iron in water containing oxygen

Ferrous hydroxide ($Fe(OH)_2$) is much more soluble than 'ferric hydroxide' (really a partial hydroxide, FeO.(OH) or $Fe_2O_3.H_2O$). If pure iron is placed in pure water under stagnant conditions, the two ions, Fe^{2+} and OH^-, are formed, and are converted to ferric hydroxide (rust) on diffusing out to a region where more oxygen is available; if the oxygen-supply is limited, magnetite (Fe_3O_4 or $FeO.Fe_2O_3$) will be produced. Material formed, at least in part, at a distance from the surface, cannot itself be protective, and the precipitation will prevent the concentration in the layer next the metal reaching a level capable of producing a protective film. Thus attack is found to continue.

In contrast, if a disc of the same pure iron is whirled in the same water, the oxygen supply at all points can be kept high. In that case, the specimen is found to remain bright and (except perhaps for the appearance of tiny dark spots) unchanged[103]; the loose rust characteristic of stagnant conditions is not formed. The exact composition of the invisible film is uncertain; it is almost certainly a cubic oxide, probably approaching Fe_2O_3.

The analogy with zinc is apparent. It should be noted, however, that this result is only obtained with pure (electrolytic) iron. Mild steel discs whirled in water, develop adherent rust, although the liquid remains clear. Probably the cell iron-cementite is involved, so that the Fe^{2+} and OH^- ions are formed at different points.

Iron covered with water drops in oxygen – nitrogen mixtures

Mears[936] ruled two sets of lines with a solution of wax on a clean iron sheet; when the solvent had evaporated, he flooded the surface with water and tilted it to let the excess run off; this left a chess-board pattern of 'square drops' on the squares between the lines. He exposed the specimen in vessels containing known mixtures of oxygen and nitrogen. Some of the drops remained absolutely clear, others produced considerable loose rust; intermediate cases (with only a trace of rust) were rare.

Typical results are shown in Fig. 3.11. In mixtures rich in oxygen, the proportion of drops developing rust was smaller than in mixtures poor in oxygen; but in those drops producing any rust at all, the corrosion was quickest in the more oxygen-rich atmospheres. This is understandable, since the nearer the position of the precipitation of rust is brought to the site of corrosion, the steeper will be the concentration gradient, the more rapid the diffusion of Fe^{2+} and OH^- outwards, and consequently the faster the corrosion. Only when the oxygen concentration in the gas-phase is high enough to bring the points of formation of the less soluble product to the metallic surface itself is stifling possible. Thus oxygen is a **stimulator** of corrosion in the sense that it increases the **conditional velocity** (the rate of corrosion attained on condition that there is any corrosion at all), but a

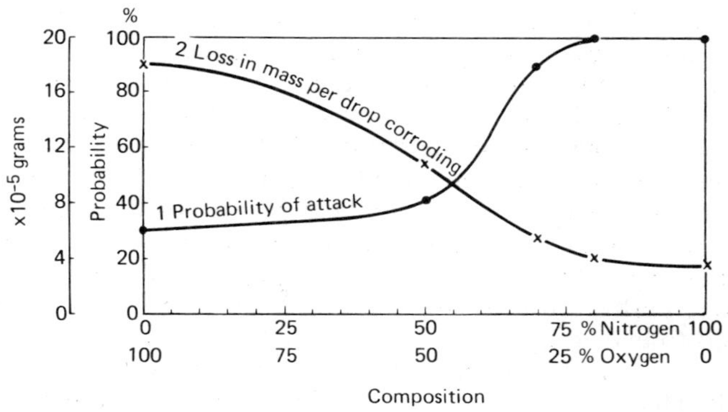

Fig. 3.11 Variation of probability and conditional velocity with oxygen-content (U.R. Evans and R. B. Mears)

deterrent of corrosion in the sense that it reduces the probability of attack starting whereas a retardant is a chemical which decreases velocity; a retardant is thus the opposite of a stimulator. This probability is conveniently measured by counting the proportion of drops which develop visible rust. The matter of corrosion-probability is further discussed on pp. 252–260.

Iron in water free from oxygen

Thermodynamical principles (p. 247) show that the liberation of hydrogen from pure water by iron is a possible reaction (i.e. one which would result in a drop of free energy). Here the cathodic reaction will probably proceed in two stages:

$$H^+ + e^- \rightarrow H$$
$$H + H \rightarrow H_2$$

The primary anodic reaction

$$Fe \rightarrow Fe^{2+} + 2e^-$$

can continue, without infringing energy considerations, until the water (in which OH^- is now in excess, H^+ is removed) becomes supersaturated with respect of solid $Fe(OH)_2$. However, a secondary anodic reaction

$$Fe^{2+} \rightarrow Fe^{3+} + e^-$$

is also possible, and energy considerations permit it to proceed until the liquid is supersaturated with respect of Fe_3O_4. Supersaturation with respect of Fe_2O_3 cannot be reached (before the necessary concentration of Fe^{3+} could be built up, the potential at anodic points would have reached the value of the potential at cathodic points, and the e.m.f. would vanish). Thus the reaction with oxygen-free water could produce either magnetite or ferrous hydroxide, according as the secondary anodic reaction does or does not come into play. This secondary reaction (although it can without doubt occur) is somewhat unpredictable and may require a catalyst such as copper. Possibly for that reason, different experimenters have obtained (1) ferrous hydroxide crystals (2) magnetite crystals or (3) a colloidal solution of ferrous hydroxide[440]; the latter observation recalls the behaviour of zinc in water containing oxygen (p. 88). Part of the apparent disagreement between results could be explained by the use of different forms of iron; some experimenters worked with iron powder. It should be noted, however, that any solid products formed do not seem to be protective, perhaps for reasons discussed on p. 87; iron powder has continued to evolve hydrogen from water for 200 days.

In contrast, at 100°C, a film is formed which rapidly brings reaction to a standstill. Thiel and Luckmann[437] found that hydrogen was first rapidly evolved by steel from boiling water – whether oxygen was present or not; but the evolution became very slow after 100 hours. Gould[437] exposed steel in boiling water in closed tubes for 76 days, and found that reaction had apparently ceased; the metallic surface had become covered with a visible film, which could be removed to plastic by methods similar to those of p. 7, and examined; the film-substance is essentially magnetite, but the composition may not be exactly Fe_3O_4.

Boiler water problems

The main large-scale use of pure water is in power boilers. Provided that oxygen and acids are absent, a protective film (loosely described as a 'magnetite film') is generally built up. If this is not broken locally through thermal or mechanical stresses, there is no serious deterioration of the steel – except at temperatures high enough for the hydrogen troubles discussed later.

Although 'purity' is a relative term, fantastically low contents of solid and gaseous solutes are achieved in feed-water at generating stations today. In large boiler plants, most of the feed-water is derived from the condensers, and this is normally of high purity; precautions are necessary to prevent it from picking up oxygen or acid gases before re-entering the boiler. There will always be some 'make-up' water needed to compensate for losses, and, where some of the steam is drawn off for process purposes or space heating, the 'make-up' may be considerable. In any case, it will require softening, conditioning and de-aerating.

The **softening** may be carried out by adding lime (to remove calcium bicarbonate as calcium carbonate) and sodium carbonate (to remove calcium sulphate in the same form); magnesium salts, if present, are converted to magnesium hydroxide, which tends to remain suspended as colloidal particles, and sodium aluminate is sometimes added to flocculate it.

Alternatively an ion-exchange resin may be used; the water is run through a bed containing suitable resins which replace Ca^{2+} and Mg^{2+} ions by Na^+ ions; thus calcium bicarbonate becomes sodium bicarbonate, which loses carbon dioxide on gentle heating, giving sodium carbonate;

sometimes this loses carbon dioxide in the boiler giving sodium hydroxide. Resins are available which effectively demineralize water by exchanging cations by H^+ and anions by OH^-.

Softened water, at least that obtained by the lime-soda or ion-exchange methods, still contains some solute, and the rapid evaporation within the boiler will produce solid. It is important that the solid shall be one normally thrown down mainly as a loose sludge, rather than one which will form on the walls an adherent scale liable to interfere with heat-transfer. To achieve this it is usual to treat the water intended for low and intermediate pressure boilers with a phosphate. Calcium carbonate or calcium phosphate are preferable to calcium sulphate or a silicate. The usual sludge is calcium phosphate, plus some silicate. The nature of the salt likely to be thrown down depends on the concentration of the various ions present; if in a water where Ca^{2+}, SO_4^{2-} and CO_3^{2-} are all present, the quantities are such that on evaporation, the solubility product of $CaCO_3$ is exceeded before that of $CaSO_4$, the former (less objectionable) compound should be obtained. This adjustment of the ion-contents to give a sludge rather than a scale is known as **conditioning**.

The removal of oxygen and carbon dioxide is also necessary. The water can be sprayed into low-pressure chambers heated by waste steam, or allowed to flow down a rack of trays, in which it encounters an ascending current of air-free steam. If **thermal de-aeration** is insufficient, **chemical de-aeration** may be used to remove the last traces of oxygen.* Sodium sulphite and hydrazine have both been used, the latter interacting with oxygen by a reaction sometimes written $N_2H_4 + O_2 = N_2 + 2H_2O$ – although it certainly proceeds in stages. Both chemicals act slowly, but the reaction is quicker in presence of traces of copper salts (often present accidentally, being derived from the condenser tubes); for sodium sulphite, cobalt is sometimes intentionally added as a catalyst to speed up the removal of oxygen. Sulphite is not suitable for very high-pressure boilers, being reduced by the iron or hydrogen to sulphide – an undesirable constituent.

If oxygen is not eliminated, pitting may result. Probably this occurs at spots shielded from oxygen replenishment by particles of debris, as in the case of zinc at ordinary temperatures (p. 89), but that is not certain. If the water is made slightly alkaline, which greatly reduces the solubility of ferrous hydroxide by the common-ion principle, more oxygen in the water can be tolerated – at least in relatively low-pressure boilers. In high-pressure boilers, where very low oxygen contents are achieved, the water is

*In 1959, Potter reported that 86 power stations in Great Britain had less than 0·043 ppm oxygen in their feed-water. (E. C. Potter, *Chem. Ind.*, 1959, 308.) At that time, most research laboratories were probably accepting a higher figure for their 'oxygen-free' water although Peers in 1952 had produced water with only 0·013 ppm. (A. M. Peers, *Chem. Ind.*, 1952, 969). Gilroy and Mayne, adapting on apparatus due to Hersch, have obtained water with 0·0005 ppm oxygen. (D. Gilroy and J. E. O. Mayne, *J. appl. chem.*, 1962, **12**, 382.)

generally kept alkaline, but this is more as a safeguard against entry of acid gases (which would greatly increase the solubility of ferrous hydroxide) than because alkali, as such, is beneficial.

In some plants, the alkalinity is supplied as phosphate; this may serve to avoid caustic cracking. Alkalinity in very high pressure boilers is supplied as ammonia.

Except in boilers working at very high pressures, reduction of oxygen to very low levels in water containing a minimum of other solutes serves to avoid corrosion trouble. In some high-pressure plants, however, **decarburization** has been caused by hydrogen entering the steel and rendering the material porous, by conversion of cementite (Fe_3C) into iron.* The evolution of hydrogen or methane (CH_4) at internal cavities may also promote cracking. There is difference of opinion as to how far the cracking sometimes met with is connected with hydrogen entering the metal, but regarding the occurrence of decarburization at certain stations there is, unfortunately, no doubt. The reason for the entry of atomic hydrogen into the steel is probably that discussed on p. 154. Under normal (healthy) conditions, the formation of molecular hydrogen occurs in two stages, and if something is present to 'poison' the second stage, a build-up of atomic hydrogen and of a concentration gradient which will cause it to diffuse into the interior is inevitable. In the hydrogen-problems of pickling, and those met with in the sour oilfields, the identity of the poison (H_2S or H_2Se) is generally known; there is still doubt about the substance – or the circumstances – which cause the poisoning phenomenon in boilers.

At one time the presence of copper in boiler-water (derived from condenser-tube material) was supposed to be the cause of trouble. This view is rarely held today, when nickel is more under suspicion. A research by Butler and Ison (1968 vol., p. 180) on a model boiler at Teddington suggested that steel containing 0·2% copper actually behaved better than steel containing little copper when the tubes were seamless, but not when the tubes were welded; in the latter case, the deepest pits were close to the weld-line. The authors were inclined to favour an electrochemical mechanism; magnetite will be cathodic towards steel, and areas covered with porous scale (presumably $CaCO_3$) may become anodic, since oxygen will not be replenished.

High-temperature boilers

Experience with boilers working at relatively low temperatures showed that if the oxygen content was kept low there was little trouble. This encouraged the belief that if only oxygen could be eliminated, troubles would be avoided in high-pressure boilers also. Those who held such an opinion were disappointed; success was achieved in reducing the oxygen concentration to a level which would once have been regarded as imposs-

*At ordinary temperatures, the volume of the iron produced would be *less* than that of the cementite destroyed in producing it. If this is true at boiler-temperature, the porosity is easily understood.

ible – but still attack continued. It is easy to be wise after the event, but the result is not really surprising. The interaction of iron with pure water with elimination of hydrogen is a reaction thermodynamically possible even at low temperature, and one which has been demonstrated experimentally. It is only prevented by the formation of a protective magnetite film and protection will cease if that film breaks away – which becomes increasingly probable as the film thickens.

There are different sorts of trouble in boilers and removal of oxygen avoids one of these. The troubles due to hydrogen embrittlement occurred when, in response to the urge to go to higher pressures and hence high heat transfer rates, it was necessary to reduce the solids content of the water. This meant 'purifying' the water and eliminating the free hydroxyl. So, because of this, and the presence of acid salts, the waters went acid. The massive tube failures occurred in those plants e.g. in the C.E.G.B. where this philosophy was adopted. A history of boiler failures according to the C.E.G.B. is discussed by C. W. Mann, *Brit. corr. J.*, 1977, **12**, 6.

Efforts in the research laboratory have been diverted from attempts to reduce oxygen concentration to a detailed study of scale structure. Much valuable information has been obtained, but agreement between different groups of experimenters has not always been complete. Work at Washington (U.S.A.) led to less porous scale than that at Leatherhead (U.K.), although the same liquid was used; the French authority, Berge (1976 vol., p. 231), performed a service in explaining the apparent discrepancy and showing that the ideas were not really irreconcileable.

It seems to be agreed that the scale formed on boiler surfaces consists of two layers, the lower one being formed below the original surface of the steel, and the upper one – generally more porous and less protective – being formed outside it. The extensive work at Leatherhead, started under Potter, and continued by Mann, with Holmes and Castle (1968 vol., pp. 173–6), has suggested reasons for the fact that, whilst over large areas the attack becomes slow as the magnetite scale thickens, there are exceptional places where it continues unabated, and pitting develops. Mann observed the presence of chloride in pits (as much as 50 mg Cl^- in each pit!) and discovered that the addition of $NiCl_2$ to pure water could increase the production of magnetite 1000 times in the first month. The formation rate (2·5 μm/hour) was of the order met with in power-boilers. The presence of nickel could be explained if at some previous time the water had been in contact with austenitic stainless steel.

If a small amount of chloride (perhaps $NiCl_2$) is present in rust-spots existing on the surface when a boiler is first put into operation, the anodic reaction will later draw further Cl^- ions into the pit, assuming that the water contains even a trace of Cl^-. The Leatherhead investigators have expressed the view that, although the presence of chlorides is necessary for the production of pits of the type encountered, in practice, pitting only occurs where some other factor comes into play – such as local overheating, which may be produced by steam-blanketing or by deposits capable of obstructing heat flow. If the hot spot is anodic, the combination of small anode surrounded by large cathode would explain the results. A more detailed study of the thermogalvanic cell constituted by two electrodes of the same metal (one hot, one cold) in a liquid would be welcome. It is generally thought that the hot electrode will be the anode, but occasionally, it seems, the current flows in the opposite direction. Russian work suggests that this occurs (1976 vol., p. 230).

Treatment of boiler water

Since early times, better water has been rendered alkaline by addition of a small amount of sodium hydroxide. So long as the concentration is low, the iron is kept in the passivity range, and the reaction builds up a magnetic scale. If the sodium hydroxide concentration is made very high, there can be corrosion giving Na_2FeO_2 (sodium ferroate) as shown by the small corrosion area above about pH12 on the Pourbaix diagram (Fig. 8.3). This diagram refers to 25°C, and it seems likely that the area is broader (extending to lower pH values) at higher temperatures. At any rate, the risk of a concentration rise increases under boiler conditions.

In the days of riveted boilers, steam used to leak out at a badly fitting rivet, so that concentrated NaOH was formed in the crevice, leading to stress-corrosion cracking. This at one time created a dangerous situation, and a treatment was put forward in the U.S.A. based on addition of sulphate; the recommendation issued was to keep the ratio of SO_4^{2-} to total alkalinity above a certain level. Different American authorities took different views regarding the efficiency of that method, and cases were established (including some in the U.K.) where cracking occurred despite the addition of sulphate. Nitrate additions proved far more effective (1960 vol., pp. 458–61). Probably the sulphate method, when it helped at all, did so by replacing OH^- by SO_4^{2-}, owing to the higher charge on the SO_4^{2-}; this replacement, being connected with transport, will depend on the geometry of the system and such a method is unlikely to be reliable. In contrast, nitrates probably oxidize the iron to the ferric condition and prevent the formation of soluble ferroate; such a change will be independent of geometry.

With the disappearance of the riveted boiler, the problem has ceased to exist in its original form, but a similar trouble, based apparently on unwanted rise in the concentration of alkali, has been met with at places where porous deposits of salts or debris exist on the heating surface. It might be thought that these are places least likely to favour a concentration rise, since steam production will occur mainly at places which are not covered with deposits. However, Bloom has pointed out that the attack on iron by the reaction

$$3Fe + 4H_2O \rightarrow Fe_3O_4 + 4H_2$$

uses up water, so that concentration is to be expected if attack is occurring at such points (which may well be anodic to the uncovered area).

Boilers at nuclear power stations

Today much of the heat used in steam-generation is derived from atomic energy. This introduces new materials (e.g. zirconium alloys), and new problems (the possible acceleration of corrosion reactions by radiation). A summary of the position, including possible problems which will arise when fusion reactors come into service – a desirable development in view of the smaller danger of atmospheric pollution – will be found elsewhere (1976 vol., pp. 241–4).

4

Influence of Environment

General

Contrast between conditions of atmospheric exposure and total immersion

The examples of attack considered in Chapter 2 were largely taken from researches carried out, at Cambridge, on specimens **partially immersed** in salt solutions with an ample supply of oxygen to the upper part and a super-abundance of water at the lower portion. In some respects, partial immersion constitutes a state of affairs intermediate between the conditions of **atmospheric exposure** studied by Vernon,[486] and the conditions of **total immersion** investigated by Bengough and his colleagues.[795] These two series of accurate research were started at the Royal School of Mines, South Kensington, and transferred later to the Chemical Research Laboratory, Teddington. In each type of corrosion, oxygen and an electrolyte usually play an important part; it is their relative supply which varies with the conditions.

Specimens freely exposed to the atmosphere will receive plenty of oxygen, but often very little moisture. In absence of electrolytes, a directly formed oxide film would stifle its own growth (p. 13). The presence of any electrolyte which can collect moisture and lead to a soluble corrosion product stimulates attack. Thus nickel remains bright in pure air, but when exposed indoors in moist air containing sulphur dioxide it becomes dull. Vernon found that this 'fogging' of nickel only takes place in air of r.h. (relative humidity) exceeding 70%; above that value, the invisible adsorbed film of sulphurous acid is oxidized, through the catalytic influence of the nickel, to sulphuric acid, which, being hygroscopic, collects water; consequently visible dew appears, even though the temperature is above dew-point. At first, the dew consists of sulphuric acid with dissolved nickel sulphate, and such a film can be wiped away with a cloth. Later, possibly during periods when the sulphur dioxide content of the air temporarily declines, solid basic sulphate appears in tenacious form, so that brightness can no longer be restored by wiping. In practice, fogging is today generally prevented by covering the nickel with a thin overlay of chromium.

Similarly iron, which in pure dry air suffers no visible change, soon becomes rusty if exposed to salt spray. Here the primary corrosion products are probably soluble ferrous chloride and sodium hydroxide – both hygroscopic bodies – while in spray of marine origin, the magnesium chloride present is also hygroscopic; the surface will remain damp even though exposure to spray ceases, and rusting, once started, is likely to continue.

On the other hand, specimens of iron fully immersed in water or salt solution, as in Bengough's experiments, will suffer attack at a rate largely dependent on the supply of oxygen to the metal. The addition to very pure water of small amounts of a salt (e.g. potassium chloride) will increase the rate of attack, but larger amounts of salt cause the corrosion velocity to decline again, because oxygen solubility falls off as salt concentration rises. If we disregard a small amount of corrosion connected with hydrogen evolution, the attack can never exceed a limiting rate determined by the arrival of oxygen at the surface, although on highly resistant materials this rate may not be reached.

Clearly the corrosion velocity is subject to a different **controlling factor** in these two cases. In **atmospheric** attack, the main controlling factor is **moisture**; under conditions of **total immersion** it is **oxygen**.

Variation of order of merit with conditions

The same substance may behave differently when brought into contact with a given liquid in two different ways. Zinc, placed in sodium chloride solution containing oxygen, is attacked more quickly than iron (other things being equal), largely because the e.m.f. of the cell Zinc | Oxygen exceeds that of the cell Iron | Oxygen. If the two metals are exposed to *fine* salt spray, the zinc is usually less attacked than the iron; here the soluble primary products (metallic chloride and sodium hydroxide) interact very close to the metal to give a largely adherent layer of hydroxide or basic chloride, which in the case of zinc provides some protection; the layer formed on iron gives less protection, probably because the ferrous hydroxide is converted to ferric hydroxide with a change of volume and interruption of continuity.

Even under fixed conditions of exposure, the order of merit of two or more substances will vary with the substances to which they are exposed. Magnesium is far more rapidly attacked by sulphuric acid, but less rapidly by hydrofluoric acid, than is lead.

Tests

Clearly the idea, once widely held, that it should be possible, by means of a 'standard corrosion test', to establish an 'order of merit' of materials which would be valid for all conditions, is wrong. The order of merit is different in different environments, and thus no one test can give the right results for all cases. Any test designed to decide the best material to withstand one particular set of conditions

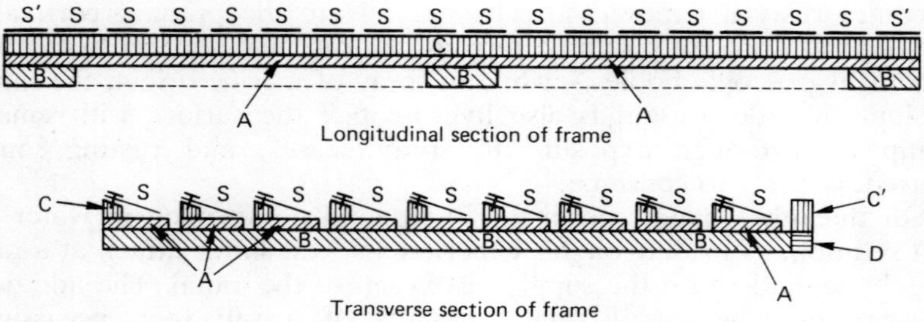

Fig. 4.1 Cambridge exposure frame for small specimens. S, steel specimens; S′, dummies; *A, B, C*, wooden strips

must be carried out in a form chosen to represent those conditions – perhaps intensified in order to give quick results, but free from any irrelevant influence.

In atmospheric exposure, behaviour varies according as the materials are washed by the rain or protected from it. For the latter conditions, Hudson[513] placed his specimens in Stevenson screens – the open-work boxes used by meterorologists to accommodate their instruments. For exposure to weather, the frame shown in Fig. 4.1 has been used in the tests organized from Cambridge; it can be placed on any flat roof – provided that there is no unequal sheltering from rain, wind or sun; the specimens, placed at 18° to the horizontal, do not screen one another. For complete immersion in the sea, specimens are often attached to rafts, whereas for intermittent immersion, they may be attached at half-tide level to a pier or other non-floating structure.

Many attempts have been made to simulate natural conditions in the laboratory, often in intensified form, so as to obtain an 'accelerated corrosion test'. At one time, it was thought that exposure in a box filled with air containing fine particles of salt spray or acid spray should serve to assess the relative resistances of materials to sea air and town air respectively. In one form of spray test, the specimens are kept wetted continuously, and the changes which in natural exposure occur to the corrosion product whenever it dries up or reabsorbs water (changes which affect protective properties) are left out of account. By applying the spray intermittently, with intervening periods of drying, an order of merit is obtained which accords better with that obtained on natural exposure to the weather. In another procedure, specimens are attached to an endless belt conveyor, so that they are alternately subjected to a spray and a drying lamp; sometimes a douche is incorporated in the cycle. Vernon improved the spray test by introducing substances which he had found to play a part in corrosion by polluted atmospheres; he proposed a solution containing sulphuric acid M/1000, sulphur dioxide (0·02% by volume), ammonium sulphate M/100 000 and carbon dioxide (nearly to saturation). For testing plated steel, 5% sodium chloride acidified with acetic acid to give pH 3·2–3·5 (the so-called CASS test) has found favour.

Laboratory testing apparatus is also available to provide, with or without agitation total partial or intermittent immersion in any liquid whose action on a series of metals is under investigation.

Criteria of destruction

As a measure of the corrosion which has occurred, the gain in mass of a specimen may be used where the corrosion product remains adherent, e.g. in high-temperature oxidation or atmospheric attack. If the product is soluble or loose, as often in immersed corrosion, the loss of mass is a better criterion; it must be determined after complete removal of the product, e.g. by acid containing enough restrainer (p. 175) to prevent attack on the metal during the cleaning.

Alternatively, the amount of metal found in the corrosion product (whether dissolved, loose or adherent) may be estimated; where the amount is small, colorimetric methods are available.

By exposing a series of specimens for different times, curves may be plotted indicating progress of attack with time; these serve to distinguish cases where attack stifles itself from those where it does not. But, since a series of nominally 'identical' specimens do not always suffer corrosion at the same rate, the points may not fall well on a single curve, and there is obvious advantage in a method providing a complete curve from a single specimen. Where a metal is attacked by acid, evolving hydrogen, the volume of the hydrogen liberated furnishes such a curve.

In corrosion of the oxygen-absorption type, the volume of oxygen disappearing was used by Bengough in the same way; since a little hydrogen is generally evolved even from neutral solutions, it was necessary periodically to burn this hydrogen by means of a heated platinum wire, the contraction of volume being noted; the specimen was a metallic disc supported horizontally on three glass points, a known distance below the surface of the liquid. The space above the liquid was filled with oxygen, the disappearance of which was followed in a burette.

A complete corrosion–time curve can be furnished from a single specimen of wire (preferably a spiral) if its electrical resistance is measured from time to time during exposure to corrosive influences. Such information will only agree with gravimetric results if the corrosion is absolutely uniform, and any discrepancy between the two methods may provide useful evidence of localized attack.

The loss of tensile strength, of elongation or of fatigue strength brought about by corrosion is of great importance to engineers. Here again, an apparent disagreement with the results of gravimetric studies points to some form of localized attack; cracking will affect mechanical properties more than pitting.

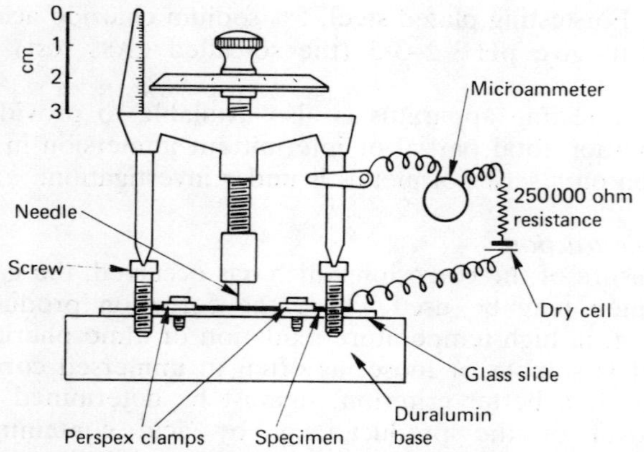

Fig. 4.2 Apparatus for measuring the depth of penetration of localized corrosion (R. S. Thornhill)

Pitting, however, possesses practical importance, since it determines the time which pipes or containers can survive before they become perforated. An electrical micrometer used by Thornhill[799] to measure the depth of penetration at pits or patches of localized corrosion is shown in Fig. 4.2. The needle is screwed down until it touches the base of a pit; when contact is made, an electric circuit is established, and a deflection of the microammeter shows that the bottom has been reached; comparison with readings obtained at uncorroded places outside the pit gives the depth. Direct reading gauges for measuring pit depths are available.

Atmospheric attack

Exposure to air containing volatile electrolytes

Early experiments carried out by the author[487] on various metals exposed to air containing the fumes of volatile reagents may serve to illustrate the factors governing atmospheric attack, although the concentrations of the corrosive substances were far higher than usually occur even in a polluted industrial atmosphere. The experiments were conducted in vessels of the form commonly used as desiccators, the specimens being placed on a false bottom in the upper part, and the volatile liquid below; hydrochloric, sulphurous and carbonic acids, hydrogen sulphide solution and ammonia were used in different vessels; one vessel contained pure water. The liquid evaporated by day and condensed on the specimens by night. Pure water and even carbonic acid solution produced little real change, beyond a little dulling in some cases, and tiny rust spots on iron. The other reagents caused appreciable attack on most metals, although lead resisted the vapours from sulphurous acid, tin underwent little change over any liquid apart from slight discoloration, whilst aluminium remained unaffected except over hydrochloric acid, which on some specimens produced a feathery growth – probably from weak spots in an otherwise protective oxide film. Iron over hydrochloric acid assumed a frosty aspect; when subsequently taken out into the ordinary damp atmosphere it absorbed moisture (ferric chloride being hygroscopic) and developed rust; hydrochloric acid vapour produced interference colours on lead, but the attack was only superficial.

The most serious action was observed where the surface absorbed moisture, and remained visibly wet even by day when, in the absence of hygroscopic bodies, water would have evaporated. Zinc exposed over hydrochloric acid developed a colourless syrup, and was rapidly eaten through; here the hygroscopic substance was zinc chloride. Sulphurous acid caused rapid rusting on iron, presumably because the portion adsorbed on the surface became catalytically oxidized to sulphuric acid, which attracted moisture; the brown–black deposit was 0·1 mm thick after two weeks. Zinc also became damp, and after two weeks was covered with a pasty deposit 0·2 mm thick. Nickel after two days began to shed a pale green liquid and the surface became covered with a green–black layer containing more sulphite than sulphate; this was 0·5 mm thick after two weeks. Ammonia, which left iron bright and unchanged, owing to the alkaline reaction, produced a marked change on copper; a violet liquid containing the ammine $[Cu(NH_3)_4](NO_2)_2$ ran off the surface; here again catalytic activity will be noticed, ammonia being oxidized to the state of nitrite.

Outdoor tests

Some results provided by the extensive tests carried out by Hudson[513] on specimens either fully exposed to the rain or placed in Stevenson screens are summarized in Tables 4.1, 4.2, 4.3 and 4.4. Among non-ferrous materials, the least resistant (zinc and $\alpha\beta$-brass) corrode nearly five times as fast as the most resistant material (80/20 nickel chromium); corrosion was more rapid in the urban atmosphere of Birmingham (Post Office roof) and the marine atmosphere of Southport (golf course) than in the rural atmosphere of Cardington.

The numbers suggest that under normal conditions non-ferrous materials resist corrosion remarkably well, but it is well to point out that special circumstances exist where attack is more rapid. If rain falls on a sloping roof of slate or glass exposed to coal-smoke or acid fumes, it will pick up much acid as it descends, and if it runs off at a limited number of points on to zinc or galvanized (zinc-coated) iron, there may be rapid attack at these points, which receive, as it were, far more than their proper 'ration' of acid. Moreover, although, as explained on p. 99, zinc or zinc-coated iron resists fine salt spray, the state of affairs is very different if large drops of salt water lodge between sheets of galvanized iron, piled near the coast or in the hold of a ship. In such a situation, the mechanism described on p. 36 pro-

Table 4.1 Average thickness corroded at Birmingham expressed in millionths of an inch per year (J. C. Hudson)

	Mass increase of wire in Stevenson screens	Electrical resistance tests on wire completely exposed	Mass loss of plates completely exposed
80/20 nickel–chromium alloy	111	311	85
Lead	—	—	145
Arsenical copper	126	353	154
High-conductivity copper	116	302	158
70/30 nickel–copper alloy	376	656	243
Cadmium–copper (0·8% Cd)	120	350	179
Nickel	245	565	230
Aluminium bronze (3·5% Al)	170	410	229
Silicon bronze	—	342	228
80/20 copper–nickel	204	361	213
Tin bronze (6·3% Sn)	151	305	243
70/30 brass	172	594	316
High-purity zinc	280	—	376
Ordinary zinc	—	—	388
60/40 brass	—	—	408
Compo wire	719	721	—

duces anodic attack in the centre of the drop, leading to the formation of zinc chloride, whilst sodium hydroxide is formed at the cathodic zone around the periphery: where they meet, zinc hydroxide will certainly be precipitated, but in a membranous form growing out at right angles to the surface and incapable of stifling attack.

Hudson's results[575] on steel specimens are shown in Table 4.3, along with corresponding results for zinc. Except in the tunnel, steel suffers attack much more rapidly than zinc (and hence than other non-ferrous materials), but here again it is quickest in industrial atmospheres containing oxides of sulphur. Table 4.4 indicates how copper in steel increases the resistance. The reason is discussed on p. 67.

Table 4.2 Average thickness corroded on the 16 non-ferrous materials shown in Table 4.1 expressed in millionths of an inch per year (J. C. Hudson)

	Mass increase of wire in Stevenson screens	Electrical resistance tests on wire completely exposed	Mass loss of plates completely exposed
Cardington (rural)	32	156	74
Bourneville (suburban)	67	248	119
Wakefield (industrial	90	368	170
Birmingham (urban)	190	395	221
Southport (marine)	71	345	144

Table 4.3 Corrosion in one year expressed in millionths of an inch (J. C. Hudson)

	Ingot iron	Zinc
Khartoum	28	22
Abisko, Sweden	158	36
Aro, Nigeria	339	57
Basrah	400	28
Singapore	517	46
Apapa, Nigeria	654	38
Llanwrtyd Wells	1888	118
Calshot	2104	121
Dove Holes Tunnel (up side)	2841	3609
Dove Holes Tunnel (down side)	3066	3449
Motherwell	3137	180
Woolwich	3472	146
Sheffield	3888	576

If, in addition to copper, chromium is present, the life is further prolonged. Much interest is today being taken in the use of **low-alloy steels**, which although not possessing the resistance of 18/8 stainless steel (p. 108) are capable of affording great economies, especially if paint is applied. Some of them contain expensive ingredients like nickel and molybdenum, but the tests of Hudson and Stanners[506] show that a steel with 2·8% Cr, 1·5% Al and 0·8% Si is almost as good as any tested, and corrodes at only one quarter of the rate of unalloyed steel.

Low-alloy steels have been used unpainted on many buildings in the middle west of the U.S.A. with satisfactory results, since the rust, which is darker and more adherent than that formed on ordinary steel, provides considerable protection. Whether such materials could be used

Table 4.4 Relative corrosion in five years' exposure tests (J. C. Hudson)

Material	Calshot (marine)		Dove Holes Tunnel (highly polluted)	Sheffield (industrial)	
	As rolled	De-scaled	as rolled	As rolled	De-scaled
Steel X (0·02% Cu)	100	100	100	100	100
Steel Y (0·2% Cu)	81	92	—	80	83
Steel Z (0·5% Cu)	78	87	—	79	82
Slagless wrought Iron (no copper added)	—	—	98	155	131
Slagless wrought iron (0·6% Cu)	—	—	92	68	67
Staffs. wrought iron (no copper added)	113	95	79	77	78
Scottish wrought iron (0·12% Cu)	102	93	90	73	66
Ingot iron	148	120	96	104	106

without a protective coating of paint or tar at places nearer the sea is doubtful. Some information regarding the relative resistance of different types of steel is however available (1976 vol., pp. 265–8). Tests at Bayonne, N.J. (U.S.A.) and Sheffield (U.K.) showed Cu–Cr steels to be superior to Cu steels, which in turn were superior to unalloyed steels. At Lille (Belgium), the behaviours of Cu–Cr steels and Cu steels were not very different. French tests have shown that close to the coast Cr–Al–Ni steels were distinctly more resistant than 3·5% Cr steel, but the advantage of using the more complex material declined with increasing distance from the sea. Japanese work has emphasized the variation of behaviour with design; places where rain water is retained may fail to produce a stable, protective rust even after four years.

Indoor tests

Vernon[76] found that copper exposed indoors obeyed the parabolic law, the mass increase M in time t being given by $M^2 = Kt$. Zinc obeyed the rectilinear law, $M = Kt$, but as shown in Fig. 1.9, the value of K varied between experiments started on different dates, although in each experiment it remained remarkably constant, despite the fact that atmospheric conditions were varying. That suggests that *the course of attack is determined by the conditions prevailing at the commencement of exposure*; observations by American and German investigators confirm this important principle.

Vernon's observations on iron indicate that the spots of rust which appear on exposure in a room are connected with dust particles settling on the surface. Various components of dust behave differently; silica is harmless, ammonium sulphate promotes attack, whilst carbonaceous particles (soot) stimulate corrosion if they carry adsorbed sulphur compounds. He exposed iron and steel specimens in a muslin cage which, whilst allowing access of air, prevented dust particles from reaching the metal, the spots of rust which would otherwise have appeared were then avoided; if after exposure for eleven months in such a sheltered situation, the iron was exposed to ordinary air containing dust particles, the rusting was far less serious than that of iron freshly exposed to dusty air, since the invisible oxide film had reached a condition capable of providing a measure of protection.

Atmospheric behaviour of copper

Vernon[490] carried out a series of quantitative experiments on copper, which he found to be little altered by air containing either sulphur dioxide or moisture alone, but to undergo serious alteration when both were present. One set of experiments, with the sulphur dioxide concentration kept at 10% gave results represented by the curves of Fig. 4.3. It will be noticed that so long as the R.H. does not exceed 63% the corrosion is slow, but that at 75% the attack is considerable. Apparently at this value the adsorbed sulphur dioxide can collect sufficient water for catalytic oxidation to sulphuric acid, which in presence of oxygen attacks

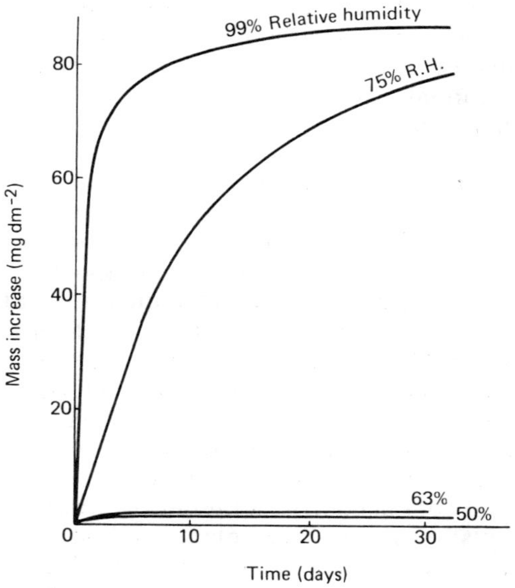

Fig. 4.3 Corrosion of copper by air containing 10% sulphur dioxide and different degrees of moisture (W. H. J. Vernon)

copper readily. Several other cases are known of this sudden rise of corrosion velocity at a certain **critical humidity**.

When the moisture content was kept constant whilst the sulphur dioxide was varied, Vernon found that 0·9% sulphur dioxide produced less corrosion on copper than either higher or lower contents; the film produced at this particular concentration was found to be normal $CuSO_4$, which provided better protection than either the basic sulphate film formed at lower sulphur dioxide concentrations or the more acidic film formed at high concentrations.

It is well known that many ancient copper roofs carry a beautiful green patina. In urban districts this appears only after some years, its development being favoured by wind and rain. Vernon and Whitby[517] found that, in the opening stages, the deposit is a dark aggregate of sulphide, oxide and soot; the final green patina is formed in part by the action of sulphuric acid in the soot, and partly by the oxidation of copper sulphide. Evidently the changes are complicated. Analyses by Vernon and Whitby of the products from ancient green roofs show that those 70—300 years old carry basic copper sulphate $CuSO_4.3Cu(OH)_2$ (**brochantite**), usually with a little basic carbonate $CuCO_3.Cu(OH)_2$ (**malachite**). The samples from newer roofs show a lower basicity, and even normal copper sulphate may be present. Near the sea coast, basic chlorides are found, their composition usually approximating to $CuCl_2.3Cu(OH)_2$ (**atacamite**) in the oldest deposits. In most coastal towns, sulphates seem to exceed chlorides in amount. On copper or bronze articles exposed for centuries in salt or alkaline desert, chloride and carbonate may become the most important constituents of the deposit formed on the surface. Copper nitrate, basic or normal – due to electric discharges – has been reported on electrical equipment.

Atmospheric behaviour of stainless steels

Whilst the low-alloy steels suffer atmospheric corrosion more slowly than ordinary mild steel, they cannot be called non-rusting. To ensure retention of brightness during ordinary atmospheric exposure, it is necessary to add **chromium** in considerable quantities. **Stainless cutlery steel** contains about 14% of chromium and 0·3% of carbon; it is usually hardened about 950°C followed by quenching (or sometimes air-cooling) and is subsequently tempered about 200°C, the structure being then essentially martensitic. A similar alloy low in carbon and consequently softer, known as **Stainless Iron**, is used in structural engineering.

Where no hardening is needed, an enhanced resistance to atmospheric attack can be obtained from austenitic (γ-iron) alloys with larger chromium contents; these have been mentioned on p. 76 in connection with acid resistance, and are sold under various trade names; a favourite composition is 18% chromium, 8% nickel – the nickel serving to ensure an austenitic condition. The austenitic state is obtained by heating to about 1100°C and cooling rapidly. If, as may happen during welding, an 18/8 alloy is heated to about 700°C, the chromium at the grain boundaries tends to separate out as carbide, leaving an impoverished network, so that the alloy becomes susceptible to intergranular attack, as explained on p. 154. This is largely overcome by adding other elements, such as titanium or niobium (colombium).

Mechanism of atmospheric attack

The early experiments just quoted gave qualitative evidence of electrochemical action in situations where two metals were in contact. Over hydrochloric acid, lead remained bright at a contact with zinc, but quickly darkened elsewhere; over ammonia, zinc developed a yellow-white excrescence at a junction with copper, although hardly changed elsewhere.But in situations where only a single metal is exposed, it was generally considered, until 1960 or later, that the mechanism is not connected with electric currents flowing between well-separated anodic and cathodic areas. The most destructive type of atmospheric attack, the rusting of iron in places where the air contains sulphur dioxide as well as moisture, was generally explained by an **acid regenerating cycle**. According to that view adsorbed SO_2, moisture and oxygen gave H_2SO_4, which attacked the iron to produce $FeSO_4$; this was oxidized to the ferric condition, with accompanying hydrolysis to give rust (largely FeO.OH or $Fe_2O_3.H_2O$) and regeneration of H_2SO_4, which could attack further iron, with further hydrolysis and regeneration – and so on. At first sight, it would seem that a small amount of adsorbed SO_2 would produce an infinite amount of destruction, but in fact the hydrolysis will not be complete, and a certain proportion of the SO_4^{2-} is removed each cycle as a sparingly soluble basic sulphate. Thus the SO_2 will produce only a finite quantity of rusting. Schikorr (1976 vol., p. 252) in his extensive tests in Germany found that one molecule of SO_2 will convert fifteen to twenty atoms of iron to rust in winter, and thirty to forty atoms in summer; since, however, the air contained four times as much SO_2 in winter as in summer, the rusting was about twice as fast.

More recent laboratory researches have shown that the reactions postulated in the acid regeneration cycle can really take place. In the early stages of exposure outdoors, they do play an important part in setting up corrosion. However, when once some ferrous sulphate and some rust have been formed, conditions become favourable to an *electrochemical cycle*, which proceeds very much more rapidly than the acid regeneration cycle. Anodic attack on the iron:

$$Fe \rightarrow Fe^{2+} + 2e^-$$

can be balanced by cathodic reduction of rust to magnetite:

$$8FeO.OH + Fe^{2+} + 2e^- \rightarrow 3Fe_3O_4 + 4H_2O$$

The magnetite can then be re-oxidized to rust by atmospheric oxygen,

$$3Fe_3O_4 + 0{\cdot}75O_2 + 4{\cdot}5H_2O \rightarrow 9FeO.OH$$

It will be noticed that now there are nine molecules of FeO.OH instead of eight; the net gain of one molecule is provided by the destruction of one atom of iron on the anodic area. Since SO_4^{2-} is not used up in the equations as printed, it would appear that a single molecule of $FeSO_4$ would produce an infinite amount of rusting; but, as mentioned above, SO_4^{2-} will gradually be removed from the system as basic salt, and, unless fresh $FeSO_4$ is formed by SO_2 present in the air, the rusting will ultimately cease.

The electric current involved in the changes has been detected and measured by Taylor (1976 vol., pp. 254–60), using a weighed plate of iron covered by filter paper soaked in M/10 $FeSO_4$ upon which rested a plate of previously rusted iron; the two plates were connected through a milliammeter (Figs. 4.4 and 4.5). The current passing was measured at different times (Fig. 4.5). It was strong at first, but as the supply of rust became exhausted, it diminished; after a time, the upper plate was turned upside down, providing a fresh supply of rust for the cathodic reaction, with marked increase in the current, whilst the exposure of the freshly formed magnetite to air allowed its re-oxidation to rust. It was found that the total number of coulombs produced after a number of these inversions predicted an amount of corrosion in reasonable accord with that indicated by the loss of mass of the lower electrode. The agreement between the electrometric and gravimetric measurements was less exact than that found by Hoar in his work on iron partly immersed in a salt solution (p. 41), but it nevertheless provides con-

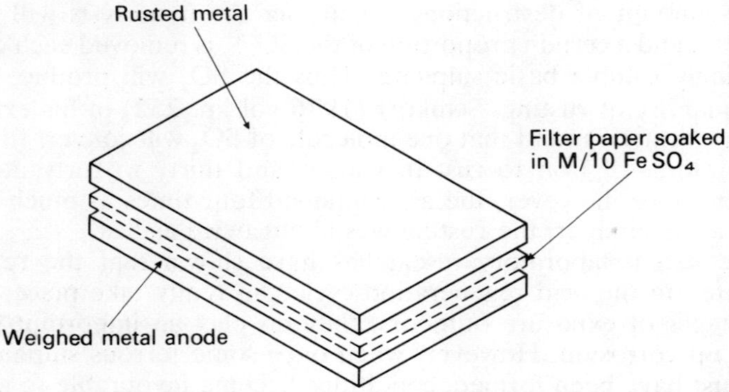

Fig. 4.4 Plate experiments designed to decide between two mechanisms. A weighed metal specimen is covered with filter paper soaked in M/10 $FeSO_4$ on which is laid a sheet of rusted iron; the rusted iron is connected to the lower specimen either directly or through a milliammeter. The specimen gains or loses mass in different cases. The results favour the electrochemical mechanism and not the acid regeneration cycle (U. R. Evans and C. A. J. Taylor)

vincing evidence of the electrochemical mechanism. On steel specimens the loss in mass was rather higher than would be expected from the electrical figures, probably because the corrosion ate down around cementite or other non-metallic particles, which were dislodged uncorroded. In contrast, where pure iron specimens were used, the discrepancy was in the opposite direction; the number of coulombs obtained was slightly too great for the loss in mass, probably because the Fe^{2+}

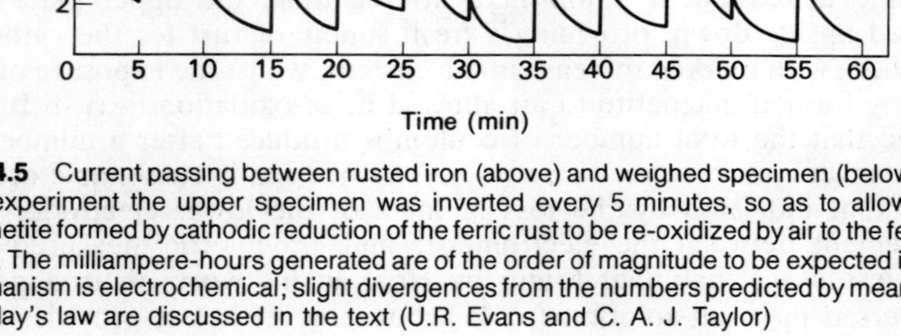

Fig. 4.5 Current passing between rusted iron (above) and weighed specimen (below). In this experiment the upper specimen was inverted every 5 minutes, so as to allow the magnetite formed by cathodic reduction of the ferric rust to be re-oxidized by air to the ferric state. The milliampere-hours generated are of the order of magnitude to be expected if the mechanism is electrochemical; slight divergences from the numbers predicted by means of Faraday's law are discussed in the text (U.R. Evans and C. A. J. Taylor)

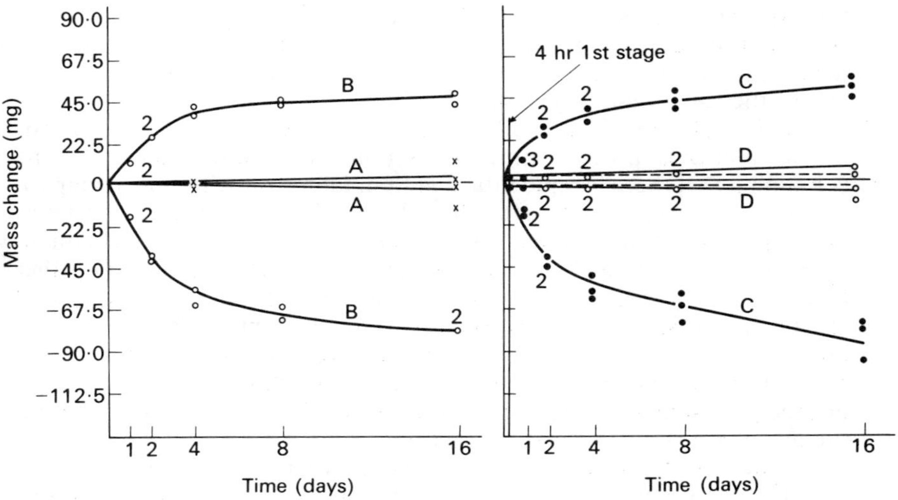

Fig. 4.6 Experiments showing influence of SO_2. Curves A relate to iron exposed to moist air without SO_2, and curves B to iron in moist air containing SO_2; both gain and loss of mass are much greater when SO_2 is present. Curves C represent two-stage experiments; after 4 hours in moist air containing SO_2, the specimens were moved to moist air without SO_2, yet the mass change is similar to B, showing that when once $FeSO_4$ has been formed, SO_2 in the gas phase is not needed for corrosion. Curves D represent three-stage experiments; after 4 hours in moist air with SO_2, the specimens were placed in water for one day and then in moist air without SO_2; the mass change is small because the $FeSO_4$ has been removed (U. R. Evans and C. A. J. Taylor)

ions were converted, by a secondary anodic reaction, to Fe^{3+} – thus providing additional electrons.*

Taylor carried out quantitative tests on weighed specimens exposed for various times in moist air containing various amounts of SO_2. Each specimen was weighed twice after the experiment; the first weighing gave the increase of mass due to rust-production, and the second weighing the loss of mass after removal of the rust; in the longer experiments, the ratio of the two numbers agreed fairly well with the theoretical value to be expected if the deposit consists solely of FeO.OH; in the short experiments there was some discrepancy, probably attributable to the presence of ferrous sulphate and other minor constituents. In some of

*This secondary anodic change will not take place under conditions of immersion in a salt solution, since the Fe^{2+}, after formation, will rapidly diffuse away; hence the agreement between results obtained by electrochemical and gravimetric methods is good (p. 41). It has, however, been observed uunder conditions less favourable to removal of the Fe^{2+} – notably in Thornhill's studies of the rusting of iron inscribed with a scratch-line and covered with filter paper soaked in a $NaHCO_3$ solution of a concentration chosen to produce passivity on the main area but allow corrosion at the scratch where the atoms are disarrayed; here the filter paper allows only slow diffusion (1960 vol., p. 867).

the experiments, the specimens were subjected to two stages of corrosion; first there was exposure, usually for 4 hours, to moist air containing SO_2; then the specimen was moved to a tube containing air (generally of the same humidity) without SO_2.

Some results are shown in Fig. 4.6. One-stage experiments in moist air, without SO_2, produce little attack (curves A), but in the same air containing SO_2 there is rapid attack at first, falling off with time owing to exhaustion (curves B). If, after four hours, (a time found to be long enough to use up the whole of the SO_2 in forming $FeSO_4$), the specimen is moved to another tube containing moist air, without SO_2, the behaviour (curves C) is much the same as in air containing SO_2 (curves B); but if the specimen is exposed to water to remove $FeSO_4$, subsequent corrosion in moist air is slight (curve D). These experiments support the mechanism suggested above, and show that SO_2 is only needed as a source of $FeSO_4$; when once this has been formed, SO_2 in the gas-phase is not needed for rusting.

The details of rust formation probably vary with conditions of exposure. If iron is exposed out of doors and becomes covered with water during periods of rain, the rust coat may be completely reduced to magnetite (in laboratory experiments rusty specimens placed in vessels completely filled with a water or $FeSO_4$ solution became black through magnetite formation). Then when the rain ceases the magnetite will be re-oxidized, producing more rust than had existed before the rain started. Iron protected from rain, but exposed to damp air containing SO_2 is likely to suffer rusting continuously. The situation is suggested in Fig. 4.7. The anodic attack producing Fe^{2+} will occur at the level XX′, and the cathodic reduction of rust to magnetite, with re-oxidation by air to rust (a larger quantity than before), at the level YY′.

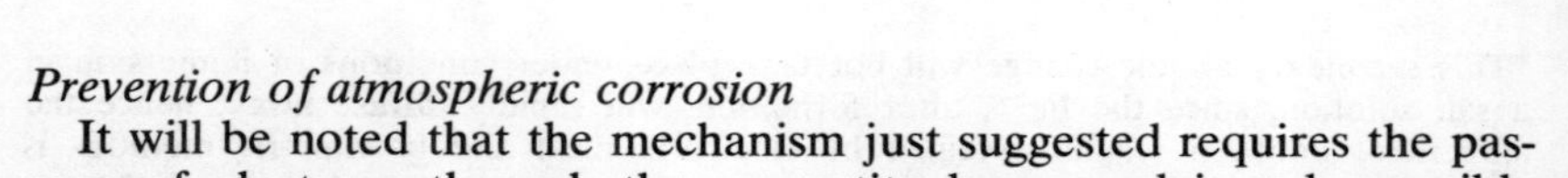

Fig. 4.7 Mechanism of atmospheric rust formation (schematic). The anodic attack on iron occurs at level of XX′ and the cathodic reduction of FeO.OH to Fe_3O_4 (later reconverted to FeO.OH, in larger amount) occurs at YY′

Prevention of atmospheric corrosion

It will be noted that the mechanism just suggested requires the passage of electrons through the magnetite layer, and is only possible because magnetite is an electronic conductor. This explains why zinc, a metal which under immersed conditions often suffers more rapid corrosion than iron, is less liable to atmospheric corrosion, since its oxide is a relatively poor electronic conductor; moreover, owing to the fact that it forms only one oxide, the cathodic action which occurs on iron cannot occur on zinc. The improved behaviour of an iron alloy containing a small quantity of Cr may be partly due to the fact that a spinel containing Cr will be a less good conductor than magnetite. (However, the instability of the divalent state in chromium probably plays a part in explaining the good behaviour of alloys containing Cr.) In these cases it would seem that atmospheric corrosion is being slowed down by an increase in the resistance of the electronic limb of the electrical circuit.

However corrosion can also be prevented or slowed down if there is interruption to the electrolytic limb, where conductivity depends on the movement of ions. The mechanism suggested above requires an electrolytic connection between XX′ and YY′ by means of $FeSO_4$ solution in the pores between the magnetite particles. If humidity is low, this solution will dry up, and corrosion will become very slow. Vernon showed (p. 98) that at r.h. values below 70% only slight attack occurred, and Taylor's results at various humidities also placed the critical humidity at 70%, although the conditions of his work were different. However, the critical value can vary appreciably with the SO_2 content of the atmosphere (1976 vol., p. 260); Barton placed it at 75% for 1 ppm SO_2, but at 67% for 10 to 100 ppm. Where the atmospheric contamination consists not of SO_2 but of sea salt, the critical value lies much lower. (U. R. Evans and C. A. J. Taylor, *Brit. corr. J.*, 1974, **9**, 26.)

Even if the r.h. of the air stands above the critical level, atmospheric corrosion may be slow if the alloy is one which will raise the resistance of the electrolytic limb. The presence of Cu in steel undoubtedly improves its behaviour, as shown in Table 4.4 and Copson has attributed this to a blocking of the pores with basic copper sulphate (1976 vol., p. 268). Such an explanation seems reasonable; any solid matter produced will decrease the cross-section available for the flow of ions, thus increasing the resistance of the electrolytic limb; apart from the geometrical changes, the removal of SO_4^{2-} from the liquid will raise its specific resistance, and the same will happen if it is removed in any other form; Shreir and his colleagues attach importance to removal as copper sulphide, and a similar view has been expressed by Wranglen.

Corrosion of buried metal work

General

Conditions existing in the soil may vary between something comparable to atmospheric exposure and what is almost equivalent to complete immersion. A metal pipe buried in very open, sandy ground has a considerable portion of its surface exposed to air. Oxygen gains sufficient access to ensure that ferric products are formed close to the metal; a film of rust may soon appear on the surface, but it frequently stifles further attack. In waterlogged soils, free oxygen is deficient, and, in the absence of those bacteria which make the oxygen of sulphates available for the corrosion reaction, there is only slow attack.*

Soils of intermediate character may produce corrosion which is localized, and accordingly intense. If air pockets are present, either as a natural feature of the soil, or produced artificially when the soil is thrown back into a trench after the laying of a pipe-line, so that spaces are left in between the individual spadefuls, differential aeration currents may flow; oxygen is taken up at the air pockets and attack is directed on the places where the soil presses on the metallic surface.

Again, cells of the type

Steel | Soil A | Soil B | Steel

may be set up where a pipe passes from one soil to another. The currents are known as **long-line currents**. There is a special risk of corrosion if a pipe is buried at a level representing the boundary between two horizontal soil-strata. Attack is sometimes met with where a steel or lead pipe passes near salt-pockets or mineral springs, whilst different constituents of the soil may themselves set up corrosion currents; thus where lead is buried in a ground containing lumps of chalk, local attack may occur, probably because the chalk precipitates lead as carbonate and thus keeps the Pb^{2+} concentration low locally, bringing about the concentration cell

Lead | Dilute Pb^{2+} | Concentrated Pb^{2+} | Lead

Steel pipes buried in cinders, especially those containing sulphur, are rapidly corroded.

Sulphate-reducing bacteria

Even more dangerous than the purely chemical or electrochemical types of attack on pipes is the action of the sulphate-reducing bacteria known usually as *Vibrio desulfuricans* (also as *Spirillum, Spirovibrio* or *Microspira desulfuricans*). These organisms can flourish only in

*Nails deeply buried about AD 87 at the Roman fortress of Inchtuthil, Perthshire, carefully covered with clean beaten earth under conditions practically excluding oxygen except from the uppermost layer were found to be almost unrusted about 1870 years later; half of them, however, showed some corrosion over half their area, whilst in a few cases there was deep pitting; the protection was partly due to scale. American studies of pilings driven into the ground under conditions which must compress the soil (in contrast with digging and replacing the soil, which will leave it loosened), show surprisingly little corrosion, even where the soil drainage and properties have been of the sort which would normally be considered favourable to the growth of sulphate-reducing bacteria. Romanoff adds the comment, 'It might be reasonable to presume that the bacteria are inoperative under the compressed soil conditions, or that the corrosion generally attributed to sulfate-reducing bacteria is overestimated. I have long been of the opinion that the latter is correct' (M. Romanoff, private communication).

anaerobic soils, like waterlogged clay, containing sulphates and organic matter. Soils containing no free oxygen are usually non-corrosive, if sterile, but the organisms mentioned enable sulphates to act as hydrogen-acceptors, with reduction to sulphides. Thus soils which are insufficiently acid to produce appreciable attack of the hydrogen-liberation type, and which contain too little oxygen for the oxygen-reduction type, nevertheless set up serious attack in presence of these bacteria. Instead of rust, a black coloration due to iron sulphide is found around iron or steel suffering from the bacterial type of destruction (but black coloration may arise in other ways, e.g. when iron is precipitated as magnetite). Both cast iron and steel pipes can be affected by bacterial action, the attack being often localized and sometimes very rapid. On cast iron, the graphite network remains unchanged, so that the pipe may appear undamaged, until tested with a knife point, when portions will be found to be very soft.

The explanation of this type of attack given by von Wolzogen Kuhr, when he first called attention to its dangers, is still largely accepted, although there are now known to be subsidiary effects (1976 vol., pp. 172–5). He showed that soils which were almost non-corrosive when sterile, since their acidity was too low for the cathodic reaction

$$2H^+ + 2e^- \rightarrow H_2$$

and the oxygen concentration too small for

$$O_2 + 2H_2O + 4e^- \rightarrow 4OH^-$$

could nevertheless cause serious attack if the bacteria mentioned above were present. The oxygen of SO_4^{2-} ions (generally present in soils) then became available for the cathodic reaction, which took the form

$$SO_4^{2-} + 8H^+ + 8e^- \rightarrow S^{2-} + 4H_2O$$

The S^{2-} produced could combine with the Fe^{2+} produced by the anodic reaction to give FeS, explaining the black coloration.

This does indeed seem to be the essential cause of the trouble, but it is now known that the FeS, once formed, can itself cause further corrosion by acting as a cathodic surface. Research at Manchester (1976 vol., p. 173) has shown that chemically prepared FeS can set up corrosion on iron; high corrosion rates are only maintained if the FeS is regularly replenished.

It is likely that differential aeration currents play a part in bacterial corrosion, since the organisms can adhere to the metallic surface, keeping the oxygen concentration very low. Also H_2S produced by bacterial processes, moving upwards to regions where oxygen is available can be converted to H_2SO_4, which is, of course, very corrosive. Furthermore the Fe^{2+} can be converted to Fe^{3+} by the so-called 'iron bacteria', producing clogging. There are thus a large number of problems, and the situation is complicated.

Prevention of corrosion of pipes

Since steel suffers little corrosion in open soils, it is possible to dig a trench, fill it with porous brick, gravel or even sand (free from clay), thus isolating the steel from the corrosive ground around; chalk in the sand is said to add to its protective character, and one authority recommends surrounding the pipe with clay to which lime has been added. In another method, usually effective but expensive, the pipe is set centrally in a long 'box' which is then filled with a cement or concrete surround. Pipes externally or internally covered with cement mixtures are available.

Another plan is to coat the pipe with compositions containing bitumen or tar (tar from horizontal retorts is superior to that from vertical retorts). For relatively mild conditions painting with a solution of tar which dries by evaporation may suffice. A better plan is to dip the whole pipe into a hot mixture giving a coat which becomes solid on cooling. For really corrosive soils, thick coatings of a bituminous or tarry base fortified with a filler such as ground limestone, silica, mica, or slate are necessary.

For many years it was customary, after applying a priming coat of tar or bitumen solution (which dries by evaporation), to wind the pipe with bandages of hessian soaked in hot tar, which hardens on cooling. Unfortunately many soils contain cellulose-destroying bacteria, which rot the hessian, so that wrapped pipes occasionally behave worse than unwrapped ones; the situation is made worse by the fact that the products of bacterial rotting, although not directly corrosive, provide food for the sulphate-reducing bacteria mentioned above. At one time, use of asbestos felt saturated with the bitumen or tar was advocated, but asbestos is dangerous to health, and the mechanical properties of the felt are not ideal. Fibreglass covered with synthetic resin may be more suitable.

By galvanizing steel pipes (coating them with zinc), considerable increase in life can be obtained, except in acid soils, for which galvanized pipes are unsuited. Logan's classical studies[279] of soil corrosion in America showed that in most soils, a zinc coating of $854{\cdot}4\ \mathrm{g\ m^{-2}}$ prevented appreciable rusting for ten years. Lead-coated pipes behaved well for a time, but when once the coating was eaten through, attack proceeded rather fast; the contrast between zinc and lead is not surprising, since zinc is anodic to iron, whereas lead is usually cathodic.

Attempts to avoid corrosion of steel pipes by altering their composition have not been very successful. Logan's tests suggest that there is comparatively little difference in the behaviour of mild steel, wrought iron and numerous low-alloy steels. High-silicon iron and austenitic chromium–nickel steels show good chemical resistance, but the former is brittle and the latter expensive.

Cathodic protection

If a carefully coated pipe could be buried without damage to the coating, it would probably resist corrosion in the average soil. In practice, the coating generally does suffer damage. Common observation of pipes laid out ready for burial at a roadside often reveals spots where the steel is already rusty at breaks in the coating; if the pipeline is one likely to be affected by stray currents or long-line currents (p. 114) the attack in the anodic regions, being concentrated on the gaps in the coat, may be more intense than if the whole had been uncoated. This danger is best avoided by making the line as a whole cathodic, and thus avoiding any anodic points. On paper it would be possible to use cathodic protection as the sole means of avoiding attack and thus dispense with coatings. A very powerful current would, however, be needed, and the cost of power would be great. It is more economical to apply a reasonably good coating and then employ a relatively small current to prevent corrosion at the inevitable points of damage. That principle is being used largely for pipelines, and also to protect the hulls of ships; in the latter case, it is necessary to avoid the use of paints which would be softened or loosened by alkali.

In a general way, cathodic control of attack can be regarded as the converse of anodic stimulation (p. 50). However, several different principles can be involved:

(1) If, by an applied e.m.f., the potential at a steel pipe surface is sufficiently depressed, the equilibrium Fe | Fe^{2+} will be established as soon as a certain concentration of Fe^{2+} has been formed in the soil-water; if the potential were then to be depressed still further, cathodic re-deposition of metallic iron would occur. The potentials needed are shown by the horizontal lines marked (19) in the Pourbaix diagram of Fig. 6.3; the lines marked 0, −2, −4 and −6 refer to Fe^{2+} concentrations (or, more accurately 'activities') of 10^0, 10^{-2}, 10^{-4} and 10^{-6} times normal. It is seen that even the lowest of the lines (−6) is only 0·18 volts below the top one (0), and at this potential level a concentration of 10^{-6}M should serve to prevent further corrosion. It is true that if the main body of the soil-water contains no iron at all, a very slow diffusion of Fe^{2+} ions outwards would occur, being replaced by fresh corrosion, but this would be negligibly slow.

(2) Although for iron it is possible to depress the potential to the value of the equilibrium $Fe \rightleftharpoons Fe^{2+} + 2e^-$, that is not possible for aluminium, since hydrogen would be evolved freely before the theoretical potential for $Al \rightleftharpoons Al^{3+} + 3e^-$ is reached. In practice, however, the cathodic protection of aluminium is possible probably because the removal of H^+ from the surface, either by movement of H^+ away from the anodic points or by evolution of hydrogen gas, produces a pH value at which the anodic product would be, not Al^{3+}, but an oxide or hydroxide film.* It is important to avoid too low a potential – at least in saline soil or water; otherwise dissolution of aluminium as sodium aluminate becomes possible.

*The pH range corresponding to passivation should extend from 4·0 to 8·6 if the solid deposited is hydrargillite, $Al_2O_3.3H_2O$, but only from 4·6 to 6·4 if it is böhmite. $Al_2O_3.H_2O$ – as shown by the Pourbaix diagrams.

(3) In soils or waters containing calcium bicarbonate, the cathodic reaction can produce a visible film of calcium carbonate; in waters containing magnesium the film will contain magnesium hydroxide. Such **chalky films** have some direct obstructive action, and any iron ions starting to pass outwards into the chalky layer will interact with the calcium or magnesium compound; the final result (which can be described as **chalky rust** if sufficient oxygen is present) may provide definite protection. One advantage of using a chalky film for protection is that, if the protecting current is turned off for a period, corrosion does not at once set in.

The protective current can be provided in two different ways (1960 vol., p. 286; 1968 vol., p. 123; 1976 vol., p. 176):

(1) An **impressed current** from a d.c. generator, or, more often, an a.c. power-source and rectifier, may be used. One terminal is joined to a relatively insoluble anode, and the other to the buried pipe or metalwork to be protected. The anode is often a ground-bed of considerable size, probably situated at a distance from the pipe system to be protected. Numerous materials are used as anodes, and their relative merits have been conveniently discussed by Berkeley (1976 vol., p. 176). Impregnated graphite is stated to be generally useful, whilst iron with 14% silicon is suitable, provided that chloride is absent; where it is present, an alloy with 3% Mo as well as 14% Si has given satisfaction. Platinized titanium (p. 79), although expensive, is much used. For marine work, anodes of lead with 1% Ag and 6% Sb perform well, but their weight is an objection.

At one time large composite ground-beds were used as anodes, consisting of carbon bars with scrap iron in contact. A current was provided from an external source, but the consumption of the scrap iron provided much of the power, so the method may be regarded as a combination of the 'impressed current' and 'sacrificial anode' methods; obviously the scrap iron required periodic renewal.

Where cathodic protection is used for the protection of a ship, a paint must be chosen for the hull which does not contain saponifiable oil; a paint containing linseed oil as vehicle would be softened or loosened by the cathodically formed alkali. That is unfortunate since the coating once regarded as most reliable for marine purposes contained red lead in linseed oil (the true inhibitor was lead azelate; see p. 226). In the Canadian navy it has been replaced successfully by a vinyl paint pigmented with aluminium (1976 vol., p. 181).

(2) **Sacrificial anodes**. If blocks of base metal are joined directly to the pipes, or other structure to be protected, the distance can be made small (avoiding the trouble known as interference, discussed later). The anodes need periodical renewal, which involves trouble and expense, direct and indirect; it is much more expensive to raise power locally by consuming magnesium or zinc than to purchase it from a generating station.

Of the metals used, the so-called **magnesium** anodes really consist of a magnesium alloy containing 6% aluminium, 3% zinc and 0·2% manganese (the manganese serves to counteract the bad effects of iron, but iron, copper, nickel, tin and lead should always be kept low). Magnesium suffers local corrosion with evolution of hydrogen and the current efficiency is low. **Zinc** anodes provide a higher current efficiency, but the current density obtained may be too low for certain purposes. Early attempts to use commercial zinc gave bad results, an obstructive film being soon formed over the whole surface, largely due to the presence of iron. If the iron content is kept very low, or, if (what may be more convenient) aluminium and silicon are added to counteract the bad effect of iron, satisfactory performance is obtained. **Aluminium** anodes gave poor results as long as unalloyed metal was used, owing to film formation; but an alloy containing 5% zinc is now being employed, whilst Russian investigators[300] report favourably on aluminium-zinc-calcium anodes.

Zinc, and often magnesium, anodes are buried in a special soil-mass of bentonite and gypsum designed to promote smooth anodic dissolution without formation of obstructive layers. The distribution of the anodes and the resistance of the circuits must be adjusted to depress the potential of steel to about −0·85 volts relative to the copper/saturated $CuSO_4.5H_2O$ electrode equivalent to −0·53 volts on the hydrogen scale (this is the reference electrode generally used to measure pipe potentials; the electrode based on silver in sea-water saturated with AgCl, much used for marine testing, has by coincidence almost the same potential). If sulphate-reducing organisms are present, the figure should be −0·95 volts. For protecting lead the appropriate level is said to be −0·7 volts. If the required depression of potential cannot be obtained with sacrificial anodes, the use of impressed current may become unavoidable.

The use of cathodic protection, especially for the protection of pipelines at the foundations of buildings, introduces other problems. The main objection to impressed current provided by a single large ground-bed, situated at a distance from the pipelines, tanks and/or other metalwork to be protected, is that some of the current, instead of passing direct between ground-bed and metalwork, may join some other pipe which does not form part of the protection scheme; two cells will be set up:

Protected pipe	Soil	Unprotected pipe	and	Unprotected pipe	Soil	Ground bed

In the first cell, the unprotected pipe will be anode, and is likely to suffer much attack. This is known as **interference**. If such troubles are to be avoided, close co-operation between the various interests involved (power, light, gas, water, telephone and transport) is advisable. It may be possible to arrive at an agreement satisfactory to all, but that will not always be easy. A permanent joint regional committee on which all interests are represented may be desirable. A Joint Committee at a national level for the Co-ordination of the Cathodic Protection of Buried Structures for Great Britain was formed in 1953.

Cathodic protection in prevention of pitting

Surprise is sometimes expressed at the fact that cathodic protection can in certain circumstances serve to stop pitting, or to arrest stress-corrosion cracking; the objection raised is that the resistance of the path joining a sacrificial anode to the bottom of a pit or to the tip of the crack, where attack (in absence of protection) would occur, will be too high, owing to the narrowness of the crack. The misunderstanding arises from the unfortunate choice of the name 'cathodic protection' to describe the method of avoidance of corrosion. The necessary condition for immunity at a point is that it shall *not be anodic*; there is no need for it to be actually cathodic. This condition can be fulfilled without the current finding its way up the crack.

Consider the case where corrosion of steel carrying a crack is occurring in absence of protection (Fig. 4.8(a)). The cathodic reaction on the large external surface will be balanced by the anodic attack at the tip of the crack (the walls will probably be protected by a film, as suggested on p. 141); it is true that the narrowness of the crack will cause a high resistance and the current will be small, but owing to the smallness of the anodic area at the tip, the current density will still be high, and the corrosion intensity will be sufficient to cause the crack to advance rapidly into the metal. Suppose now that a sacrificial anode of, say, magnesium is connected (Fig. 4.8(b)). The cathodic reduction of oxygen on the external face will now be balanced by anodic attack on the magnesium; for the e.m.f. of the cell with magnesium as anode will be higher than the other and the resistance very much lower than that involving current flowing up the crack. Thus corrosion at the crack-tip will become negligibly slow. The same is true if an impressed current system is used (Fig. 4.8(c)). There is no doubt that the so-called cathodic process is effective in arresting stress-corrosion cracking; under favourable conditions cases are known where a crack already started has been brought to a halt; if, however, there has been a faulty diagnosis and what was called stress-corrosion cracking is really hydrogen cracking,

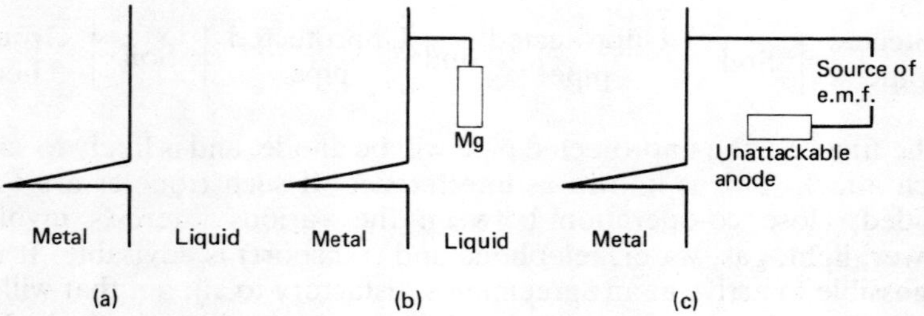

Fig. 4.8 Prevention of corrosion cracking

cathodic protection, by increasing the supply of hydrogen, will make things worse.

Stray currents

It will be understood that the burial of metal (especially a new pipeline) in a region where cathodic protection is already installed for existing metalwork carries perils. Even in regions where no cathodic protective schemes exist, stray currents may be flowing through the ground, capable of causing anodic attack. Such currents were causing damage in days before the introduction of cathodic protection; it is indeed arguable that it was the methods adopted for dealing with the menace of stray currents which led to cathodic protection as we know it today.

The most troublesome stray currents were met with in the days of electric street tramways, and, although that form of transport has disappeared the subject is worth mentioning owing to the instructive character of the preventive measures. The electric current needed to drive the trams was conveyed by overhead conduction passed through the motor in the vehicle, and was supposed to return home to the power-station (or more probably a local transformer station) by the rails over which the tram moved. Much of it, no doubt, did follow this return route, but if the rail sections were badly joined together a considerable amount of current would leave the rails, and sink into the ground, then join some steel pipe or lead-sheathed cable which chanced to lead in the right direction returning to the rails nearer home. This sets up two cells in series:

Rail/ground/pipe or cable
and Pipe or cable/ground/rail.

Corrosion would be caused to the rail in one cell and to the pipe or cable in the other. This caused great havoc by attack on lead sheaths of telephone cables. For the steel pipes with their larger diameter, the attack at the anodic areas was less intense, but there was trouble at any place where two adjacent pipe-sections were not in good electrical contact, so that the current had to leave the steel and pass through the soil. This set up attack at whichever side was anodic. The trouble was largely eliminated by bonding all the pipe-sections electrically to one another, so that there was no need for current to leave the metallic path. The telephone authorities protected their cables by a method exactly opposite to that just mentioned; they deliberately introduced gaps into the lead sheaths, which no longer provided a continuous path of low resistance, so that very little current flowed along the sheath.

A third method available for protecting either pipes or sheaths was known as **drainage**. It was indeed an early form of cathodic protection with the protecting current largely provided by the offending tramway authorities; the plan was to join the rails to the pipes or cables (through resistances or other devices) in such a way as to make the whole length of pipe or cable cathodic. The drainage required to be carefully controlled. 'Overdrainage' could cause the formation of alkali on a lead cable surface, with consequent attack.

The various preventive methods were successfully developed, and the situation appeared to be under control when the tramway system was discontinued and the problem disappeared.

Another cause of stray currents – on land and sea – was welding. If, at a ship-building yard, two ships, floating in the sea, were being 'fitted out', wires would be run out from the same generator situated on land to convey current for welding purposes. Since welding is an intermittent operation, the two ships were frequently at very different potentials, and current then flowed through the sea water causing attack on whichever ship chanced to be the anode; the attack, concentrated at breaks in the paint coats, would be intense and rapid. Here also the problem was capable of control by common sense measures.

Corrosion of immersed metals

Iron

The most extensive laboratory studies of iron and steel fully immersed in salt solutions (particularly potassium chloride), with oxygen above the liquid, have been carried out by Bengough and his colleagues, Stuart, Lee and, particularly, Wormwell. Some of the experiments lasted several years, and the iron became covered with a layer of granular magnetite itself overlaid with flocculent ferric hydroxide. During considerable periods, constant corrosion rates prevailed; changes of velocity came about comparatively suddenly – evidently through changes in the physical state of the corrosion product, which would affect the amount of oxygen reaching the metal.

Bengough's results demonstrate that the speed of corrosion is limited by the rate at which oxygen reaches the corroding system. The replacement of air by oxygen increases the rate of attack, whilst the raising of the oxygen pressure above the atmospheric value causes a further increase in corrosion, unless the specimen is exposed to the high-pressure oxygen *before* the salt solution is introduced; in the latter case, the iron becomes partially passive, and the rate of corrosion is slowed down.

By increasing the cross-section of the containing cell (keeping the specimen-size constant), a greater rate of entry of oxygen becomes possible, so that the rate of corrosion is speeded up. On the other hand, an increase in the size of the specimen (under a certain range of conditions) hardly increases the total corrosion. The intensity of attack is actually diminished; this is easily understood, since, if a small specimen can consume all the oxygen reaching the appropriate level in the vessel, a large specimen will obtain little more.

Where the upper face of the specimen is placed only a short distance below the liquid surface, the corrosion rate diminishes as this distance is

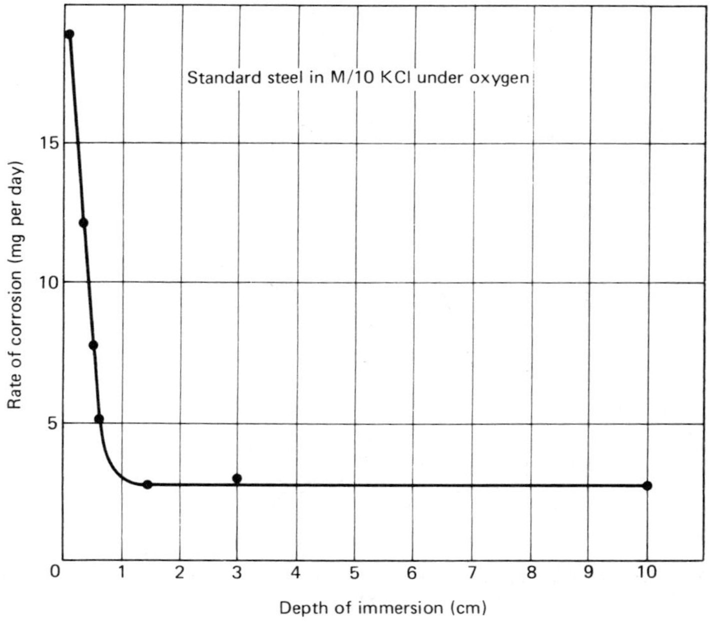

Fig. 4.9 Influence of depth of immersion on corrosion rate (G. D. Bengough, A. R. Lee and F. Wormwell)

increased (Fig. 4.9). But when the depth of immersion becomes great the corrosion rate ceases to fall off further and it is clear that the mechanism of supply of oxygen to the metal must have altered. At shallow immersion, oxygen probably reaches the metal by **diffusion** downwards across the liquid layer separating gas from metal, so that the corrosion rate diminishes with increasing depth, whereas at deep immersion, oxygen travels down by **convection currents**, which are almost independent of depth. There is still some doubt regarding the origin of the convection currents in Bengough's experiments. Apparently they are influenced by the nature of the corrosion product, since, notwithstanding that iron can use up, in fairly concentrated salt solutions, all the oxygen reaching it, yet zinc is corroded more quickly than iron (Fig. 4.10), suggesting that it receives a more ample supply of oxygen; the fact that on iron some of the oxygen is used up in converting ferrous to ferric compounds cannot fully account for the discrepancy. Copper and aluminium corrode far more slowly, and probably do not attain the maximum rate permitted by considerations of oxygen supply; the resistance of copper is connected with its intrinsically noble character, but that of aluminium is due to the protective character of the alumina film.

In the actual corrosion of immersed steel work, the oxygen will usu-

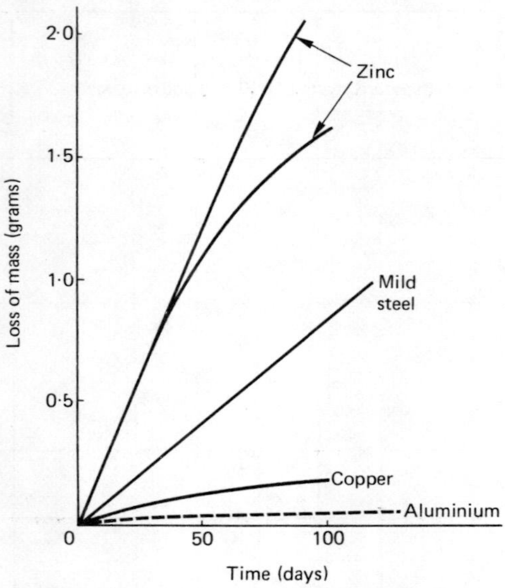

Fig. 4.10 Corrosion of metals fully immersed in 0·5M sodium chloride in presence of oxygen (G. D. Bengough and F. Wormwell)

ally reach the metal through the motion of the water (e.g. where it runs through a tank or a pipe). In other cases, there may be convection currents set up by temperature differences. In the rare event of the complete absence of mechanical stirring and of thermal convection, convection currents could still be set up by the difference between the specific gravities of the corrosion products and the corroding medium respectively (which difference, however, is not so large as might be supposed, since rust and other hydroxides contain a great deal of water).

Zinc

The attack upon horizontal totally immersed zinc was studied in detail by Bengough. In high-purity water the attack falls off with time; however, the corrosion, being concentrated at pits, may cause serious damage. The retardation occurs more quickly on annealed zinc than on abraded or turned material, where residual stresses may cause intermittent breakdown of the films, with speeding up of the attack; such occurrences are well brought out by the irregularities of Bengough's curves. In potassium chloride solutions, the retardation becomes less pronounced as the concentration increases, so that in M/10 solution the corrosion velocity remains constant with time. In view of the mechanism suggested for zinc in salt solution (p. 39) and in pure water (p. 89), the differences in Bengough's curves at various concentrations are easily understood.

Metal subjected to rapidly moving water

General

Where water containing dissolved oxygen (and possibly entrained air bubbles) moves over a metallic surface, the rate of attack is not likely to be limited by shortage of oxygen. At places where some small foreign particle rests on the surface, a small area may be *locally* shielded from oxygen, and, under certain conditions, differential aeration may set up attack, with the unaerated spot as a small anode, and the aerated area as a large cathode – the combination which is always likely to produce intense attack (p. 59). On bare steel, this localized corrosion is most probable if the water contains just sufficient inhibitor to prevent attack at the free surface, but not at the shielded spot (where replenishment of inhibitor will be hindered). Intense attack on steel by water not containing an inhibitor is most probable if the whole surface is covered with oxide (mill-scale) except for small breaks, which form the small anodes in the large cathodic area.

A different form of intense localized attack may occur if the water carries suspended matter which, impinging at some particular point, determined by the flow of the water, keeps damaging a film that would otherwise stifle attack, and leads once more to intense localized corrosion. Cases of corrosion trouble due to sand suspended in water have been reported from industrial plants – notably from beet sugar works. Impingement of air bubbles or vacuum bubbles constitutes a somewhat special case, and is considered on pp. 128, 130.

In the case of copper, another type of local attack may be set up if water passes rapidly over the surface at one point, but is relatively stagnant over the rest of the surface. The removal of copper ions from the point of rapid motion keeps the ionic concentration locally low, and this point will be permanently anodic to places where copper ions can accumulate. Intense corrosion may thus arise, leading to perforation, e.g. on externally cooled copper condenser tubes at a point opposite the water-inlet where the velocity is high. It might be thought that this effect should be obtainable with all metals. Actually ion-concentration cells – set up by differential stirring – can be demonstrated on lead. But on iron and zinc stirring has another effect, since the place where water is moving most rapidly receives the best supply of oxygen, and becomes cathodic rather than anodic. In general, the differential aeration effect will become more important, and the ion-concentration cell less important, as we pass to base metals. For various reasons the ion-concentration cell is rarely dangerous except on copper and its alloys.

Corrosion connected with rapid movement of liquid may be serious on brass condenser tubes and bronze propellers. Before this can be understood, some general remarks on the **dezincification** of brass are necessary.

Dezincification of brass

The brass in common use contains about 30% zinc with 70% copper, and consists of a single α-phase; since the attackable zinc is dispersed in excess of the nobler copper, it would hardly be expected that liquids which attack zinc but not copper would cause more than superficial corrosion. General experience has shown that the two constituents of a binary alloy can be 'parted' in acid or similar reagent, but only if the content of nobler constituent falls below a certain **parting limit**; from alloys with a smaller content, the baser constituent can be dissolved out fairly completely, leaving the nobler metal as a porous skeleton or sludge. If the content of the nobler constituent is above the parting limit, there is only superficial attack. Thus a gold–silver alloy, sufficiently rich in silver, may be treated in nitric or hot sulphuric acid, or alternately made an anode with an external e.m.f., so as to dissolve out the silver, leaving the gold unattacked. On annealed material, a sharp parting limit is usually found at a composition where one half of the atoms belong to the noble species. (In a less corrosive liquid, it may occur where only one quarter belong to the noble species). A sharp limit only occurs if annealing has allowed the two atomic species to take up an ordered arrangement; on cold-worked material, with the arrangement disturbed, the parting limit is less sharp.

This 'parting-limit principle' suggests that 70/30 brass should suffer only superficial dezincification in liquids capable of dissolving zinc but not copper. Actually brass, if free from arsenic, is found to suffer considerable deterioration in moving sea-water – a serious matter in brass condenser tubes and marine fittings generally.

The attack often seems to start at places which are relatively rich in zinc (such as the grain-boundaries on insufficiently annealed material); here the zinc passes into solution leaving the copper behind. When once a small amount of residual copper has been formed by local dezincification, and when once a small deposit of zinc and copper compounds (usually basic chlorides) has been formed locally by corrosion of *both* constituents of brass, then a cell can be set up which, once started, may produce serious local attack; this cell is

Brass | Sodium chloride solution containing copper compounds | Copper

The brass will be the anode, and both zinc and copper will pass into solution:

$$Cu \rightarrow Cu^{2+} + 2e^-$$
$$Zn \rightarrow Zn^{2+} + 2e^-$$

At the cathode, copper will be deposited in amounts equivalent to the copper entering the solution at the anode,

$$Cu^{2+} + 2e^- \rightarrow Cu$$

whilst the passage of zinc at the anode is balanced by the reduction of oxygen at the cathode

$$O + H_2O + 2e^- \rightarrow 2(OH)^-$$

The final result is, therefore, the same as though only the zinc were attacked, and the e.m.f. driving the cell should be the same as for dezincification; but owing to the fact that there is no need to remove zinc atoms from the interior of a brass mass, the resistance would be expected to be lower, and the action, once started, will proceed smoothly. The metallic copper thus formed will not be residual but redeposited; this smooth-running type of attack can only occur at a place where a reservoir of copper compounds has already been formed, and in this respect is autocatalytic; having started at a certain point, the action will continue there in preference to elsewhere.

Some years ago this explanation of dezincification was generally accepted, but today the situation is known to be more complicated. (1976 vol., p. 239.) Rothenbacher has found two layers of copper on dezincified brass. The structure of the outer one bears no relation to the structure of the brass, and here the copper would seem to be formed by redeposition. But it is believed that it is formed, not by cathodic deposition of copper from Cu^{2+} ions formed by anodic attack on the alloy, but by a 'disproportionation' reaction; the anodic attack on the copper produces Cu^+ ions, which on emergence, interact to give Cu^{2+} and metallic copper. This explanation of redeposition was originally proposed by Lucey. The inner layer of copper has a different structure, and appears to be residual; it is thought to be copper left behind when the zinc moves outwards. The difficulty of accepting a residual origin for the copper has always been the high resistance which would oppose the movement of zinc outwards through an alloy consisting mainly of copper. However, the difficulty has been largely disposed of by a suggestion of Pickering and Wagner, who point out that the outward movement of cations is equivalent to the inward movement of vacancies, and that this mechanism should provide paths for the preferential extraction of zinc.

For more than fifty years, it had been known that certain batches of brass underwent 'complete corrosion' (leading to zinc and copper compounds in the proportions in which the metals occurred in the brass), whilst others of similar composition underwent dezincification, (leading to metallic copper). Bengough and his colleagues were the first to discover that the principal factor determining behaviour was arsenic. If the brass contained a small quantity of arsenic, it resisted dezincification, although, if subjected to conditions sufficiently drastic to produce attack on the copper as well as the zinc, it would suffer complete corrosion. Arsenic is often added to brass intended to resist dezincification by sea water, if it is not already present as an impurity in the copper employed for making the alloy. Alternatively the heat-treatment may be adjusted so as to avoid the zinc-rich areas where – it is thought – the action starts. The subject is discussed in the 1960 vol., p. 74.

Condenser tube corrosion

In boiler plants, the chambers used to condense the steam contain rows of tubes through which cold water is run (sea water is used in the case of marine boilers and coastal power stations); the steam condenses outside. For the tubes, 70/30 brass, sometimes containing 1% of tin, was commonly used in the first two decades of the century, although a two-phase alloy with about 40% zinc was also in favour, mainly on the European Continent. The majority of the tubes enjoyed lives of many years, but occasionally a tube would become perforated in a few months – or even weeks. Failures were found by Bengough and his colleagues to be of two main types:

(1) *Deposit attack* was most frequent if the water speed was low – as in many land condensers and some ships; it occurred at places where debris or some foreign body (sand, mud, coke, wood, shells or seaweed) had settled. The corrosion set up was due, in the first instance, to differential aeration; the portion screened from oxygen by the debris was a small anode surrounded by a large cathode, but, when once corrosion product had accumulated, it could act as a screen and cause the attack to continue even after the original foreign body (which started the trouble) had been swept away. If the brass was of the 'dezincing' type, 'plugs' of copper were sometimes formed which passed right through the tube walls, owing to the conversion of brass to copper. These were often inconspicuous, and the tube might appear undamaged, until tested with the knife, when the plugs were found to be quite soft.

These troubles could generally be prevented by keeping the tubes free from debris, using special brushes which did not scratch the tube (surface defects made by scratching might themselves be the starting points of attack). To prevent the attack (if it started) developing into dezincification, brass containing arsenic was often recommended.

(2) *Impingement attack* occurred most often when the water speed was high. The turbulent flow of the water caused bubbles of low-pressure air to impinge on the tube surface near the inlet end of the tube. At points where *large* bubbles kept impinging on the tube, rapid pitting was often produced. In general the pits were free from corrosion product, and were undercut at the forward end; sometimes the deflection of bubbles by an adhering particle led to trenches of horseshoe form.

A succession of blows at a single spot (whether due to bubble-impingement or anything else) may merely cause elastic deformation to the metal, which springs back after each blow, but may cause permanent damage to the less ductile film-substance; corrosion is almost certain to set in, unless the material is one which can quickly develop new protective material, healing the gap.

Considerable advances towards the avoidance of impingement attack were achieved by designing the condenser so as to prevent the impingement of large bubbles (small ones were relatively harmless). The real solution, however, came through the development of special alloys.

These might be brasses containing aluminium, which give protective films (probably alumina) not easily damaged by impingement and capable of rapid self-repair if damage does occur; the composition favoured was 75–76% copper, 22–23% zinc and 2% aluminium. Another valuable alloy consisted of 70% copper and 30% nickel. Both gave excellent resistance to impingement attack; the nickel alloy had the advantage of suffering no attack analogous to dezincification – since it contained no zinc. In general, the 70/30 copper–nickel alloy should contain iron and manganese; for many purposes, less than 1% of each suffices; by raising the content of each to 2%, resistance to suspended sand is improved (an important point at power stations drawing waters from estuaries), but the resistance to polluted waters deteriorates. Alloys with only 10% nickel and 0·7 – 2·0% iron have given good results.

In more recent times stainless steels have been competing with 70/30 copper–nickel and aluminium brass as tube materials. The Cu–Ni alloy seems to be preferred to aluminium brass where the cooling water is likely to be polluted. Kenworthy states that 70/30 Cu–Ni with 0·4 – 1·0% Fe and 0·5% – 1·5% Mn has given excellent results in the (British) Royal Navy. The only real trouble occurs with ships in reserve, where there may be sluggish circulation, mud settlement and sometimes water-pollution. It is important to drain condensers when they are being closed down, using clean water to flush them out (1968 vol., p. 180).

The performance of brass tubes can be improved by the addition of ferrous sulphate to the water. Tests on a model condenser at a Japanese power station showed that aluminium brass tubes became covered with a smooth brown film when 0·01 ppm of Fe^{2+} was present in sea water which provided complete protection at 3·7 m s^{-1} flow rate; in absence of Fe^{2+}, there was serious trouble at the inlet end.

A series of tests on a model boiler was carried out in order to decide upon the best material for a new power station in Australia; copper alloys, nickel alloys, stainless steel and titanium were included; the results led to the selection of aluminium brass with plans for dosage of the water with ferrous sulphate. (*Brit. corr. J.*, 1976, **11**, 6.)

Even with the best materials, there is still risk of corrosion to condenser tubes if the water used contains hydrogen sulphide or *cystine*

$$\begin{array}{l} S{\cdot}CH_2{\cdot}CH(NH_2){\cdot}COOH \\ | \\ S{\cdot}CH_2{\cdot}CH(NH_2){\cdot}COOH \end{array}$$

a substance derived from seaweed. It cannot be claimed that this trouble has been entirely overcome, and it is best to arrange that the condensers shall not contain polluted port water during periods when the engines are idle. Despite some unsolved problems, however, condenser trouble is at the moment under control. In 1961 Rogers reported the absence of serious condenser corrosion in the Royal Canadian Navy since the early 1950's. (T. H. Rogers, First International Congress on Metallic Corrosion, London, 1961; report (Butterworth) 1962, p. 605).

Cavitation

Impingement attack in condenser tubes is closely allied to the cavitation which causes trouble on bronze ships' propellers, water turbines and hydraulic and hydroelectric systems generally. If water flows rapidly through a locally constricted tube, it becomes milky at and beyond a certain point, owing to the formation of cavities caused by the sudden change in flow conditions. They are often called 'vacuum bubbles', but must always contain water vapour, and usually (if there is air dissolved in the water) low-pressure air. Where they impinge on metal, serious damage may be produced. This is often attributed to the 'water-hammer action' caused by the blow delivered when a vacuum bubble 'collapses' on coming into contact with a metallic surface. Undoubtedly mechanical damage due to such a cause can be severe, even in cases where corrosion is impossible; non-metallic substances, like bakelite, are affected, whilst metals can suffer cavitation damage in non-corrosive liquids like paraffin. Nevertheless, it seems probable that the main damage caused in service by vacuum cavities is due to disruption of the protective film; if this damage is always produced at the same point (as is likely if the cavities always impinge at one place), it may be rapid and severe. The destruction is an example of conjoint mechanical–chemical action; without the mechanical factors, the chemical changes would soon become slow (cf. p. 132).

Those who hold that the trouble to propellers and condenser tubes is purely mechanical should consider the case of $\alpha\beta$-brass described by Campbell and Carter, where the hard β-phase was removed, leaving needles of the softer α-phase standing proud; this was clearly electrochemical action, the β-phase being anodic to the α-phase as cathode. (H. S. Campbell and V. E. Carter, *J. Inst. Met.*, 1961–2, **90**, 363.) Or again they should read Callis's description of deep narrow cavities (1960 vol., pp. 752–6). He writes, 'It is difficult to imagine that a cavitation void would penetrate to the bottom of one of the holes, to implode finally on the very bottom of so tortuous a channel.' Callis attaches importance to 'irregular flow', and this seems reasonable, as it will favour ion concentration cells (p. 125), which are unlikely with lamellar flow. Campbell and Carter also attach importance to turbulence; the reason why impingement by large bubbles is more dangerous than that by small bubbles is that the former causes turbulence. Turbulence in the absence of bubbles can cause attack on copper or 70/30 brass; on alloys like aluminium brass, which possess film-repairing properties, occasional periods of turbulence may lead to no permanent damage, although persistent turbulence may cause trouble.

Cavitation damage cannot be overcome simply by selecting very hard materials, nor, in general, by choosing alloys resistant to corrosion; much can be done by combining good mechanical resistance with high chemical stability. Often it is still more important to obtain a smooth surface, free from porosity, on or just below the surface layer; this requirement suggests that the damage may be partly due to bubbles formed within such pores, which, under rapidly fluctuating pressure, expand and contract, thus loosening metal and protective films, and causing conjoint mechanical–chemical damage. But the main method advocated today for the avoidance of cavitation troubles depends on designing the shape of the moving parts in such a way that any vacuum bubbles formed impinge at points where they will do no harm; if this is impossible, a more resistant material, or a special surface condition, must be provided at points of impingement.

5

Effect of Stress, Strain and Structure

General

Conjoint action

This book does not deal with purely mechanical failure. However, since mechanical and chemical influences acting together can cause fracture in cases where mechanical influences acting alone would allow an indefinitely long life and where chemical influences acting alone would produce only slow thinning, the response of metal to mechanical stress has to be considered. It is assumed that the reader has already studied the mechanical properties of metals, or has access to other books dealing with them. The present section, however, may serve to refresh the memory, and summarizes the principles in a form suitable as a starting-point for considering conjoint (mechanical–chemical) action.

Crystal structure of metals

In general, metals are crystalline solids, the atoms being arranged on a three-dimensional pattern or **lattice**. Each grain represents a crystal, grown from a single nucleus; where two grains meet, the two lattices do not normally fit, and a looser structure is inevitable. Preferential penetration of sulphur or oxygen along grain-boundaries, mentioned on p. 23, is easily understood.

On ideal metals with faultless atomic arrangement within grains, metals would display a strength far greater, and a ductility far lower, than metals as we know them. These contain defects, the most interesting being various types of **dislocations**. Around a 'screw dislocation' the atoms are arranged as on the screw-thread, forming a very flat microscopic corkscrew staircase; the corkscrew structure can sometimes be shown up by etching.

Of greater practical importance are the 'edge dislocations'.[374] On certain planes of atomic packing, there are places where the atoms, instead of maintaining the ideal distances characteristic of a perfect crystal, are spaced slightly too loosely on one side of the plane, and slightly too closely on the other. The result is that when a stress is applied, it is possible for the two parts of the crystal separated by the plane to glide relatively to one another without serious expenditure of energy. The position of the dislocation moves along the 'gliding-plane' as gliding proceeds. At any moment, a proportion of the atoms next to the plane are in a state of high energy, being out of register with those on the other side, but the proportion is not increased by the movement, so that the work needed to produce gliding is moderate. Had the structure been faultless, all the atoms along the plane would have been brought to high-energy positions at the same moment, and the work needed to cause gliding along the same plane would have been enormous. Thus actual metals are far less strong than ideal metals. But they have another feature. At parts of the plane between the dislocation lines the atoms are in almost perfect register with their 'opposite numbers' across the plane, and the continuity of structure is not disturbed; thus the two parts remain adherent – notwithstanding the gliding; the effect of the stress is to produce deformation without separation of the two parts – surely the most valuable property possessed by metals as we know them!

The effect of gliding is to produce microscopic steps at the free surface of a metallic block, but in the interior dislocations will become piled up near a grain-boundary – a result directly confirmed by electron diffraction. The distance which dislocations can easily move will be shortest when the grain size is smallest, and for this and other reasons fine-grained metal is stronger than coarse-grained metal. After deformation of a metallic article, particularly if there has been gliding on several sets of planes, the disorganization of atomic structure will oppose gliding when forces are imposed from new directions; the material is said to be **work-hardened**. The disarrayed structure, however, represents increased energy, and on heating the atoms may rearrange themselves in orderly manner, forming new grains – generally with straighter boundaries than those obtained by solidification from the molten state. The **annealed** material thus produced is less strong but usually more ductile than cold-worked metal; the number of nuclei from which the new grains grow depends on the amount of deformation, and on the annealing temperature; by adjusting the conditions, the grain size of the annealed material can be made coarse or fine, as required. After certain heat-treatments of cold-worked metal, the internal stress is relieved by the grains forming blocks (or **sub-grains**) with their atomic rows set at very slight angles to one another. Certain reagents can produce microscopic etch pits along sub-grain boundaries; Vogel,[381] studying such pits formed on germanium, found them to be spaced out at distances which agreed well with calculations from crystal-geometry.

When a rod of metal is subjected to tensional stress, its behaviour varies with the stress applied. If this is low (within the **elastic range**) there is no gliding. The inter-atomic distance in the axial direction becomes slightly greater than in unstressed metal, but when the applied tension is removed, the atoms spring back to the original spacing. A greater stress will cause gliding along planes of dense packing, and now there will be a permanent deformation remaining after the tension is removed. At a still greater stress, the T.S. (**tensile strength**) is reached, and the rod thins out at some point and fractures.

Residual stresses left by deformation

If metal is subjected to a complicated deforming operation, such as rolling, drawing or extrusion, internal stress will remain; sometimes tensional regions (with the atoms spaced abnormally far apart in the direction parallel to the surface) will overlie compressional regions; sometimes (as after peening, or bombarding with small hard particles) compression overlies tension. Obviously the two forces must be in equilibrium (if they were not, the metal would change shape until equilibrium was established). If, however, the outer part of the metal is converted to oxide or removed by an etching agent, the equilibrium is upset. Such considerations have importance in deciding the starting-points of corrosion (p. 45).

If a piece of sheet-metal is bent around a rod, the deformation in the interior may be insufficient to produce gliding, and only elastic stresses are produced; near the two surfaces, gliding occurs, relieving the situation, but when the bending forces are removed, the elastic stresses in the interior cause a slight spring back, until equilibrium is established; this equilibrium will occur when the structure is in *compression* on the *convex* side of the bend and *tension* on the *concave* side – the opposite of what might at first sight be expected.

Effect of structure on distribution of attack

General

Crystal structure may determine the incidence of corrosion in two ways:

(1) Where planes of atoms cut the free surface of a metallic article, the work needed to dislodge atoms from an incomplete layer is less than to start removal from a new layer, so that corrosion produces facets related to crystal orientation.

(2) Less work is needed to dislodge an atom at the boundary of two differently orientated crystals than to dislodge an atom in a crystal interior having all its neighbours similarly orientated; thus corrosion often starts at intergranular boundaries.

Production of crystal facets

Gwathmey and Benton[383] ground a single crystal of copper so as to form a sphere, and subjected it to anodic attack in phosphoric acid (using a hollow concentric sphere of larger radius as cathode). They found that the destruction of metal did not proceed uniformly (which would have left the spherical form unchanged, merely diminishing the radius), but occurred more slowly in some directions than in others; facets gradually appeared, so that the sphere became a polyhedron. The facets developed were those perpendicular to the direction in which corrosion is slowest; (in general, these represent crystal faces along which the atoms are densely packed and which therefore offer resistance to the attack). Glauner[385] made quantitative measurements of the corro-

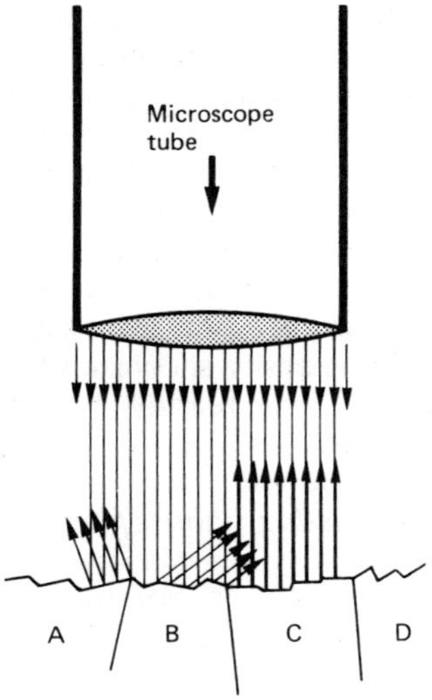

Fig. 5.1 The reason why some grains in a microsection appear bright and others dark

sion velocity in different crystal directions on copper immersed (without current) in acids containing hydrogen peroxide; in acetic acid the corrosion was slower on the (110) than on the (111) and (100 faces), whereas in propanoic acid the (110) face was the most rapidly attacked.

When a prepared section (such as is used for the microscopic examination of metals) is subjected to an etching agent, a system of little facets is produced on each grain; since the crystal orientation will vary from grain to grain, the inclination of the facets will also vary. If the section is examined under the microscope by vertical illumination, some grains will appear bright (those that chance to reflect light up the tube) and others dark (Fig. 5.1). If oblique illumination is used, it is possible, by rotating the stage supporting the specimen, to make bright grains dark and dark grains bright.

Intergranular trenches

On many specimens of metal, a second type of etching can be produced, resulting in the appearance of dark lines along the boundaries of the grains. These may arise in different ways. Sometimes two grains may be attacked to different depths, so that a difference of level arises, resulting in a shadow when the illumination is not vertical. On impure metal, the difference of composition between grain interiors and the borders often leads to a difference in corrosion rate; if the boundary zones corrode the more slowly, they will constitute a network standing as relief, but if they are corroded preferentially, they will be engraved as trenches. Probably a common cause of special attack along the grain boundaries is the presence of small amounts of eutectic – not always visible in microsections. A eutectic involves a large number of closely spaced grain boundaries and thus forms a zone of perceptible thickness, particularly susceptible to attack. Apart from this, the metal in a eutectic (or a eutectoid, the analogous structure formed from the solid state) is probably in a state of strain, the crystal-planes being bent, so that the material is especially susceptible to attack.

In unalloyed metal, the dark-boundary type of structure is difficult to develop unless impurities are present. For instance, on pure cadmium the grain boundaries are not easily shown up by etching, whereas cadmium containing a trace of zinc readily develops a network of lines marking the grain boundaries.

Factors determining the crystallographic effects developed during corrosion

In their detailed study of the corrosion of zinc in potassium chloride solution, Bengough, Lee and Wormwell[379] observed that, when very pure metal was employed, crystal facets appeared; zinc crystallizes in the hexagonal system, and accordingly beautiful six-sided pits were cut into the grains. When small amounts of impurities were present, the attack was directed mainly on the grain-boundaries, where impurities would be segregated, and geometrical figures were usually absent.

The facets developed by etching depend upon the corrosive agent. The six-sided pits developed on zinc placed in sodium hydroxide have sloping sides, representing pyramidal faces, whereas when hydrochloric acid is used, prism faces appear.

It is often the case that impure materials tend to be attacked – in the early stages – along the grain-boundaries, whilst pure ones develop crystal facets in the grain-interiors; nevertheless, exceptions are known. Cases have been reported of intergranular attack on aluminium, magnesium or zinc of high purity. At first sight, it would seem that attack should commence at the grain-boundary even on absolutely pure metal, since, as explained above, the atoms are here least firmly held. But in a perfect crystal, the layer of easily removable atoms along the grain-boundaries is a very tenuous one, and it seems unlikely that, in general, corrosion will dig in deeply unless other favourable factors are present. If mechanical work or even cooling stresses have produced disorganization, with dislocations piled up near the grain-boundaries, the strip susceptible to attack is likely to be wider, and intercrystalline corrosion may penetrate to a greater depth; possibly cases of intergranular attack in pure materials are attributable to such causes.

But even where the layers of suceptible matter separating the grains are relatively broad, several factors can retard the penetration of attack inwards from the surface along the grain boundaries. For instance, if the reaction requires the presence of a reagent (e.g. an acid) at the actual site of attack, the rate of renewal of that reagent at the bottom of an intergranular fissure must become continually slower as the fissure becomes deeper. Even where the presence of the reagent is not necessary at the site of attack (e.g. if oxygen present at a cathodic area outside a fissure can provoke attack at the anodic area at the bottom of the fissure), the electrical resistance of the liquid path joining anode and cathode must increase as the fissure deepens. If the metallic salts formed anodically in the fissure interact with alkali formed cathodically to produce metallic hydroxides just inside the mouth of the fissure, this may further increase the resistance. All these causes of obstruction will be lessened if tensional forces operate (due either to internal or applied stresses) tending to open out the fissures, making them wider as they become deeper. Hence, stresses often favour intergranular attack – as pointed out below.

Another factor militating against intergranular attack on relatively noble metals is the increase of the concentration of cations in the fissure, which would shift the potential in the direction unfavourable to the continuance of anodic attack. The presence of substances which lock up the metal as complex ions will prevent this factor from coming into operation. Thus brass, which suffers general attack in most acids, undergoes intergranular cracking in ammoniacal atmospheres, provided that internal stresses are present; the ammonia probably locks up the copper as the complex ions $[Cu(NH_3)_2]^+$, preventing the accumulation of Cu^+ ions, which would otherwise bring the action to a standstill. This was once important in causing the 'season-cracking' of cold-worked cartridges stored in an atmosphere containing traces of ammonium carbonate, as was not unusual in warm climates (uncombined ammonia acts much more slowly, if at all). Mattsson finds cracking most rapid in the pH range where CuO, Cu_2O and $[Cu(NH_3)_2]^+$ can co-exist (E. Mattsson, views reported by H. S. Campbell, *Chem. Ind.*, 1962, 688; esp. pp. 689, 690.) Season-cracking generally follows intergranular boundaries, but can follow slip planes (indeed single crystals sometimes show the phenomenon). Careful work by Nutting shows that cracking in brass which appears to be intercrystalline is really connected with attack along gliding planes but confined to regions close to the grain boundaries, where dislocations are piled up. (J. Nutting, *Chem. Ind.*, 1962, 1560.) The trouble – once a serious problem – can be avoided by suitable annealing, so as to remove the internal stresses without destroying the work-hardness. The correct conditions were found by Moore and Beckinsale.[691]

Another example is the crumbling of lead foil in an originally acid solution containing nitrate, which becomes reduced to nitrite. This is probably true intergranular attack, due to penetration along the boundaries separating the grains, which fall apart from one another; single crystals of lead suffer no decay in the same solution. It is believed that the solution locks up the lead as complex nitrite anions, thus preventing the accumulation of Pb^{2+} ions, and allowing the penetration to continue unchecked; the action of nitric acid on metals proceeds most readily at crevices for reasons explained on p. 71.

Internal stresses

Neutral liquids

In some of the author's laboratory experiments, preliminary bending of sheet specimens has drastically altered the distribution of attack, both on the compressional and tensional sides. The total corrosion of iron partially immersed in potassium (or sodium) chloride (or sulphate) is usually determined by the uptake of oxygen at the cathodic zone along the water line: if the specimen is bent about a vertical axis before being immersed, the oxygen supply is not likely to be altered, but the corrosion is found to rise to a higher level along the bend axis than elsewhere.

Disturbance of the ideal distribution (p. 38) by bending may be due to cracking of the air-formed film. Film-rupture was certainly operative in the author's experiments on copper in 0·04M silver nitrate. Freshly abraded copper, bent locally and at once immersed, showed general blackening, due to deposition of finely divided silver, without any special phenomenon at the bend. Similar specimens, kept nine days in a desiccator after abrasion, then bent and placed in silver nitrate, showed marked deposition of silver all round the bend and little elsewhere. Specimens which were abraded, then at once bent and subsequently exposed to air for nine days before being tested in silver nitrate, showed no special deposition at the bend; the exposure had 'healed' the damage.

Acid liquids

In acid liquids where oxide films will usually be absent, it is the strain in the metal, rather than damage to a film, which determines the distribution of attack. Interesting patterns are developed when locally deformed iron, after heating at 200°C, is etched with acid cupric chloride (**Fry's reagent**). The more rapid attack upon the strained metal is usually attributed to the energy stored up in the distorted material. Fry's reagent has been used in criminology to detect cases where identification numbers have been obliterated by thieves; the strain left by the punching remains even after the external form has gone, and application of the reagent shows up the figures once more.

In general, acid attacks cold-worked iron more rapidly than the annealed material; in contrast, straining makes little difference to the rate of attack by sea water, probably because the corrosion velocity is here governed by the supply of oxygen, which will not be renewed more quickly on strained iron than on unstrained.

Welded iron containers used for transporting mixtures of sulphuric and nitric acid sometimes undergo intergranular corrosion a few centimetres from the weldline, due to tensional stresses; annealing about 600°C, which removes the stresses, usually prevents this trouble. But, whereas tensional stresses promote attack, surface treatment of a kind calculated to close up fissures and leave the metal in compression parallel to the surface may increase resistance; hammering of the copper columns once used in the distillation of acetic acid was said to lengthen the life.

Cold-working accelerates the action of hydrochloric acid on aluminium and magnesium, whilst annealing restores their resistance; the rate of attack upon tin, on the other hand, is often diminished by rolling, probably because tin recrystallizes spontaneously after rolling even at low temperatures, so that the lattice structure of tin may be more perfect after rolling than before it.

Damage by voluminous corrosion products

General

Most solid compounds formed by corrosion occupy (if unconstrained) a larger volume than the metal destroyed in producing them. Rust, for instance, normally occupies a larger volume than the iron contained in it. The formation of such a product within a cavity opening on to the metallic surface must set up a compressional stress, and will often have one of two results: (1) it may seal the cavity and stop further attack within it, or (2) it may disrupt the metal. The decision between these two contrary results depends upon the strength of the material, the geometrical relation of the cavity to the surface and the existence of stresses, internal or applied, which may assist or oppose disruption.

Iron and steel

Numerous cases are known where voluminous rust has developed forces strong enough to cause breakage. The snapping of rivets on bridges by rust formed in crevices between plates was mentioned on p. 63. Pipes at a sulphuric acid works are reported to have burst through the expansion caused by iron salts formed within pores. Near acid works the author has found strips of iron swollen to several times their original thickness. These had become quite rotten, but where, owing to external support, the swelling had been prevented, the internal attack had not developed, evidently owing to sealing by the corrosion product. In such situations a bolt passing through the material may locally prevent swelling and disintegration. In the inspection of a marine structure, it was once noticed that the attachment of a ladder (not intended to give structural support) had arrested attack.

In rolled material, the grains tend to be elongated, the structure becomes pastry-like, and the majority of grain-boundaries run parallel to the surface. Thus attack following the grain-boundaries, which in cast metal showing columnar (or even equiaxed) structure would soon stifle itself, can on rolled material follow a course just below the surface and parallel to it. The surface layers are forced up by the voluminous corrosion product formed below them, and this disrupts the material and prevents sealing.

Corrosion parallel to the surface is met with on wrought iron and laminated steel exposed to the atmosphere, and is sometimes the cause of swelling – especially on railings near the coast. Cast iron and unlaminated steels are less liable to be disrupted by the internal pressures, because the grain-boundaries do not systematically run parallel to the surface; the forces are unlikely to be powerful enough to break up the metal, although they may expel liquid from pores. Cast iron which has remained in the sea for long periods, and is then taken out and dried, may appear unchanged. It usually contains, however, hygroscopic magnesium or ferrous chloride in pores, and after being kept indoors for a few days exudes little globules of moisture, each covered with membranous rust. This is useful as evidence that the iron or steel has been exposed to sea water.

Similar drops may appear on objects which have been buried in salt ground; they usually contain chlorides. Archaeological treasures from salt deserts, brought to a museum, should be washed free from chloride, dried and placed in cases containing lime or caustic soda.

Non-ferrous metals

Attack along planes parallel to the surface, causing detachment of flakes on bending, is met with on rolled zinc – as mentioned on p. 39 (see Fig. 2.6(d)). A more important example, however is the **layer corrosion**[678] of certain aluminium alloys. As explained on p. 152, aluminium alloys containing copper develop on slight under-ageing an intergranular network of material which is anodic to the main body. Under immersed conditions, this would lead to slow intergranular corrosion, producing no serious damage; in presence of stress, it may cause cracking along intergranular paths. Under marine atmospheric conditions, the soluble anodic and cathodic products would interact to give bulky aluminium hydroxide within the metal, and this might stifle further attack before the penetration deepens. If, however, the alloy has a banded structure a little below the surface, the attack can turn sideways and proceed parallel to the surface, pushing out the surface layers and causing general disintegration into a mass of thin loosely adherent flakes. The phenomenon is known as **exfoliation**.

Stress intensification

If a hole exists on the surface of a metallic rod or strip to which a tensile stress is applied, **stress intensification** is present, for reasons well understood by engineers. The same intensification can occur at a pit produced by corrosion; a crack may develop even when the load applied is below the value needed to produce cracking in absence of a pit. In general, the crack will not extend far, being held up when a region of sound material is reached. If, however, corrosion is able to demolish the sound material, the cracking may be resumed and will then continue until another sound region is reached. We thus get the cracking proceeding by jerks, with a small extension in length whenever cracking is resumed and also a sudden jump in electrical potential, since the cracking will expose fresh film-free material and in some materials may diminish the activation energy needed for cations to enter the liquid. These mechanical jerks and electrical jerks have long been known, and it has been suspected that the two sorts of jerks occur at the same moment. Their simultaneous occurrence was established by van Rooyen who applied a sound-amplification device in following the progress of cracking of certain aluminium alloys (96 Al/4 Cu and 93 Al/7 Mg) stressed in sodium chloride containing bicarbonate – a liquid in which Farmery had observed the jerks previously. (D. van Rooyen, *Corrosion*, 1960, **16**, 421t.) Van Rooyen found that sudden potential changes and the emission of sound (caused presumably by the cracking) occurred at the same instant. In contrast, stainless steel, and also a magnesium alloy (6·5 Al/1 Zn/0·2 Mn/rest Mg), suffered cracking which advanced slowly and steadily, without jerks; mild steel was intermediate. In general, no sound was heard in cases where there was no sudden potential movement.

It would seem that there are two types of stress corrosion cracking. One is essentially a mechanical fracture with corrosion serving to demolish the obstructions which would otherwise arrest the advance of the crack. The other is essentially steady electrochemical destruction of metal, with the mechanical stress serving to concentrate the corrosion on the tip of the crack, so that, instead of a rounded pit, a narrow fissure is eaten into the metal until fracture occurs. The two types of advance – continuous and jerky – have also been studied in Fontana's laboratory. (R. W. Staehle, F. H. Beck and M. G. Fontana, *Corrosion* (Houston), 1959, **15**, 373t). In the case of the transgranular corrosion of stainless steel, Eckel has reported mechanical bursts alternating with periods of electrochemical corrosion. (J. F. Eckel, *Corrosion* (Houston), 1962, **18**, 270t.)

Intergranular attack due to depletion

An alloy which consists of a single phase at high temperatures but of two phases at low temperatures will, on cooling from a high temperature, tend to deposit the minor phase preferentially at grain-boundaries, where the loose structure favours nucleation; this will also produce a change of composition in that part of the major phase which lies close to a grain-boundary. In stainless steel, the precipitation of carbide particles rich in Cr at a grain-boundary may sometimes leave an adjacent band depleted in Cr. Similarly an Al alloy containing Cu, deposits an intermetallic compound containing Al and Cu along the grain-boundaries, and leaves a band depleted in Cu. If a material in such a condition is later exposed to corrosive conditions, the attack may sometimes penetrate preferentially along grain-boundaries. This will happen, for instance, if the depleted material is anodic to the unchanged material present elsewhere in the grain.

There is no doubt that the depleted material in stainless steel will suffer anodic attack because the Cr content will be insufficient to maintain a resistant oxide film. But depleted material may not always be present; if the temperature is sufficiently high, diffusion may occur with sufficient rapidity to prevent the Cr content declining to an unfavourable level.

The second phase (chromium carbide or an intermetallic compound) may also play a direct part in promoting the attack. If it is cathodic, that will increase the anodic attack on the neighbouring material; if it is anodic, it will be attacked itself.

Where a grain-boundary runs at right angles to the metallic surface, corrosion along an intergranular path is likely to become slower as it advances owing to the increasing resistance of the path between the anodic area at the tip of the fissure and the cathodic area on the face outside. But if the material is held under the tensile stress, the fissures will be opened more widely and the increase of resistance with time will be less, so that the slowing down will not occur to the same extent as in the absence of stress. Furthermore, the stress, increasing the distance between atoms, may stimulate the anodic reaction at the tips of cracks by diminishing the energy needed for an atom to pass into the liquid as an ion; Farmery (1960 vol., p. 389) has demonstrated that a current can be obtained from a differential stress cell, consisting of two plates of the same material, one unstressed and the other subjected to tensile stress, in the same solution.

Another effect of tensile stress is that it can keep rupturing any protective film which would be formed by anodic action at the tip of the crack; no rupturing will be produced on the walls, so that the attack is concentrated on the tips of the crack. This will clearly favour the formation of narrow crevices following those grain-boundaries which lie roughly perpendicular to the surface of the metal.

It is easy to see that, in material where the grain-boundary material is anodic to the unchanged material of the grain-bodies, corrosion and tensile stress acting together may sometimes produce rapid cracking in material which would suffer little change from corrosion in the absence of stress, and no change at all from stress in the absence of corrosion.

Relation between intergranular corrosion and stress corrosion

The argument set forth above suggests that stress corrosion is essentially an extreme form of intergranular corrosion. It is, however, important to notice that not all alloys which exhibit special attack along grain-boundaries develop cracking on the application of stress in a corrosive environment. There are also materials which can suffer stress corrosion without being apparently susceptible to intergranular attack in the absence of stress. Such cases have caused some investigators to doubt whether the two types of deterioration are really different forms of the same phenomenon.

An interesting case was examined at Swansea (1976 vol., p. 351). An aluminium alloy containing 4·3% Zn and 2·4% Mg was known to be susceptible to cracking under stress in a corrosive environment, but it did not normally show intergranular corrosion in absence of stress. It was, however, shown by Eurof Davies and his colleagues that if a suitable anodic potential was applied, intergranular corrosion was obtained in absence of stress. (D. Eurof Davies, *Corrosion* (Houston), 1971, **27**, 371.) This would seem to clear up the anomaly. An atom can only pass into a liquid as an ion if it possesses sufficient energy to surmount the energy barrier. Applied tensile stress will increase the distance between atoms and this will increase the proportion of atoms which possess the required energy; an applied anodic potential will provide a field which will reduce the energy barrier to be surmounted; they are just alternative ways of increasing the proportion of atoms capable of escaping.

Apparatus for studying stress corrosion

In early studies of stress corrosion it was considered important to work either at constant load or at constant strain throughout an experiment. Figs. 5.2 and 5.3 give examples of simple apparatus

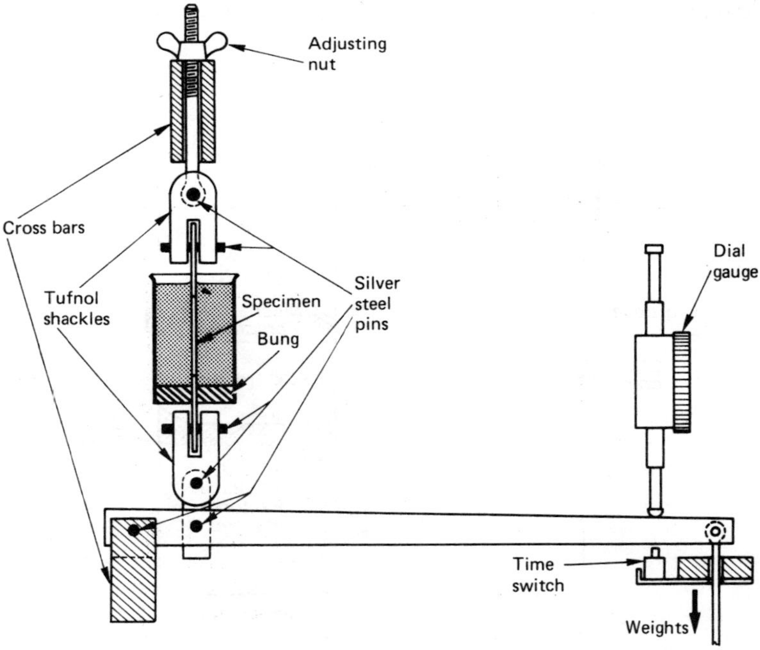

Fig. 5.2 Apparatus for studying stress-corrosion cracking – an example of a constant-load method (H. K. Farmery and U. R. Evans)

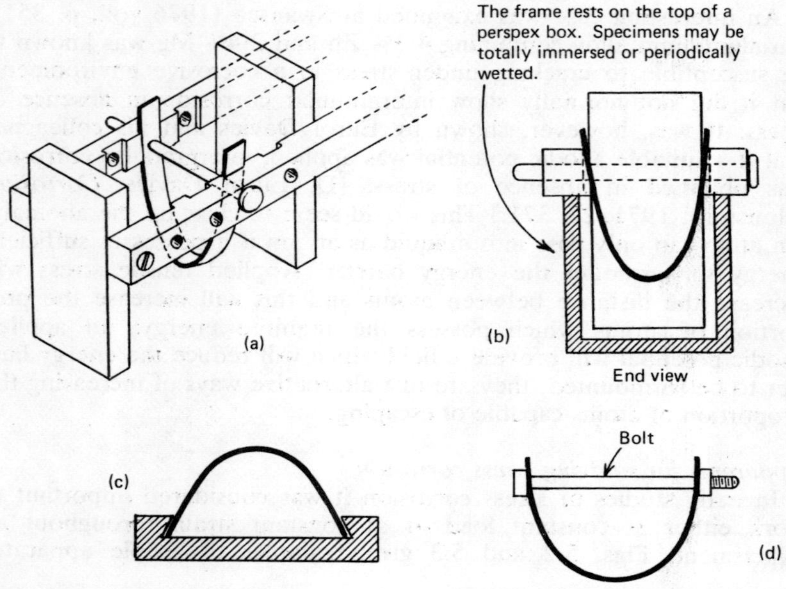

Fig. 5.3 Apparatus for studying stress-corrosion cracking by constant-strain method; (a) and (b) represent apparatus used by H. K. Farmery and U. R. Evans

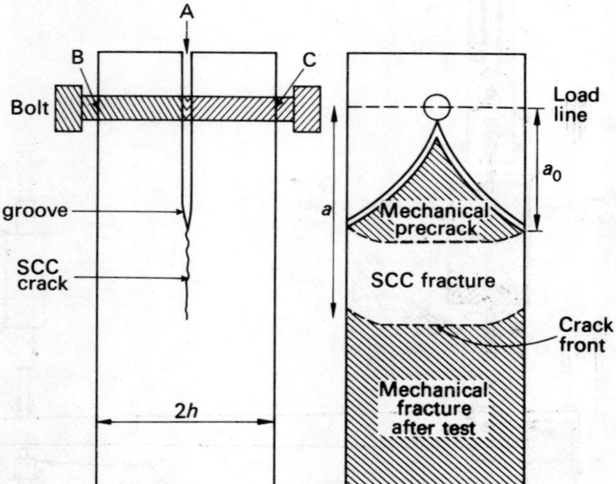

Fig. 5.4 Double cantilever-beam specimen and stress intensity calibration (M. O. Speidel)

which proved useful. However, for most of the problems of today, something different is required, since it is desirable to measure the rate of advance of a crack under different conditions. Fig. 5.4 shows a method used by Speidel, who applied stress by means of a bolt, and used a pre-cracked specimen. The adoption of a specimen on which a crack is already present at the moment when the corrosive influence is first applied is due to the fact that in material as supplied for practical engineering purposes, tiny cracks are usually present; a laboratory test carried out with a carefully prepared specimen with a smooth fault-free surface will give misleading results, since an incubation period will be needed before a crack starts to form. Controlled pre-cracking provides better simulation.

Proof of the electochemical mechanism of corrosion cracking

If the advance of the crack is of the continuous type, depending on anodic attack at the tip (with the walls immune owing to a protective film which escapes rupture), there should be a relation between the crack-penetration rate and the current passing. A series of careful experiments carried out in Hoar's laboratory and collected in Table 5.1 shows satisfactory agreement between the rate as calculated from the current measurements and that observed under the microscope. Such agreement can only be expected if there are no periods when the applied stress (intensified at the crack-tip in accordance with well-established principles) causes the crack to advance without corrosion. Engell has quoted the case of Al–Zn–Mg, where, he states, 'any explanation of these crack velocities cannot be given by any reasonable electrochemical hypothesis alone'. This may be readily accepted, but the word 'alone' is all-important (1976 vol., p. 325).

However, the mathematical basis of stress corrosion is far from simple and is being developed today with the help of the fracture mechanics which has been used with success in the elucidation of purely mechanical failures (1976 vol., pp. 334–7).

Classification of stress-corrosion phenomena

Although for convenience, cases of corrosion cracking have been considered above as falling into two groups, continuous and jerky, this may be an over-simplification. Parkins (1976 vol., pp. 331–4) has suggested that there is a spectrum of phenomena passing gradually from cases where the dominant factor is corrosion to those where it is stress (Fig. 5.5). Yet, however gradual the passage may be, there are sharp distinctions between certain types; Parkins' picture of the different situations arising according as the substance produced at the grain-boundary is anodic or cathodic is reproduced in Fig. 5.6.

An important feature of stress-corrosion cracking is that it occurs at the borderline between corrosive and non-corrosive conditions. Scully's view of the inception of the action is suggested schematically in his diagram reproduced in Fig. 5.7. If metal carrying an oxide film

Table 5.1 Comparison of crack penetration rates[1]. (a) Estimated from maximum current density on bared metal and (b) Measured microscopically (T. P. Hoar)

Material	Solution	Temperature	Potential	Bared metal CD	(a) Crack pen. rate (from CD)	(b) Crack pen. rate (micro-scopic)	Ref.
			V(SHE)	A/cm^2	mm/h	mm/h	
18/8	42% $MgCl_2$	154°C	−0·14	0·3	0·4		West[2]
Stainless			−0·14	0·16	0·2	0·5–1·0	Scully[3]
steel			0·065	0·6	0·7		Lees[4]
			0·15	0·3	0·4		
Carbon	4M $NaNo_3$	Boiling	−0·22	0·25	0·3	0·2	
steel	pH 4·8		−0·18	0·8	1·1	0·9	Galvele[5]
			−0·14	1·5	2·0	1·5	
	4M $NaNO_3$	Boiling	−0·05	0·8	1·1	1·1	
	pH 3·2		0·00	2·2	2·6	1·6	Lees[6]
			+0·75	5·0	6·2	6·9	
Carbon	10M NaOH	121°C	−0·90	0	0	0	
steel	pH 14·5		−0·80	0·15	0·20	0·14	
			−0·75	0·12	0·16	0·12	
			−0·70	0·09	0·12	0·13	Jones[7]
			−0·65	0·07	0·10	0·12	
			−0·6	0·08	0·10	0·10	
			−0·55	0·03	0·04	0·09	

Material	Solution	Temperature	Potential	Bared metal CD	(a) Crack pen. rate (from CD)	(b) Crack pen. rate (micro-scopic)	Ref.
α-brass	M NH_4Cl, 0·05M $CuSO_4$ + NH_3 to give pH 7·3	25°C	+0·25 +0·26	0·05 0·20	0·07 0·3	0·3	Podesta[8] Lees[9]
Al–7% Mg	M NaCl pH 2–5·5	25°C	−0·80 −0·75	0·1 up to 80	0·11 up to 90	0·25 very high	Ford[10]

[1]T. P. Hoar, Lecture at Firminy (France).
[2]T. P. Hoar and J. M. West, *Proc. Roy. Soc.* (*A*) 1962, **268**, 304.
[3]T. P. Hoar and J. C. Scully, *J. electrochem. Soc.* 1964, **111**, 348.
[4]D. J. Lees and T. P. Hoar.
[5]T. P. Hoar and J. R. Galvele, *Corr. Sci.* 1970, **10**, 211.
[6]D. J. Lees and T. P. Hoar.
[7]T. P. Hoar and R. W. Jones, *Corr. Sci.* 1973, **13**, 723.
[8]T. P. Hoar, J. J. Podesta and G. P. Rothwell, *Corr. Sci.* 1971, **11**, 241.
[9]D. J. Lees and T. P. Hoar.
[10]T. P. Hoar and F. P. Ford, *J. electrochem. Soc.* 1973, **120**, 1013.

	Corrosion dominant						Stress dominant	
Weld decay	Carbon steel in NO_3	Al-Zn-Mg in Cl	Brass in NH_3	Austenitic steel in Cl	Mg-Al in CrO_4-Cl	Titanium in Methanol	High strength steel in water	Brittle fracture

← Pre-existing active paths →
← Strain generated active paths →
← Specific adsorption at sub-critically stressed sites →

Fig. 5.5 The stress corrosion spectrum (R. N. Parkins)

is stressed so that gliding produces an area free from oxide, healing may take place at such a rate that only the tip is left bare, and anodic attack concentrated on this small area will result in a fissure advancing rapidly into the metal. If healing takes place too slowly, we shall, instead of a crack, obtain a pit or even general corrosion. If it occurs too quickly, we shall get no attack at all. For cracking, borderline conditions are needed. Staehle's ideas of crack-propagation are presented in Fig. 5.8 and those of Parkins in Fig. 5.9.

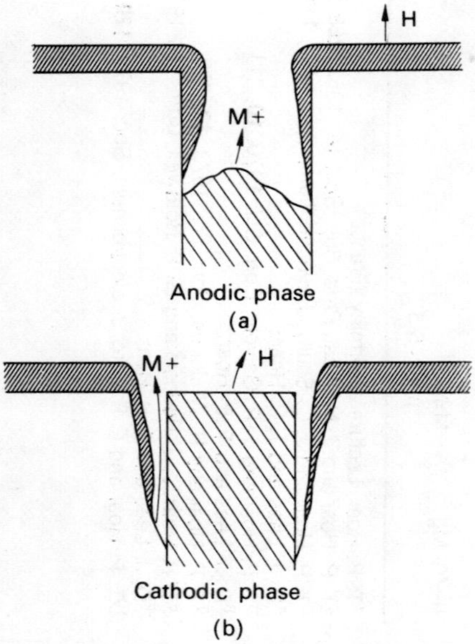

Fig. 5.6 Galvanic cell mechanism of intergranular attack (R. N. Parkins)

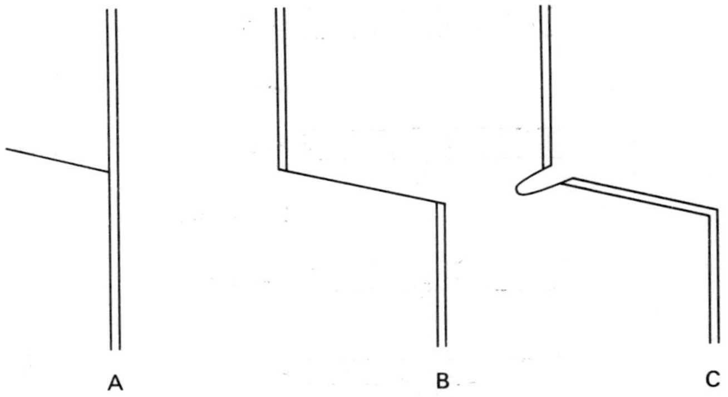

Fig. 5.7 Schematic diagram suggesting the nature of the attack that occurs on the emergent slip-step surface in alloys that develop wide slip-steps when exposed to a passivating environment. (a) Alloy surface (b) Same surface after plastic deformation and the production of a wide slip-step (c) Subsequent re-passivation of nearly all the surface; only a small area remains active (J. C. Scully)

Undoubtedly when once corrosion cracking has started, the production of acid in the fissure will play an important part in causing it to continue. This has long been recognized as an important factor in pitting. In the case of stress corrosion, an exact method for determining the pH value existing in the cavity has been provided by Floyd Brown (1976 vol., p. 327). It depends on the use of liquid nitrogen to freeze

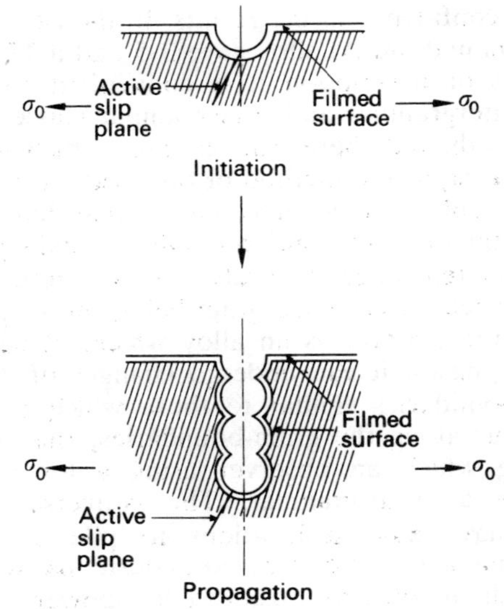

Fig. 5.8 Schematic aspects of crack propagation by successive step emergence and dissolution (R. W. Staehle)

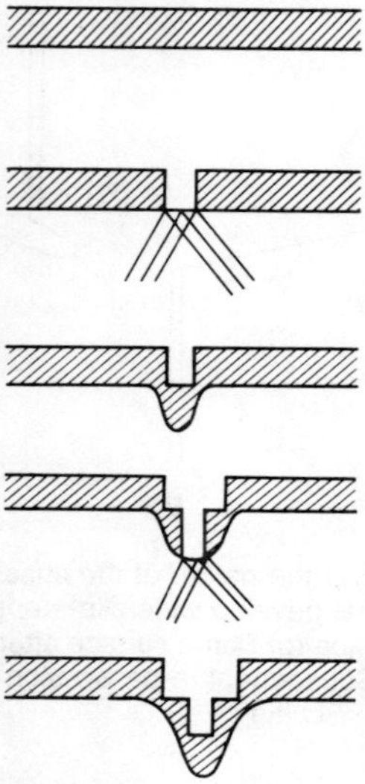

Fig. 5.9 Oxide film rupture mechanism of stress-corrosion cracking (R. N. Parkins)

the moisture in the crack and thus allow examination at leisure after the specimen has been broken open. Paper carrying a pH indicator (or $K_3Fe(CN)_6$ to detect Fe^{2+}) can be pressed on to the walls of the crack; the agreement between values obtained from two opposing walls increases confidence in the results. In the case of 0·45% C steel, pH 3·8 was obtained; an Al alloy gave 3·5 and a Ti alloy 1·7.

For purposes of description it is convenient to divide corrosion cracking into intergranular and transgranular classes according to the nature of the path, but there can be cases which would be variably described according to the method of observation. For instance, metal which has been subjected to deformation may have the structure of its grains slightly distorted and on mild annealing the atoms of a single grain may rearrange themselves as a number of sub-grains of more perfect structure, each sub-grain being set at a small angle to its neighbours. If the material is an alloy which, at equilibrium, should consist of two phases, it can undergo changes of composition along the sub-grain-boundaries similar to those which might in other circumstances occur along the grain-boundaries; this could lead to corrosion cracking which an observer using a low-power microscope would describe as transgranular. The converse case is possible. Deformation may cause dislocations to pile up near the grain-boundaries giving layers of disarrayed material particularly susceptible to attack, which an observer using a low-power microscope would describe as intergranular attack, although it is directed along slip planes near the boundaries. An example is provided by Nutting's work on brass.

Effect of planar arrangement of tunnels (1968 vol., pp. 249–59)

Nielsen, studying the early stages of cracking, has observed that pits are formed at susceptible spots, which may extend inwards along paths of susceptible matter, forming tunnels; under certain circumstances, a number of tunnels, if set in the same plane, can unite to form a crack – thus creating a dangerous situation. Tromans and Nutting have found that the pits must be close together if they are to unite to form a crack. Generally the tunnels will only merge if they lie in the same plane; tunnels which keep changing course, producing irregular tangles, are less likely to develop into cracks. Cross-slipping is a likely cause of tangles, and materials which cross-slip are generally not susceptible to transgranular failure. However, other factors besides the tendency to tangle can be important in deciding whether or not an alloy will be susceptible to corrosion cracking; the power to form protective films is important, but since cracking will occur at borderline conditions, it is not possible to state simply that the presence of a certain element is safe or dangerous; the amount is all-important, and the concentration most likely to cause susceptibility to cracking will depend on the concentration of other elements, and on the composition of the corrosive liquid. The subject is complicated.

Stress corrosion of pure metals

It was at one time thought that pure metal was not susceptible to corrosion cracking, but a research at Baltimore (1968 vol., p. 265; 1976 vol., p. 319) has shown that unalloyed copper can undergo stress-corrosion cracking in a liquid containing $Cu(NH_3)_5^{2-}$. This is easily understood. Stress intensification at the tip of a crack, by increasing slightly distances between atoms, will shift potential in an anodic direction, establishing a cell with small anode and large cathode. However, if the liquid at the outset contains no copper, a small amount of Cu^{2+} formed by anodic action will shift the potential in the cathodic direction and the e.m.f. will vanish. If the liquid is one in which copper already exists in the form of complex ions, then this concentration polarization is avoided, and the action will continue.

Nevertheless, those who argue that for unalloyed metals, or even mechanically weak alloys, stress-corrosion cracking is not a serious menace, are probably justified. A weak material subjected to tensile stress, will start to yield before any large alteration in the interatomic distance has occurred. A strong material, capable of withstanding large stresses with only elastic deformation, is able to develop a differential stress cell with high e.m.f. and rapid cracking may develop; the disastrous consequences of possible failure in a structure where very large forces are present are easy to imagine. Stress corrosion is essentially a danger connected with the use of extremely strong materials.

Behaviour of aluminium alloys

The classical researches on aluminium alloys which led to the concept of the depleted layer were carried out between 1932 and 1940 by Akimow and Tomashow in Russia and by Dix in the U.S.A. An alloy containing copper was found to undergo changes near the grain-boundaries producing a layer depleted in copper and anodic to the unchanged part of the material. Recently, the interpretation provided by these early investigators has been regarded in some quarters as insufficient. The reader may care to study the various opinions put forward (1960 vol., pp. 670–2; 1976 vol., pp. 349–51); probably in the end he will accept the conclusions of Kaesche, who considers that 'the Dix concept of electrochemical corrosion has been shown to be correct'.

The use in aircraft of aluminium alloys which attain high strength by 'ageing' (long continued storage without heating) or 'temper-hardening' (brief heating under carefully controlled conditions of time and temperature) has led to fears that the treatment designed to confer the greatest strength may introduce susceptibility to stress-corrosion cracking. The various alloys behave differently, and in some cases it is possible to attain high strength without danger of susceptibility, if the time and temperature laid down in the specification are accurately adhered to. If, however, the treatment is continued too long, the strength will decline. There is a natural tendency to avoid this decline by stopping the treatment too soon – at a time when the changes have occurred at the grain-boundaries but not elsewhere; that leaves the material susceptible to cracking. Today, most authorities consider that it is safest to 'over-age' – that is, to continue treatment for a time longer than that which gives the greatest strength. Cole, for instance, advocates the slight sacrifice of mechanical strength involved in over-ageing, and suggests that the risk of mechanical fracture can be avoided if compressional stresses are introduced by controlled machining, peening or blasting. Protection can also be provided by anodizing, followed by coating with a paint containing a chromate pigment (strontium chromate is preferred to zinc chromate). This should provide reasonable assurance against corrosion cracking (1968 vol., p. 262).

It would seem that the attack on strong aluminium alloys is not a case of simple anodic attack helped by stress, but is essentially the pulling apart of grain from grain, helped by the electrochemical destruction at places which would resist the disruption. Some of Farmery's specimens examined after being broken under stress-corrosion conditions exhibited bright reflecting surfaces along the fracture planes without signs of corrosion product. There is other evidence for that view; 'mechanical jerks' (sudden extensions) have been noticed in stress-corrosion tests on aluminium alloys, whilst 'electrochemical jerks' (sudden alterations in readings on the electrical meters) were noticed in early work by Gilbert and Perryman.[261]

Behaviour of steel

The caustic cracking mentioned on p. 96 as a serious danger in the days of riveted boilers is a type of stress-corrosion cracking; treatment of the boiler water with nitrate provided effective cure. It may seem surprising to find that another serious type of corrosion cracking occurs in chemical processes involving nitrite solutions, but the conditions are quite different. In caustic cracking, the boiler water is strongly alkaline at the points where trouble occurs, whereas nitrate cracking occurs mainly under acidic conditions. It is the nitrates of weak bases (such as NH_4NO_3 or $Ca(NO_3)_2$), which cause trouble; they produce acidity by hydrolysis; $NaNO_3$ is relatively harmless unless HNO_3 has been added; addition of alkali prevents cracking. Presumably the cathodic reaction is the reduction of NO_3^- – (discussed on p. 70). Parkins and Henthorne state that the rates of crack-propagation actually observed can be explained on electrochemical principles 'without recourse to arguments involving bursts of mechanical rupture' (1968 vol., p. 258). This brings the phenomenon into line with the results of Hoar and his colleagues (Table 5.1).

Parkins has also found corrosion cracking in carbonate solutions and thinks that it may be responsible for several industrial failures, reported but hitherto unexplained (1976 vol., p. 343).

Behaviour of stainless steels

In contrast with aluminium alloys, where we meet with damage which is essentially mechanical cracking aided by electrochemical attack, stainless steel seems to exhibit a continuous advance of a crack without periods of mechanical rupture. In many liquids, the advance is due to steady anodic attack on zones depleted of chromium as a result of the precipitation of chromium carbide. If the alloy contained no carbon, this depletion would not occur. Low carbon (0·03C) austenitic stainless steels are in current production but the production of material with a very small carbon content is expensive, and the usual method of combating the trouble is to introduce into the alloy a stabilizer, such as Nb or Ti, which possesses a high affinity for carbon. Either of these elements are effective in preventing trouble in most corrosive liquids (including a $CuSO_4$–H_2SO_4 solution which is often used in testing), but in HNO_3 (which also is used), they behave differently. Nb confers resistance, but Ti leaves a carbonitride which is itself dissolved by 85% HNO_3 and protection fails.

There was at one time considerable doubt regarding the correctness of the depletion mechanism, but research by several independent investigators has proved that the depleted zones do really exist, although the Cr content is not reduced to zero, and that depletion does play an important part in the special attack (1968 vol., pp. 247–9; 1976 vol., pp. 344–8).

The precipitation of chromium carbide (and consequently depletion of Cr) occurs when the stainless steel is held for an appreciable time at about 700°C, and great care must be taken to avoid the material remaining long in that temperature range. An important case to be considered is that of welding, where clearly a range of temperatures is produced on each side of the weld-line; there will be two zones where the temperature is about 700°C, and in these zones the formation of susceptible material is to be expected. It may sometimes be necessary, after welding stainless steel, to heat the whole article to 1100°C, so as to bring all carbides into solution, and then to quench (1960 vol., p. 214).

Hydrogen troubles

General

It was stated on p. 139 that voluminous corrosion products in cavities cause disruption. Hydrogen is the most voluminous product known; it is a cathodic product, whereas most of the others are anodic products resulting from the precipitation of the product by alkali from the cathodic reaction.

Passage of hydrogen through iron

It is common knowledge that Fe or Zn, placed in HCl or dilute H_2SO_4, evolves H_2 as a gas; Zn, although lower in the potential series, is attacked very much less quickly than Fe, provided that it is relatively pure; impurities, if present, may be redeposited on the metal and attack, slow at first, gradually becomes faster.

In the case of iron, there is another possibility. In addition to evolution as a gas, hydrogen may pass into the metal as atomic H. Indeed if the iron used is thin foil, the hydrogen may pass right through it, being evolved as gas on the far side. Passage of hydrogen through iron is most easily demonstrated if the iron is made a cathode – an effect published in 1930 by Aten; but Morris in 1935 showed that it could be obtained without an applied e.m.f. (1960 vol., pp. 396–400). For systematic measurement, methods involving an applied cathodic current are convenient. Devanathan and Stachurski, working in Bockris laboratory, used an iron diaphragm separating two vessels containing M/10 KOH with two electrodes arranged so that one side of the diaphragm was cathodic and the other side (which carried a thin layer of Pd) was anodic. The current was adjusted to such a value that the hydrogen-coverage on the anodic side was zero, which meant that hydrogen was being removed on that side by anodic action as quickly as it was being introduced on the cathodic side. The current measurement provided knowledge of the rate of passage through the foil, which was found to be inversely proportional to the thickness – indicating a simple diffusion process (1968 vol., p. 154).

Effect of adsorbed substances on uptake of atomic hydrogen

It was stated above that the uptake of hydrogen from H_2SO_4 is greatest if certain impurities such as S and Sc are present in the acid. Recent work has shown that many substances which are readily adsorbed on the iron surface promote entry of H into the metal, even when they retard the evolution of H_2 as gas bubbles in the acid; work at Ferrara (1968 vol., p. 158) has shown that *o*-tolyl thiourea, although diminishing the total amount of hydrogen formed, increases the proportion which enters the iron*. Bockris, Beck and their colleagues found I^-, CN^- and naphthalene in the liquid to increase entry of hydrogen into iron, whilst benzonitrile and other nitriles diminish it. They attribute the action of I^- and CN^- to a lowering of the strength of the bond between hydrogen and metal. Another explanation might be suggested. The formation of a H_2 molecule requires the cathodic discharge of two H^+ ions on two adjacent sites; if a large proportion of the sites are put out of action owing to the attachment of I^- or CN^-, a H atom produced by discharge at one still uncovered site may find no partner, since all adjacent sites are occupied by I^- or CN^-. Since evolution as H_2 gas is largely prevented, the few H atoms formed either remain attached to the surface or diffuse into the metal; according to this view, the promotion of entry is due to interference with the only alternative method of escape (as H_2 gas into the liquid). The fact that the large nitrile molecules diminish entry is easily understood. Their adsorption will cover up a large fraction of the surface area, but their edges will not exactly fit together where they come nearest to one another; spaces will be left sufficiently large to provide plenty of pairs of adjacent sites suited for the discharge of pairs of H^+ ions and the formation of H_2 molecules. Evolution of gas will proceed readily, and the amount produced *per unit area* will probably be the same as if the whole surface were clean and uncovered. However, the absolute surface available will be much less, and since the hydrogen actually produced will escape into the liquid as H_2 gas, the amount entering the metal as H will be small.

If the promotion of entry is due to covering up most, but not all, of the sites where H^+ could be discharged, it would be expected that substances which increase entry when in small amounts, might prevent it altogether when present in large amounts. It is a fact that As_2O_3 can either promote or suppress entry under different conditions (1968 vol., p. 157), and this may perhaps be explained in the manner just suggested. But the situation is complicated; under some

*This is a serious matter since thiourea derivatives have been used in acidic pickling to prevent undue attack on the steel during the removal of the scale. Certain polyamines (1976 vol., p. 223) provide inhibition without embrittlement.

circumstances a film of metallic As is formed and under other circumstances molecules of AsH_3. Also, as already stated, As can remove S, which itself promotes entry. The action of H_2S in promoting entry seems to depend on a different principle. There is no doubt that the total production of hydrogen is accelerated when H_2S is present in the acid; some catalytic process seems to favour the discharge of H^+. If the proportion of the total hydrogen which diffuses into the metal as atomic H is not greatly changed, the absolute amount of the entry will be increased if H_2S is present. H_2S is produced by the action of acid on sulphide particles in the steel, and is influenced by the presence of other elements. Steel containing Cu produces less H_2 when placed in citric acid than unalloyed steel; the relationship of Cu and S is important, as shown by Hoar and Havenhand; it seems that the function of the Cu is to remove S, probably as Cu_2S (1960 vol., pp. 402, 948).

Types of deterioration produced by hydrogen-uptake

Atomic hydrogen entering steel may produce trouble in more than one way. It may diffuse inwards until it reaches some microscopic hole, on the walls of which the atoms can combine to form molecular H_2; if the hole has 'clean' walls, there will be no difficulty in finding pairs of neighbouring sites, and gas at very high pressure can be formed. Simple electrochemical calculations (p. 282) suggest that potentials likely to be produced during the attack of acid on the external surface of the iron would be capable of evolving hydrogen at enormous pressure; experiments with hollow electrodes have indeed indicated the actual formation of gas at pressures up to 0·2–0·3 MN m^{-2} (1960 vol., p. 403); probably, in the microscopic holes present in commercial steel, much higher pressures will be reached. The blistering of steel mentioned on the next page is a more convincing demonstration than any reading on a gauge; it requires a huge pressure to blow bubbles in solid steel at room temperature.

Over a long period, Zapffe used the idea of high-pressure hydrogen in internal cavities to explain various types of mechanical breakdown; considerable scepticism was felt at the time, but in recent years authoritative opinion has become more favourable to that sort of explanation. Truman, for instance, discussing a paper by Farrell, has stated his belief that high-pressure hydrogen often plays a role in propagating a crack. Details of the various views must be studied elsewhere (1960 vol., pp. 416–24; 1968 vol., pp. 155–7; 1976 vol., pp. 218–22).

If early doubts about Zapffe's views were due to absence of evidence of the existence of internal voids in which high-pressure hydrogen could accumulate, a more recent research should reassure the doubters (B. F. Dyson, M. S. Loveday and M. J. Rodgers, *Proc. roy. Soc.* A), 1976, 349, 245.) In the case of a nimonic alloy, a profuse generation of cavities at grain-boundaries during deformation has been demonstrated; the diameter of the cavities (mostly less than 0·3 μm) is found to be independent of the effective plastic strain and is the same whether strain is applied by tension, compression or torsion. The number of cavities per unit area, which is proportional to the effective strain, is the same for all three modes of deformation. It is unlikely that the nimonic alloy is exceptional in this respect. Similar cavities would probably be formed by deformation on materials capable of liberating hydrogen from a corrosive liquid.

But even if no holes are present in the metal, it is likely that atomic hydrogen, uniformly distributed throughout the whole mass, will change mechanical properties for the worse. The production of even a small crack demands the provision of energy, because the walls of the crack constitute an increase in the surface area, and this represents an increase in the total interfacial energy of the system; unless energy can be provided from an external source there will be no cracking. Now the adsorption of hydrogen on a clean surface represents a diminution of interfacial energy. Thus, in the production of a crack in iron containing hydrogen, where the new surface formed will not be clean iron carrying hydrogen, less energy is required than would be the case if no hydrogen were present. It would seem that the existence of hydrogen atoms should render the iron 'brittle' and liable to sudden fracture on administration of a shock which would be insufficient to cause damage to hydrogen-free iron.

Thus 'embrittlement' of iron by hydrogen would be expected even if no holes were present, but many cases of breakage are occasioned by high-pressure hydrogen in holes. Only if the holes are close to the surface will the pressure suffice to force up the iron as a blister. Holes situated in the interior may produce no change in the absence of stress, but if there is either internal stress due to previous deformation, or applied stress (too low to produce fracture in the absence of hydrogen), the combination with the stress due to the high-pressure hydrogen in a hole may start a crack. Extension may be slow, since any growth of the crack increases the volume available for the gas, and pressure will fall unless there is replenishment by hydrogen diffusing from the solid iron around. However, there is a risk that ultimately the crack may extend far enough for a sudden catastrophe. The danger is the more insidious because the advance of the crack is slow and may be overlooked.

Hydrogen troubles in pickling

In early days, attention was called to the effect of hydrogen by troubles arising during the pickling of steel to remove mill-scale – a treatment necessary before electroplating and advisable before painting. Edwards[417] studied blistering produced during the pickling of sheet steel, and found that the blisters occurred along lines of inclusions (of oxide or perhaps sulphide) situated a short way below the surface. It may well be that the material of the inclusion, less malleable than the metal, failed to deform sufficiently during rolling, so that true cavities were left; even if an inclusion can be deformed sufficiently to maintain contact at all points with the metal around it, it provides a place where the formation of a hydrogen bubble demands less energy than is needed at the average point in the steel, since in general the adhesional energy will be less between steel and inclusion than between two steel surfaces.

The case has practical importance, because cathodic protection has sometimes been recommended for steel which is being pickled, so as to reduce the attack of the acid on the metal; cathodic treatment may indeed accelerate removal of a thin scale, since reductive dissolution is faster than direct dissolution (p. 55). However, if cathodic conditions produce blistering or leave the steel in a state where there is a danger of slow-cracking, cathodic treatment may become inadvisable. The use of an oxidizing acid or a mixture of acid and oxidizing agent avoids risk of hydrogen trouble and has provided satisfactory removal of scale in some cases.[408]

Hydrogen troubles caused by plating

Even if no hydrogen is introduced during pickling, there may be entry of atomic H during electroplating. The embrittlement caused by cadmium plating can be reduced if, instead of a cyanide bath, a bath containing di-amino-*n*-butyric acid is used; its superiority is attributed to lower adsorption. The subject is discussed by Beck. (W. Beck, A. L. Glass and E. Taylor, *Plating*, July 1968, 232.)

The availability of processes involving no cathodic deposition (such as metal-spraying) may perhaps provide a more general manner of dealing with the problem (p. 221).

Hydrogen troubles in welding

(See 1960 vol., p. 421; 1976 vol., p. 222.) The presence of hydrogen is a probable cause of certain phenomena, known as 'hair-line cracks' and 'flakes', met with near welds, especially in austenitic steel. Opinions on the question have varied, but an interpretation offered by Huddle in 1942 may be worth summarizing. He pointed out that hydrogen is an austenitic stabilizer and that if it is present, the transformation of austenite to martensite which normally occurs on cooling to a certain temperature may not occur completely. If subsequently part of the hydrogen charge is lost, the change to martensite takes place; hydrogen is less soluble in martensite than in austenite, so that the remainder will be evolved as gas, probably quite suddenly, producing a high pressure, and causing the defects mentioned.

Hydrogen troubles in the oilfields (See 1960 vol., p. 422.)

In the so-called sour oilfields, the water, which is both acid and contains hydrogen sulphide, is apt to produce hydrogen cracking. This may be wrongly diagnosed as stress-corrosion cracking; if attempts are then made to combat stress corrosion by cathodic protection, the disease is likely to be made worse. In some materials, the action of these waters may produce intergranular layers of a sulphide, and sulphide cracking may be the result. A simple means of distinguishing at sight between these three forms of deterioration would be very welcome.

Clearly any decision to use cathodic protection in a case of cracking must only be reached after careful consideration; cathodic protection was largely developed as a result of admirable early work in the oilfields, and perhaps there may sometimes be a temptation to adopt it where it might be better to use another method. Clearly the decision must be left to the 'man on the spot'; it is important that there is someone available possessing the appropriate knowledge of the scientific principles involved, combined with technical experience.

Hydrogen troubles in nuclear power generation (See 1976 vol., pp. 226, 242.)

At nuclear plants, materials are used which suffer hydrogen troubles of a different character from those met with in steel. Zirconium, for instance, forms a hydride as a separate phase, and its production has been demonstrated in zircaloy exposed to concentrated LiOH. Evidence that the embrittlement of titanium subjected to stress is due to hydrogen has been brought forward by Scully; the important step is the nucleation of hydride on an operative slip-plane, which restricts the ductility of the grain. More work on this matter seems to be desirable.

Corrosion fatigue

General

If a specimen is subjected to alternating stress (tension and compression in turn) over a range insufficient to cause immediate fracture, gliding may occur within some of the grains, but when the dislocations reach a grain-boundary they are halted, retracing their movement along the gliding-plane when the stress is reversed. If the material were ideal, it might be hoped that the dislocations would merely move to and fro along the plane, and that no damage would result. In practice a large number of cycles can be withstood without apparent damage, but in material as we know it, slight irregularities will prevent smooth gliding indefinitely, and roughening along the original gliding-plane will make movement difficult, so that gliding will then start on another parallel plane. In the end, bands of material will have become disorganized, and ultimately one of two things must happen: (1) if the stress range is low, gliding will cease altogether, the only changes still produced by the alternating stress being elastic, (2) if it exceeds a certain level (the **fatigue limit**) the gliding will become

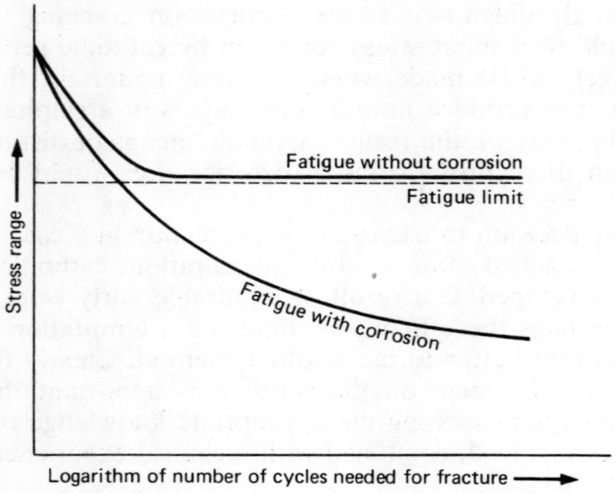

Fig. 5.10 Effect of alternating stresses with and without corrosion

so irregular, as to cause separation between the moving surfaces, first locally, producing gaps, which later will join up into cracks. Thus above the fatigue limit (after a time which is shorter at high stress ranges), there will be failure; below the fatigue limit, the life, in absence of corrosion, should be indefinitely long (Fig. 5.10).

In the presence of a corrosive liquid, the situation will be different. Disorganized atoms along a gliding-plane may require less activation energy to pass into a liquid than more perfectly arrayed atoms elsewhere; certainly, whilst the atoms are in motion along a gliding-plane, preferential attack may reasonably be expected even below the fatigue limit. This means that there is no 'safe stress range' within which the life should be infinite. It is, however, convenient to determine an **endurance limit** – namely the stress range below which the material will endure some specified number of cycles (the number must be stated). The endurance limits of a few materials – based on the classical work of McAdam[712] are shown in Table 5.2; in air, the limit usually falls well below the T.S., whilst in water it is lower still, except on copper, where corrosion would be slight (here the cooling action of the water seems sometimes to *raise* the endurance).

It should be noted that, although stress-corrosion cracking is often intergranular, corrosion-fatigue cracks are usually transgranular, following gliding-planes inclined at such an angle as to provide high resolved shear stress. There are exceptions to both rules. Whitwham,[702] studying corrosion-fatigue cracks on steel, found that, although mainly transgranular, they followed grain-boundaries for short distances, where such boundaries chanced to run in a convenient direction.

Table 5.2 Effect of corrosion on fatigue-strength (S. F. Dorey, based on D. J. McAdam)

Material	Ultimate tensile strength. Tons per sq. in.	Endurance limit ± tons per sq. in. Approximately 5 × 10^7 cycles		
		Air	Fresh water	Salt water
0·16% Carbon steel (hardened and tempered)	29·3	16·0	8·9	4·0
0·24% Carbon steel	24·8	10·3	7·6	—
Copper steel (0·98 Cu, 0·14 C)	27·5	14·3	8·9	3·5
1·09% Carbon steel	46·2	17·85	9·4	—
Ni steel (3·7 Ni, 0·26 Cr, 0·28 C)	40·5	22·0	10·1	7·1
Cr-V steel (0·88 Cr, 0·14 V)	67·2	30·0	8·0	—
Ni-Cr steel (1·5 Ni, 0·73 Cr, 0·28 C)	62·0	30·3	7·2	6·2
Stainless iron (12·9 Cr, 0·11 C)	40·0	24·5	17·0	13·0
Stainless steel (14·5 Cr, 0·23 Ni, 0·38 C)	42·0	23·2	16·0	16·0
Ni-Si steel (3·1 Ni, 1·6 Si, 0·5 C)	112·0	48·7	7·5	—
Monel metal (fully annealed)	36·5	16·0	11·6	12·5
Pure nickel	34·0	14·8	10·4	—
Nickel (cold-rolled)	58·7	22·3	13·0	10·7
Duralumin	31·0	7·8	4·5	3·6
Aluminium bronze (7·5 Al)	40·2	14·5	11·2	9·8
Pure copper (annealed)	13·6	4·2	4·5	—

Notes. The 'fresh water' was a well water containing calcium carbonate. The 'salt water' was a river-water with a salinity about one-sixth that of sea water. Speed of tests = 1450 cycles per min.

Types of corrosion-fatigue cracks

Forsyth and Stubbington (1968 vol., p. 270), studying an Al alloy containing 7·5% Zn and 2·5% Mg, have established two distinct modes of cracking:

(1) **a shear mode** with fracture occurring in a plane 45° to the applied stress (a direction which receives the greatest resolved shear stress)

(2) **a tensile mode** with fracture roughly normal to the stress direction.

They suggest an interpretation which deserves study. In the absence of corrosion, there will be no continued damage so long as the stress does not exceed the fatigue limit. The first few thousand cycles may indeed produce fine slip (not always easy to observe), often following two sets of planes, so that dislocations become locked at intersections, forming 'sessile groups'; the material thus becomes hardened and there is no further deterioration. Corrosion, if present, can release the locked dislocations, probably by the destruction of material at the intersection points and damage may continue.

Measurements at the Fulmer Institute have shown that, where the initial mode of cracking is at 45°, cathodic protection is effective in preventing it; if it is at 90°, cathodic protection is ineffective. If the 45° cracking is connected with anodic attack, and the 90° cracking with cathodic production of hydrogen, the facts are easily understood. This emphasizes the importance of deciding, in any practical case of cracking, whether it is the result of corrosion or hydrogen embrittlement – a matter mentioned in connection with stress corrosion (p. 120).

Although the most striking development of corrosion-fatigue cracking occurs in liquid water (especially salt water), the presence of water vapour can itself allow cracking to develop at a stress range below the fatigue limit which would be obtained under dry conditions. Gough and Sopwith[704] showed that tests carried out in dry, neutral gas, free from oxygen, gave endurance values higher than those obtained without precautions; on 70/30 brass the improvement was 26%. Later workers obtained similar results on other materials. It is likely that moisture is more important than oxygen. Many materials, like aluminium, which would resist water vapour under static conditions, by developing protective films, will continue to react if there is persistent rubbing. Probably many cases of failure commonly described simply as 'fatigue' are really 'corrosion fatigue' in the sense that they would not have occurred if it had been possible to exclude moisture.

Where the presence of liquid water is unavoidable, water-soluble inhibitors can be considered. Gould[726] prolonged the lives of steel wires subjected to alternating stress in water by adding potassium chromate; the life diminished if the stress-range was increased, or if chlorides were present. In the U.S.A., zinc yellow ($4ZnO.K_2O.4CrO_3.3H_2O$) has been used to control the corrosion fatigue normally caused by salt water dripping from railway refrigerator cars; here the chromate and zinc salts may provide anodic and cathodic inhibition respectively.

It is interesting to ask whether it is the disorganization of the material or the motion which promotes corrosion along gliding-planes. To answer this question, Whitwham[708] carried out experiments on two sets of specimens. One set was subjected, first to dry fatigue and then to fatigue under corrosive conditions. The other underwent corrosion fatigue only (the dry stage being omitted). There was no significant difference between the corrosion-fatigue lives of the two sets, and it was concluded that the disorganized material produced by dry fatigue was not exceptionally sensitive to attack, and that the material must, as it were, be 'caught on the move'. Probably the atoms, when they are at positions of high energy (half-way between positions of low energy), require relatively little activation energy to enter the liquid.

Methods of measuring resistance to fatigue and corrosion fatigue

Alternating stresses can be applied in several ways. Rod specimens

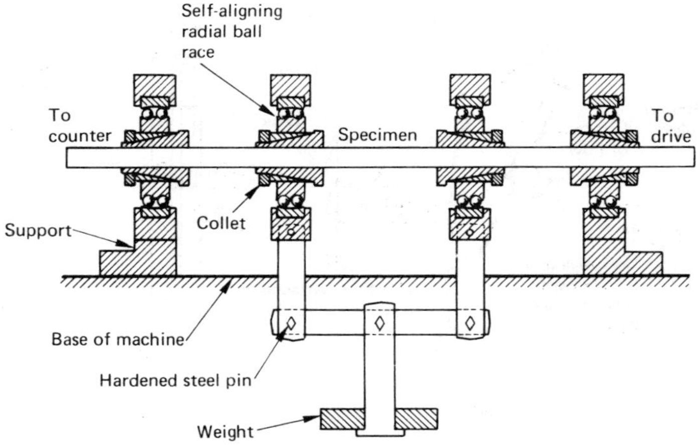

Fig. 5.11 Fatigue-testing machine giving uniform stress between the two loading points

may be alternately stretched and compressed in a push–pull machine, conveniently worked by an electromagnet; alternatively they can be subjected to torsion, being twisted first in one direction and then in the other. A commoner procedure is to use a horizontal rotating rod, carrying a weight at one or more places which has the effect of bringing each point on the surface alternately into tension and compression; in a convenient form of rotating bar test shown in Fig. 5.11, the weight is supported at two points, so that the stressing is uniform over the length between the points of support. For if the weight applied at each point is W, and the distance between them is S, the effect at a distance x from the left-hand support will be $Wx + W(S - x) = WS$, and so is independent of the value of x.

For testing a wire specimen, it may be kept bent into a bow form and then rotated, so that each point is alternately stretched and compressed. In the Haigh–Robertson fatigue machine, the wire is bowed in a horizontal plane; in the Kenyon machine, the plane is vertical and, the bending forces being applied at four points, a uniform stress is maintained over a length of wire, for the reason just explained.

In most of the testing methods indicated above, the stress fluctuates about zero stress as the mean value. In service, it may sometimes fluctuate about a mean which is not zero; probably the severest conditions will generally be those where the mean value is tensional; it is possible to carry out tests which introduce such conditions.

Some of the machines hitherto described were originally designed as pure fatigue machines but have been adapted for corrosion-fatigue testing by directing salt spray or a stream of water on the specimen. In the Kenyon machine[715] which was designed specially for corrosion

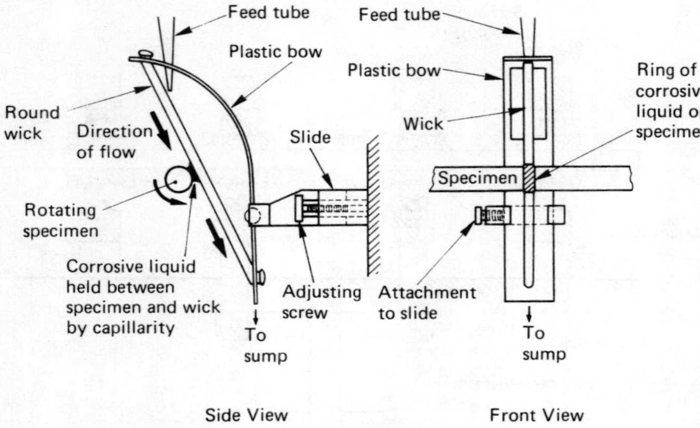

Fig. 5.12 Feed for corrosive liquid (A. U. Huddle and U. R. Evans)

fatigue, it was arranged for the lower part of the bent wire to rotate below the surface of the corrosive liquid. There are certain special features called for in a reliable corrosion-fatigue machine which the dry-fatigue machine does not provide; one is that the machine itself shall be built of relatively corrosion-resisting material; Huddle designed a machine for use at Cambridge in which an aluminium alloy replaced steel. He also designed a special method of feeding corrosive liquid on to a rotating rod specimen (Fig. 5.12). The liquid runs down a wick, held taut in a curved plastic bow attached to a slide, which, by means of a screw, can be made to approach the surface of the rod until the distance is such that a film of liquid is held between wick and rod; the rotation carries a narrow ring of liquid right round the periphery, and this is constantly renewed, thus providing a well-defined and reproducible wetted area without any contact between metal and wick. Tchorabdji-Simnad,[717] also at Cambridge, adopted the principle for his work on wires stressed in the Haigh–Robertson machine, but, in place of a wick, he ran the liquid down a glass rod coated with wax except for a pathway on the side nearest the wire; the liquid followed this unwaxed pathway.

Procedures for corrosion-fatigue tests

Two main procedures are available:

(1) **One-stage tests**. Here the corrosion fatigue is *continued until breakage*. The logarithm of the number of cycles needed to produce breakage is generally plotted against the stress range, as in the curves of Fig. 5.13 selected by Gough from McAdam's experimental data.

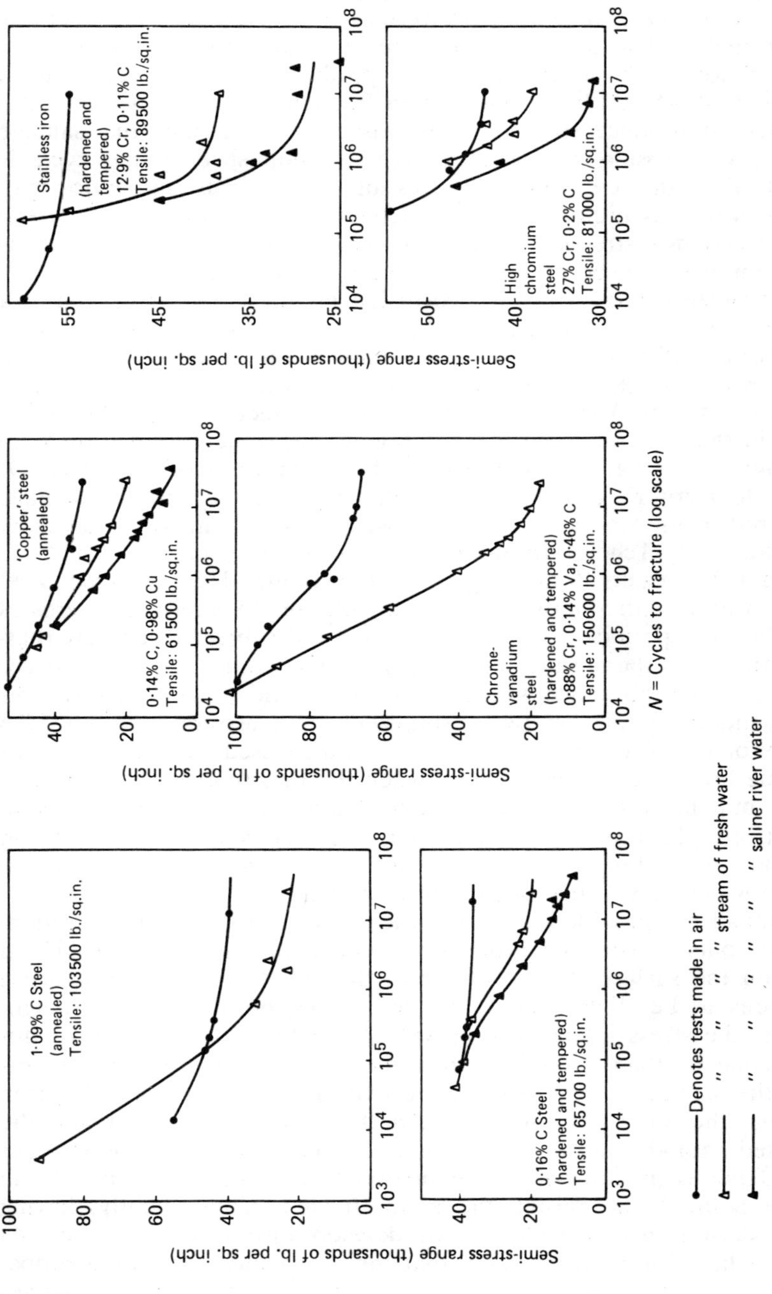

Fig. 5.13 Typical curves showing the number of cycles needed to produce fracture at different stress ranges in absence and presence of corrosion (H. J. Gough, based on data by D. J. McAdam)

The effect of **frequency** deserves attention, since, in order to obtain quick results laboratory tests are commonly carried out at frequencies much higher than those normally met with under service conditions. But although the time needed to produce fracture, measured in hours, is shortened by testing at high frequency, the number of cycles needed to produce fracture is increased thereby, since the damage done by corrosion during each cycle is diminished if that cycle is short. Thus the low-frequency stressing which occurs in service may be regarded as either more destructive or less destructive than the high-frequency stressing adopted in the testing-room, according as duration is expressed in number of cycles or in days and hours.

(2) **Two-stage tests**. Here the corrosion fatigue is interrupted after a definite number of cycles, and the residual strength is estimated by measuring either (a) the endurance limit in the absence of corrosive influences (i.e. the stress which can be withstood for some definite number of cycles) (b) the number of cycles needed to produce fracture in the absence of corrosive influences at some definite stress (c) the tensile strength, or (d) the shock resistance (Izod number). Of these four criteria of residual strength, chosen to measure the damage suffered in the first stage, (a) was used by McAdam, (a) and (c) by Gould, (b) by Tchorabdji-Simnad, (c) and (d) by Huddle.

Simnad's results[717] are especially interesting. He subjected steel wire wetted with potassium chloride solution to alternating stresses for different periods, then dried the wire and subjected it to alternating stresses in the absence of corrosive agency until fracture occurred; it made little difference whether – in the second stage – the wire was surrounded by air or by a solution of potassium chromate, an inhibitor of corrosion. Since the stress range used was below the air fatigue limit, the life was infinite when the first (corrosion) stage was zero, but the total life (both stages) became very short when a comparatively short period of corrosion was allowed. If the corrosion period was prolonged further, the total life became longer again – a result which may at first sight appear strange.

Simnad has provided an explanation based on stress concentration. At the outset, corrosion produces a number of roughly hemispherical pits on the surface. If one of these (through some local peculiarity) happens to be slightly more pointed at the bottom than its neighbours, the stress intensification will be greater at the bottom of this particular pit than at the bottom of the others; since stress, by distorting the lattice or rupturing protective films, will render the iron anodic, the e.m.f. working between the bottom of this pit and the aerated cathode outside will be greater than that operating at other pits. Thus corrosion will be stimulated at its bottom, making it still more pointed and raising the e.m.f. further. Consequently, having once started, this particular pit will develop downwards as a crack. At first sight it might seem likely that the pit in question would propagate itself right through the metal. However, when the crack becomes very sharp, the increase in resistance as the crack deepens will outweigh any further rise in e.m.f. Sooner or later, therefore, the rate of propagation of this first crack will slow down, and other pits will begin to develop into cracks. If it happens that the corrosion stage ceases when only isolated pits have developed into cracks, the concentration of stress at the tips of the cracks will be very high, so that if the alternating stressing is continued under dry conditions, without further application of corrosive substance, failure will soon occur. If corrosion continues until there are numerous cracks close together, the stress concentration is less, and the rate of propagation of the cracks under dry conditions is slower. Consequently the total life up to fracture is longer when corrosion is continued to failure than when it is discontinued at the stage of high stress concentration.

Simnad's experiments are important, not only in revealing the probable mechanism of corrosion fatigue, but in suggesting that, in service, protective measures, if they are to be effective, must be continuous; if the precautions taken to exclude corrosive influences are relaxed for a short period there is a risk that the life of the member affected by corrosion fatigue may actually be shorter than if no protective measures had been applied at all. Obviously this will only occur under certain unfortunate conditions, but the danger is a real one.

Corrosion-fatigue failures in service

There was a period when corrosion-fatigue breakdowns were becoming alarmingly common in engineering practice owing to the introduction of new materials with enhanced tensile strength and even high fatigue strength, as judged by tests in air. The use of such materials appeared to justify reductions in cross-sections, but when these were made without consideration of the possibility of corrosion fatigue, trouble was only to be expected.

The earliest example of corrosion fatigue was the failure of paravane towing ropes, during the war of 1914–18, which Haigh rightly attributed to the chemical action of the aerated salt water acting on the vibrating wires. Since that time, numerous examples of premature failure due to corrosion fatigue have been reported from ships (propeller shafts, rudders and dredger pins), aircraft, railways (axles exposed to dripping from carriages, also steel sleepers), pumps (shafts, rods and bodies), boilers (tubes and superheaters), rock drills (at the central hole through which water is forced to wash out debris), automobiles (steering arms and axles) and even pedal cycles, where the breaking of spokes has been ascribed by Huddle to corrosion fatigue.

An important case is the corrosion of off-shore structures. It must be recognized that a structure designed to withstand the highest stress expected and thus enjoy an indefinitely long life in the absence of corrosion, may fail quickly if corrosive conditions are present, whilst a protective scheme which would promise a long life in absence of stress will allow rapid failure if at intervals high stresses are applied. Many different opinions have been expressed and deserve study. (See *Brit. corr. J.*, 1976, **11**, 62 and *The Metallurgist*, Dec. 1976.) It is thought that coatings should be anodic to the basis metal, not cathodic, and that duplex systems (e.g. galvanizing followed by painting) deserve special consideration. An important point is that a large complicated structure will almost certainly contain defects, so that the initiation stage will not occur; crack-development will start at once, and laboratory testing methods of a type which include the initiation stage will be misleading as a basis for the choice of materials.

An alarming feature of the present situation is the fact that some erudite discussions of dry fatigue hardly mention corrosion fatigue, although any calculation of design could be disastrously misleading unless satisfactory protective measures are adopted. No doubt the writers would acknowledge that where corrosion pits are present on a metallic surface, the effect of stress intensification must be taken into account; but this does not allow for the fact that when two factors (fluctuating stress and exposure to corrosion conditions) are supplied simultaneously they may produce different results from those expected when they are supplied at different times.

Some of the laboratory work on corrosion fatigue may have been carried out under conditions unsuited for the calculation of the service life of a structure. This would account for the mistrust felt by some engineers in laboratory experimentation. But if it is true that the existing methods are based on an imperfect simulation of service conditions, that is a reason not for neglecting the corrosion factor but rather for carrying out fresh work using a procedure which provides a satisfactory model. In particular, the principles of dimensional analysis must be kept in mind when designing the method.

Prevention by choice of materials

Although some of the new 'strong' alloys exhibit poor resistance to corrosion fatigue, materials exist which combine resistance to corrosion fatigue with good mechanical strength. The austenitic stainless steels, which show excellent resistance to stressless corrosion, behave well even in presence of alternating stresses.

The numerous measurements of corrosion fatigue on different materials made by McAdam and others – of which a few are collected in Table 5.2 – may be helpful in guiding choice of materials, but several factors should be borne in mind in predicting liability both to pure fatigue and corrosion fatigue. On certain materials, the presence of notches (whether produced intentionally, as part of a design, or unintentionally as the result of a rough surface finish) will gravely prejudice performance as a result of stress concentration; other materials, like grey cast iron (where fissures exist following graphite flakes), are already 'notch-saturated', and the addition of further notches has little effect. Again, if a member of a machine or structure is exposed to vibration at a frequency close to its 'natural period', resonance may set up oscillations which are very likely to cause failure, unless the material is one of high 'damping capacity', which will dissipate the energy as heat and prevent dangerous amplitudes being reached. Stainless iron combines good corrosion fatigue endurance with outstanding damping capacity.

A test capable of predicting whether a given combination of metal and liquid will produce corrosion fatigue or merely dry fatigue would be very welcome: in the latter case fairly accurate calculations of life expectation are possible. Work at Coventry gives hope that such a test may be obtainable. The metal specimen immersed in the liquid with potential controlled by means of a potentiostat, is subjected to strain-cycling, the strain being changed from tension to compression (or from compression to tension) at equal intervals of time. (C. Patel, T. Pyle and V. Rollins, *Nature*, 1977, 266, 517; *Metal Science*, June 1977.) After each reversal a transient current is registered. With some combinations, the strength of current attained increases on each successive reversal, and these are combinations with which corrosion fatigue must be feared. With other combinations, the current value reached diminishes after each successive reversal, suggesting that repassivation is predominating over film breakdown; here there is hope that corrosion fatigue will be absent. The situation is not simple, but there is hope that a reliable method of assessing inhibitors may emerge on these lines.

Importance of design in avoiding fatigue breakdowns

Quite as important as correct choice of materials is the careful design of machines and structures so as to avoid stress concentration. In marine practice, fatigue failure has been known to start from places where the stress has been raised locally owing to the faulty shape of some stressed member; in different cases, stress concentration has originated in an oil hole, in a small shoulder, in marks left by tools, or in unsuitable surface condition generally; small flaws or grooves must be scrupulously avoided. Pits or notches left by previous corrosion (produced in absence of stress) are also known to have been starting places of trouble.

Prevention by peening

Whilst microscopic notches in the surface layers and tensional stresses, which help to keep them open, tend to promote fatigue and corrosion fatigue, compressional stresses have the opposite effect. Early experiments at Brunswick proved that treatment in a special machine designed to leave the surface in compression enhanced fatigue-resistance. In practice, compressional stresses are usually applied by peening, that is bombardment with round shot. Under correct conditions, peening confers enhanced resistance to fatigue and corrosion fatigue, but much depends on the right choice of pressure and material; grooves produced by bombardment with jagged particles might act as stress-raisers. Gould[727] studies the fatigue behaviour of steel in contact with very dilute acid after peening with different sizes of shot directed against the surface at different pressures; favourable results were obtained with large shot at low pressure, or small shot at high pressure; over a fairly wide stress range, the 'best' procedure gave nearly ten times the life of the 'worst'.

In applying laboratory results to service problems, it must be remembered that a long exposure to low-frequency stressing gives time for the chemical destruction of the compressed layer. A rather thick compressed layer appears desirable.

Prevention by nitriding

Similar benefits may be brought about on ferrous materials by producing in the surface layers certain nitrides which are not only hard but also voluminous. Special steels are employed containing elements like chromium and molybdenum possessing an affinity for nitrogen; aluminium and vanadium are often present. Some of these materials have a high pure fatigue strength even before nitriding, but their resistance to corrosion fatigue is greatly enhanced by heating in anhydrous ammonia. The effect is ascribed to the volume increase accompanying the formation of the nitrides, which produces a state of compression in the surface layers. The process has been called 'chemical peening'. The improvement is illustrated by the figures of Inglis and Lake reproduced in Table 5.3, which also brings out the good performance of austenitic steels.

Table 5.3 Effect of nitriding and other factors on fatigue (N. P. Inglis and G. F. Lake)

Material	Fatigue limit (air)	Corrosion-fatigue limit (Tees Water)
Mild steel	17·0	2
Mild steel, nickel plated	11·2	not more than 3
Nitralloy steel (1·58% Cr, 0·87% Al, 0·33% Mo, 0·26% carbon)	33·0	5·0
Nitralloy steel, nitrided	37·0	25·0
18/8/1 Cr/Ni/W Steel, fully softened	17·6	11·1
18/8/1 Cr/Ni/W Steel, air-cooled from 650°C.	17·6	6·5

Prevention by coatings

The electroplating of steel to prevent fatigue raises the question of internal stresses in the coating; if there is strong torsional stress, plating might make air-fatigue behaviour worse and corrosion behaviour little better. Most nickel coatings contain tensional stresses, but Williams and Hammond[725] have found that by addition of sodium naphthalene trisulphonate to a plating bath of the Watts type (p. 214), tension can be reduced to zero or even replaced by compression. The fatigue strength of a weak steel may be raised above that in the unplated condition, although a strong steel will still lose strength. Chromium deposits also carry tensional stresses and contain hydrogen; by heating steel above 440°C, the stress may become compressional with improvement of fatigue behaviour. Another method is to shot-peen after plating, which may lead to a fatigue strength exceeding that of unplated steel.

The coating of aluminium alloys to prevent corrosion fatigue has been the subject of much research. Cole states that 'treatments and paints which give good protection against static corrosion are equally effective against corrosion fatigue (H. G. Cole, First International Conference on Metallic Corrosion, London 1961; report (Butterworth), 1962, p. 642). There is some evidence, however, that sacrificial metal coatings, including even cladding, are much less effective against corrosion fatigue under some conditions of fatigue loading. It is certainly wise to supplement these coatings with a good paint scheme.'

Wear

Dry friction

The extensive work of Bowden and his colleagues[736] showed that when two unlubricated metal surfaces slide over one another, the motion is often jerky and consists of alternate **slipping** and **sticking**. During periods of slip much heat is evolved, and the temperature may in some circumstances approach or reach the melting point; then the surfaces stick together and movement ceases until the force accumulated bursts the bridge and a slip once more occurs. Friction is in some cases less when the two surfaces are composed of two different materials than when they are composed of the same material; in the latter case, however, it may often be diminished if the material contains two phases; thus ordinary steel causes less friction than austenitic steel. The presence of a film of oxide, sulphide, chloride or iodide will diminish the risk of welding, and thus reduce friction; consequently metal carefully cleaned *in vacuo* furnishes unusually high friction numbers, which diminish as soon as a trace of oxygen gains access to the metal. In a properly lubricated bearing surface, seizing should not occur, but other troubles may sometimes be encountered.

Corrosion by lubricants

The use of lubricating oil to diminish friction and wear is well known. If a lubricating oil contains fatty acids, it may be corrosive at the outset, but much of the corrosion suffered by bearing surfaces is probably due to acids formed by oxidation, during service; certain metals, notably copper, catalyse the oxidation.

Frettage

Whilst, at bearings where relative motion is desired, wear can be largely avoided by suitable lubrication, a different type of damage occurs at places where slight slipping occurs between two steel surfaces which are *intended* to be firmly clamped together. If the small relative movement is within the range of elastic distortion, no change in the material is produced; more extensive movement leads to debris. Hard materials suffer more than soft ones, which easily seize, and escape the undesired sliding. If it is impossible to avoid movement by increasing the pressure between the surfaces, the interposition of soft material between the two hard surfaces may solve the problem; or V-serrations may be substituted for flat faces. A method adopted with some success is the interposition of a layer of soft, fusible metal, such as lead or indium; cadmium has been used on aircraft to prevent frettage and the fatigue cracking that might follow.

The product of this type of wastage is essentially oxide. On a stationary surface exposed to air, oxidation will, of course, be retarded as the film thickens and will soon, at ordinary temperatures, become negligibly slow. Where there is relative motion, the temperature will be raised by friction and the film may be rubbed off periodically, thus exposing fresh metal, so that the retardation will never be realized and the mean rate of destruction may be serious. The fact that interference colours have been noticed on such surfaces supports this cause of the wastage. If, however, the trouble were solely due to oxidation, wastage would cease if oxygen were completely excluded. Many attempts have been made to discover experimentally whether frettage ceases in the absence of oxygen, and different experimenters, working under different conditions, have obtained contradictory results. There does, however, appear to be less destruction when oxygen and moisture are both excluded, and the products are different. Feng and Uhlig[743] found the damage to steel in nitrogen to be only one-sixth that in oxygen; in nitrogen the debris was metallic iron, whereas, in presence of air, oxide was formed. It does not appear that frettage trouble can be stopped entirely by excluding oxygen; stick–slip action, localized at a single point, could produce a quantity of metallic debris, which would turn to oxide when the surface was exposed for examination.

Some authorities regard frettage merely as frictional wear concentrated on a few points, and hence intensified. That is probably an over-simplification. In many cases, the oxide is a hard body, which, if it accumulates, acts as an abrasive. Waterhouse (1960 vol., p. 747; 1968 vol., p. 279; 1976 vol., p. 371) has made a careful study of the mechanism of frettage. He points out that the form of the damage may vary with the geometry. If this is such that the debris cannot escape, deep holes will be produced by the abrasive action of the hard oxide. If it can escape, there will usually be only shallow dish-like depressions. He carried out electrochemical studies, which showed up the effect of film-breakage. On Al, Cr and Ta frettage produced a substantial fall of potential, but on noble metals, like Ag and Cu, there was no change of potential. Like Field, he found that fretting damage which is barely visible to the eye can lead to very serious fatigue deterioration.

This has long been suspected by engineers, but the experiments of Field and Fenner[747] have now provided definite proof. They placed a flat specimen in a push-pull fatigue machine, frettage by bridge-shaped clamps bearing on the surface. When the clamps were present, the fatigue life was greatly reduced, showing that frettage can enhance fatigue damage. Removal of the clamps after they had been in position for a time exceeding one-fifth of the fatigue life obtained in absence of a clamp, failed to increase the longevity; it would seem that here the initiation of the fatigue crack occurs in the early stage before debris has accumulated.

6

Passivity and Inhibition

Nomenclature

General

The disagreements commonly expressed during discussions of passivity and inhibition are at least partly due to the fact that the words are used in many different senses. Attempts to reach universally acceptable definitions have met with limited success. The statement which follows may serve to indicate how the words will be used in the present book.

Passivity

Early investigators, such as Keir, Faraday and Schönbein, were interested in the fact that iron, after treatment in concentrated nitric acid, became altered in properties; it now failed to react with nitric acid of somewhat lower concentration, which violently attacked iron not pre-treated in concentrated acid. Schönbein suggested that the altered iron should be called 'passive', and the name became generally adopted.

It was found that the passivity (the passive condition) could be produced by anodic treatment. Iron subjected to anodic treatment in dilute sulphuric acid at low current density dissolves as ferrous sulphate at a rate decided by Faraday's law. At high current density, dissolution takes place for a time and then virtually ceases; thereafter, the current passing is devoted to the production of oxygen. The iron is now altered in properties, but the alteration may only survive for a few seconds in the acid after the current is turned off. This is clearly the same change as occurs on treatment in nitric acid without current, and the iron is again described as 'passive'. The change from the 'active' to the 'passive' condition can be detected electrically. In an experiment at constant potential, the current falls off sharply when the iron becomes passive: in an experiment at constant current, the potential rises sharply; in the latter case there is an abrupt fall of potential when the iron becomes active again. The alteration from the active to the passive state is known as 'passivation'.

Inhibition

A substance which, added to a liquid (generally water or an aqueous solution), diminishes the corrosion produced, is known as an **inhibitor**, but inhibition is of two kinds. The substance can reduce the probability that any attack at all will occur on a small area; such a substance is known as a **deterrent**. Alternatively, the substance may reduce the velocity at which corrosion takes place; such a substance can be called a **retardant**. The same substance can be both a deterrent and retardant; an alkali added to water generally reduces the probability of corrosion starting on a small area, and generally reduces the velocity of attack if any corrosion takes place at all. In contrast, Mears' experiments on drops of water placed on iron in oxygen–nitrogen mixtures showed oxygen to be a deterrent, but not a retardant (p. 90). The 'restrainers' added to pickling baths to limit the attack on iron which is being treated in acid for the removal of scale are certainly retardants, but they do not prevent all attack and do not seem to be deterrents; it is convenient to use the word 'restrainer' for the retardants used in pickling.

Anodic passivation

Alternative anodic reactions

If a cell consisting of two zinc electrodes in dilute H_2SO_4 is joined to an external battery or other source of e.m.f. through a resistance to regulate the current, one electrode (the anode) will suffer anodic attack; at the other (the cathode) H_2 gas will be evolved (unless an oxidizing agent is present, in which case it will be reduced by the cathodic reaction). If the two electrodes consist of iron, anodic attack will occur, provided that the current density is kept low; if it is high, the anode will become 'passive', covered with an oxide film (invisible whilst in optical contact with the metal) and the current will be devoted to evolution of O_2 gas; corrosion will almost cease. If the liquid is a NaOH solution, an iron anode will become passive even at the lowest current density, because Fe^{2+} and OH^- are incompatible (a precipitate is formed when NaOH and $FeSO_4$ solutions are mixed). It might be expected that the film which stops attack on the metal would be a hydroxide, but in fact an oxide seems generally to be formed, probably by the mechanism suggested in Fig. 2.12 (a); OH^- ions are adsorbed in an oriented array, and the Fe^{2+} ions (which in H_2SO_4 would have passed out into the liquid) stay in the oxygen zone (giving oxide) whilst the rejected H atoms join on to OH^- further out to produce water; this mechanism involves the same consumption of electricity as the passage of Fe^{2+} into the liquid and avoids the demands for energy which would be necessary for an Fe^{2+} ion to pass from the negative O zone to the (relatively) positive H zone. When a single layer of oxide has been formed, it can thicken by the mechanism suggested in Fig. 2.12 (b), but before the film is thick enough to be visible, passage of metal into it ceases and current is devoted to O_2 production. In pure water, a similar change can take place as soon as the liquid water has become saturated with the (not very soluble) $Fe(OH)_2$, and it is possible to obtain either oxide, as in Fig. 2.12 (c), or hydroxide, as in Fig. 2.12 (d).

In a solution of sodium sulphate or nitrate, anodic corrosion will lead to a soluble ferrous salt at low current density, but passivity will be produced at high current density by the mechanism suggested (for NO_3^-) in Fig. 2.12 (e). The current density needed for passivity is lower in nitrate solution than in sulphate, whilst in chloride solution it is difficult to obtain passivity at all, unless alkali is also present; at certain values of the ratio Cl^-/OH^- we may get corrosion localized at places, as suggested by Fig. 2.12(f). If the liquid is concentrated HNO_3, passivity can be produced without applying a current from an external source; the reduction of HNO_3 provides a powerful cathodic reaction at some parts of an iron specimen, which must be balanced by anodic reaction elsewhere; in dilute HNO_3 this leads to rapid attack on the metal, but in concentrated HNO_3, where still higher current might be expected, there may be insufficient Fe^{2+} ions possessing sufficient energy to pass from the O zone to the NO_2 zone in (Fig. 2.12(e)); thus an oxide film is formed, the metal becoming passive.

Metal cannot be rendered passive by anodic treatment unless it is raised into the passivation region of the Pourbaix diagram (p. 248). On iron (for reasons believed to be connected with the existence of vacant electron sites in the atom), a relatively low current density will be sufficient to cause the necessary rise in the potential; on Zn a much larger current density is needed, so that Zn becomes passive much less easily than Fe.

The anodic passivation of iron, zinc and other metals was studied in detail in the classical work of W. J. Müller mentioned on p. 53. Unfortunately, Müller's otherwise excellent quantitative studies of anodic corrosion were conducted under conditions which allowed variation both of current and potential. This complicated the interpretations of his accurate results. Later, methods were worked out for working either under **galvanostatic** conditions (with the current kept constant) or under potentiostatic conditions, with the potential kept constant). A third procedure has sometimes been used, known as **potentiokinetic**, in which the potential is moved up or down at a definite rate, the current corresponding to various values being noted. This method, which has been criticized by Herbsleb, can lead to misleading results; a hysteresis effect is observed and curves obtained with rising potential may not coincide with those obtained with falling potential. The other two methods are more satisfactory, and, if properly carried out, lead to results concordant with one another (G. Herbsleb and W. Schenck, *Corrosion Science*, 1969, **9**, 613.) The galvanostatic procedure requires no expensive apparatus but the potentiostatic procedure is generally preferred; the best plan is to set the instrument at some fixed potential, wait until the current has become constant and record the value; then another potential value is taken and so on; generally the changes of potential should be arranged at equal intervals of time and potential. An early example of valuable results obtained by the potentiostatic method was provided by the work of Olivier.

Iron in sulphuric acid

The curves for an iron anode in 10% sulphuric acid, shown in Fig. 6.1 are due to Olivier.[230] In the absence of any applied e.m.f. (point 0), the iron is attacked with hydrogen evolution, but as the potential is moved in the positive direction, hydrogen evolution slackens and the rate of attack increases as the applied anodic current increases. In region (1) current density rises with positive movement of potential, but suddenly it starts to fall off as we enter region (2). This occurs when the liquid layer next to the surface has become so supersaturated with ferrous sulphate that crystals separate on the metal surface, causing a slight dullness; they are identifiable under the polarizing microscope. If the rate of potential increase is rapid, (curve a), a higher current density is reached than if it is slow (curve d), since with a rapid increase there is less time for nucleation; curves (b) and (c) refer to intermediate rates. When the solid sulphate separates it will cover up much of the surface, and although the true current density on the still uncovered metal is higher than at lower potentials, the current read on the meter is lower, the drop being greatest on curve (a) owing to the greater super-saturation,

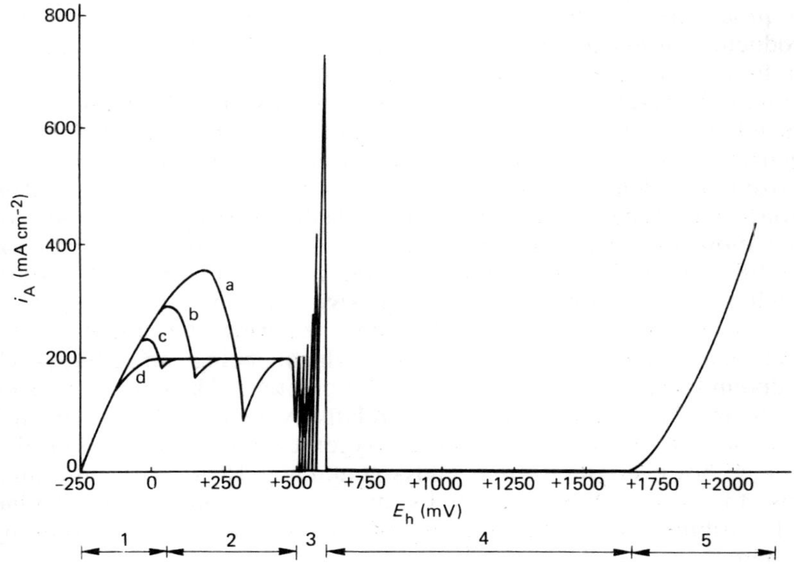

Fig. 6.1 Anodic polarization curves for iron in 10% sulphuric acid (R. Olivier)

ultimately, curves (a) (b) (c) and (d) all come together, and there follows (region (3)) a violent oscillation of current with time (the cause of which is discussed later). At a potential of about 0·58 volt, energy conditions for the first time permit of the maintenance of a continuous film of oxide, and since this, when once formed, will hinder the passage of Fe^{2+} ions into the liquid, the current becomes very low, and the ferrous sulphate crystals dissolve, with diffusion into the main body of the liquid. The small current passing in region (4) is generally believed to represent the amount needed to replenish any destruction of the film through slow dissolution in the acid, but since this dissolution is a chemical process, the small current passing in this region is almost independent of potential. In region (5) energy considerations make possible the evolution of oxygen gas, and the current accordingly rises again.

The composition of the film formed anodically in sulphuric acid or sulphate solutions varies with conditions of formation. It is generally held to be a cubic oxide, a member of the series of which magnetite (Fe_3O_4) and γ-Fe_2O_3 represent limiting compositions. However, hydroxide is sometimes found, and work in Cohen's laboratory has revealed the reason (1976 vol., p. 141). He showed that it is possible to deposit films on a platinum anode from a $FeSO_4$ solution; the films are largely γ-FeO.OH with some Fe^{2+}, SO_4^{2-} and water. Now an experiment on the passivation of an iron anode in H_2SO_4 usually commences with a period during which the iron is active and produces $FeSO_4$ in solution; it is likely that hydroxide will be deposited after passivation has set in. Thus anodic behaviour depends on procedure. A slow or step-wise rise of potential, allowing the production of much Fe^{2+} in the liquid, will modify the composition of the film and probably its protective power.

Fig. 6.2 presents Olivier's curves for various iron–chromium alloys (including 'stainless irons' of 12% and 14% chromium). Here a logarithmic scale is used, a device which serves to separate the horizontal middle parts of the curves for the different alloys; on an undistorted scale these would be indistinguishable. The zone of oscillation, shown by broken vertical lines, occurs on the curves for the 2·8% and 6·7% chromium alloys, but not on the others. A feature peculiar to the alloys is the **trans-passive** zone in region (3) where there is definite attack, the chromium entering the liquid in the hexavalent state, as chromic acid, which, being an energy-rich compound, can only be formed at high potentials. The current passing in region (3) is greatest on alloys relatively rich in chromium, and continues up to potentials where oxygen-evolution starts; Pražák's work,[232] however, shows that alloys containing 18 to 30% chromium cease to dissolve above +1·8 volts; he calls this **secondary passivation** and attributes it to the blocking of channels through the film by oxygen.

Frankenthal, studying anodic behaviour of an alloy containing 24%

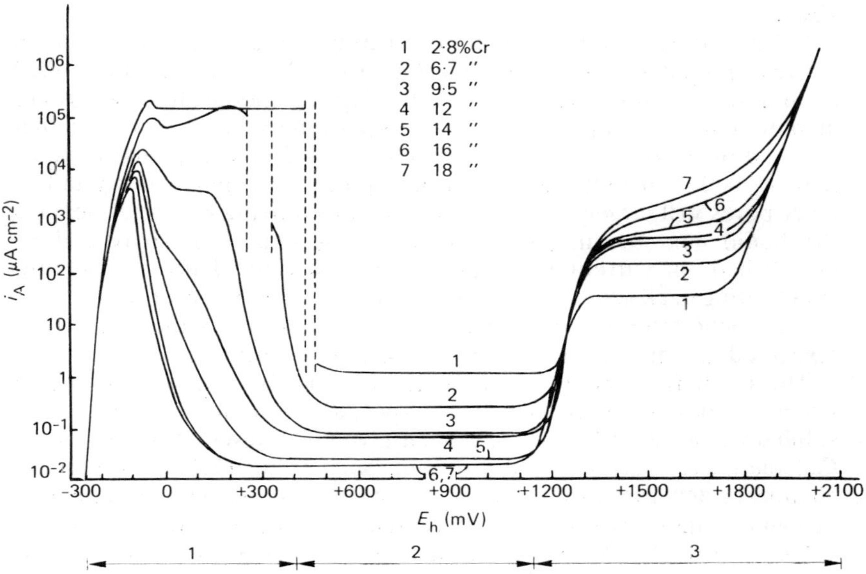

Fig. 6.2 Anodic polarization curves for iron–chromium alloys in 10% sulphuric acid (R. Olivier)

Cr, has shown that there are two kinds of films (R. P. Frankenthal, *J. electrochem. Soc.*, 1967, **114**, 542; 1969, **116**, 580.) The primary film is stable only within a few millivolts of the potential at which the change between activation and passivation would occur under thermodynamically reversible conditions; its formation or destruction is a reversible process. The average thickness is small; there is less than one O atom for each atom on the surface layer of the metal. The secondary film is slowly formed at more positive potentials and grows to a thickness exceeding 10^{-9} m; it does not show reversible behaviour and can be very resistant to reductive dissolution at potentials where, under reversible conditions, its destruction would be expected. The existence of these two types of film shows that both the schools of thought which ascribed passivity to adsorption or film-production respectively may, in this instance, be right; the primary film can best be regarded as due to adsorption, whilst the secondary film must be attributed to a three-dimensional oxide.

Curves similar to Figs. 6.1 and 6.2 have been constructed for 18/8 Cr–Ni stainless steel. Japanese work (1976 vol., p. 146) supplies information regarding the composition. In the potential range analogous to region (4) of Fig. 6.1, the oxide film contains Cr, Ni and Fe, but at lower potentials the matter deposited contains more Cr, and at higher potentials it consists mainly of iron oxide.

Causes of oscillation

Olivier noted violent oscillations on iron at the borderline between activity and passivity (region (3) in Fig. 6.1). The explanation provided by him was as follows. Although in 10% H_2SO_4, anodic attack will occur at potentials just below 0·58 volt, passivity may be expected in that region if the acidity drops. Now when ferrous sulphate crystals are being deposited the current passing through pores threading the crystalline layer pores will largely be carried by H^+ ions moving outwards, leaving the liquid next the metal less acid; in due course passivity will be established, the current dropping to a very low value. Diffusion can now slowly bring back acid, and dissolution of the oxide will occur (as Fe^{2+} not as Fe^{3+}, since the former is indicated by the Pourbaix diagram as the 'expected' product); thus anodic attack can be resumed.

The oscillations between the active and passive states which occur when stainless steel or nickel is exposed to anodic treatment in a sulphate–chloride mixture have been given a different explanation by Galvele (J. R. Galvele, *J. electrochem. Soc.*, 1976, **123**, 464.). Electrochemical calculations show that if pitting attack starts at a point, the ions of higher valency will move preferentially into the pit. When a certain value of SO_4^{2-}–Cl^- has been reached, repassivation will occur, and the current will fall to a very low value; slow diffusion of SO_4^{2-} out of the pit will then occur, and after a time fresh anodic attack will start.

Electropolishing

The small constant current passing in region (4) of Fig. 6.1 is necessary for the maintenance of the film, and is generally believed to compensate for the slow chemical dissolution. If the film is continually cracking, this current may serve also to provide the oxide needed for repair of the gaps. If a liquid is chosen which dissolves the film-substance quickly, a much higher current density will be needed to maintain the film. At microprominences on the surface, the removal of metal from the film-surface into the liquid by diffusion, convection or stirring will occur more readily than at the intervening depressions, so that the film covering the metal will be kept slightly thinner at prominences than elsewhere. Metal will pass into the thin film here faster than elsewhere, the potential gradient being steepest where the film is thinnest; also cavities between film and metal, left by cations diffusing out into the film, will develop more quickly below prominences than elsewhere. In a reasonably short time, preferred attack will destroy the metal at prominences and leave the surface bright. Some smoothing would occur even on a film-free surface since for reasons of potential distribution in the liquid, the current density will be slightly higher at prominences than elsewhere; but this difference is insufficient to give real brightness. The work of Hoar, Mowat and Farthing[253] indicates that for true polishing a solid film is needed; they found that when a drop of mercury was brought into contact with a copper anode in phosphoric acid, it 'wetted' the copper under conditions favourable to etching, but refused to wet it under conditions favourable to polishing. This view, however, does not seem to be universally accepted.

The solutions used in electropolishing are curiously varied. Jacquet[258] the talented inventor of the process, used a mixture of perchloric acid and acetic anhydride for aluminium alloys, steel, tin and titanium. That mixture has occasioned serious explosions, probably caused by departure from instructions; a safer bath, also due to Jacquet, contains acetic *acid* (instead of anhydride) along with perchloric acid; others recommend a mixture of alcohol and perchloric acid. Much simpler baths are available for certain metals. Copper and its alloys are commonly polished in phosphoric acid, whilst Vernon and Stroud used 25% potassium hydroxide for zinc. (W. H. J. Vernon and E. G. Stroud, *Nature*, 1938, **142**, 477, 1151).

Electropolishing was adopted in industry more quickly than some other innovations. In the laboratory, it has proved useful for preparing sections of metals for microscopic examination. The method shows up detail which is missed after mechanical polishing, since it depends essentially on the removal of disorganized matter, whereas mechanical polishing, to some extent at least, produces a smooth surface by causing metal (generally containing oxide and grains of the polishing agent) to flow over the surface, filling up, or bridging over, the depressions.

Anodizing of aluminium

It is possible to generate oxygen on an iron anode in dilute acid for long periods without the serious thickening of the film; the optical studies of Tronstad[235] suggest some slight thickening at first. Iron oxide is an electronic conductor, so that movement of cations outwards through the film is not needed to deal with the current, thus there is only a little thickening.

Aluminium, however, forms an oxide which is not an electronic conductor, and current can only be carried by ions – probably mainly cations, moving outwards, although in a sulphate bath there is also movement of SO_4^{2-} inwards. Under potentiostatic conditions the film will thicken until the potential gradient becomes insufficient to cause the ions to jump the hurdles separating the sites of low energy (see p. 242). The thickness of the compact layer reached will be proportional to the potential drop applied across it, since the thickening will become negligibly slow below a certain potential gradient; it is sometimes stated that the thickness is proportional to the e.m.f. of the cell

cathode | acid | aluminium anode

and this is approximately true, since when the current has become small, the polarization at the cathode and the ohmic drop across the acid will be unimportant. Similar relations exist on other metals. Charlesby and Polling[788] obtained eight orders of brilliant interference colours by the anodic treatment of tantalum. The same colour recurred at intervals, and the e.m.f.s needed to cause interference of a given wavelength were found to be spaced at equal intervals.

In some **anodizing baths** (e.g. borate or tartrate) the film produced on aluminium is essentially non-porous. In sulphuric acid baths, the compact layer next to the metal becomes covered with a much thicker porous layer, found to consist of hexagonal cells each threaded by a central pore, running at right angles to the metal. The porous layer is perhaps best explained by the fact that when the non-porous layer is approaching its limiting thickness, the anodic reaction ceases to be solely the production of oxide. The formation of soluble aluminium sulphate also takes place; this is the reaction which would produce the greatest energy drop, and therefore the reaction which would be expected under 'reversible' conditions. It is, however, widely believed that the first product is oxide, and that the pores are formed by the acid dissolving it as sulphate.

Researches by Dekker, Neufeld, Leach and Bradhurst (1976 vol., p. 148) provide much information regarding the morphology of anodic films on aluminium, and also the parts played by movements of anions and cations during their formation.

Permanent passivity

The principles presented above have been applied by Hoar and Mears to the problem of finding a material likely to resist pitting indefinitely in a chloride solution. Their immediate object was to discover a safe material for surgical implants, but any such discovery, if realized, would have attractive possibilities in connection with desalination plants, pumps for brackish wells, washing machines and anodes for cathodic protective installations. They point out that water is only stable within a certain range of potential, bounded by the two lines labelled (a) and (b) on Fig. 8.3. Outside that range, water will be decomposed, evolving O_2 above (b), or H_2 below (a). If a material could be found with its active range well below (a) and its transpassive range well above (b), no serious attack or pitting need be feared, although slow general corrosion connected with passage of cations through the intact oxide film must be expected. Certain titanium alloys appear to fulfil these conditions; some alloys today used for implants fail to answer requirements.

Corrosive and inhibitive solutions

Effect of oxidizing agents

If, instead of making a piece of metal anodic by means of an external e.m.f., we merely immerse it in a liquid containing an oxidizing agent (which may be oxygen), the anodic current density, and hence the behaviour, depends on the cathodic reaction. For iron in concentrated nitric acid (p. 72), this is so rapid that the necessary supply of cations cannot be found to pass into solution, and the anodic reaction builds a protective film, producing passivity; in dilute acid, there is violent attack. The same general explanation could be given in other cases. Kolotyrkin, obtained under potentiostatic conditions the current–potential curve for a nickel anode in 0·5M sulphuric acid shown in Fig. 6.3 (the resemblance to the curves of Figs. 6.1 and 6.2 will be noticed). (Y. M. Kolotyrkin, First International Congress on Metallic Corrosion, London, 1961; report (Butterworth), 1962, p.10.) He then measured the dissolution-rates of nickel immersed (without current) in 0·5M sulphuric acid containing various oxidizing agents – $Fe_2(SO_4)_3$, $K_2Cr_2O_7$, H_2O_2, $KMnO_4$, $Ce(SO_4)_2$ and O_2 – at different concentrations. He converted those rates into electrical current values (in the sense of Faraday's law), and plotted the current values against the measured potentials.The points fell almost exactly on the anodic curve; when the potential was between + 0·4 and + 1·1 volts, corresponding to region (2) of Fig. 6.2 the dissolution rate was very low;

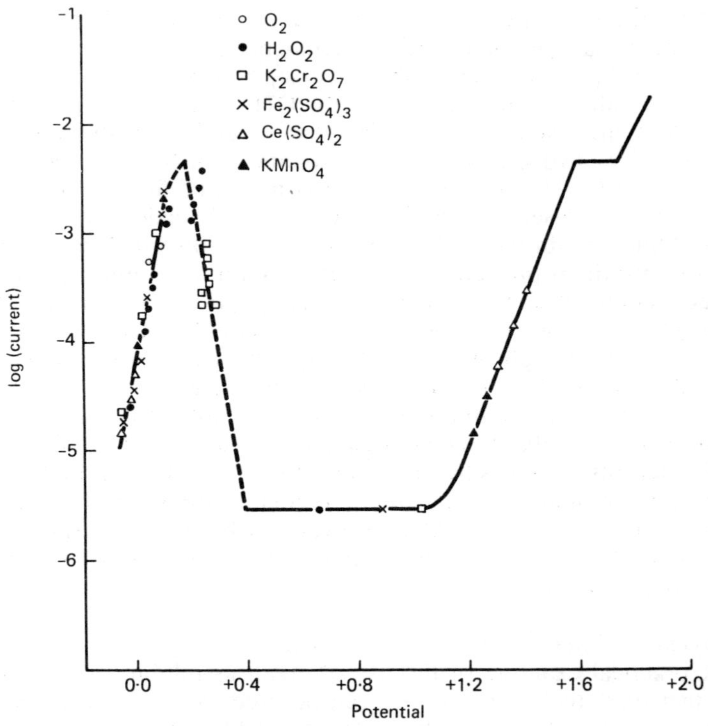

Fig. 6.3 Behaviour of nickel in 0·5M H_2SO_4 with applied e.m.f. (curve) or with addition of oxidizing agents (points). The ordinates represent the current on a logarithmic scale or, in the case of the points, the current equivalent to the measured corrosion rate (Y. M. Kolotyrkin)

below and above that range it was faster. The results show clearly that the addition of an oxidizing agent is equivalent to applying an anodic current from an external battery.

Alkaline inhibitors

The best known inorganic inhibitors, such as sodium carbonate, hydroxide, phosphate and silicate, yield alkaline solutions. These prevent corrosion of iron in water containing oxygen, and the question has been raised as to whether, in all cases, the inhibition is not due to a phalanx of oriented OH^-. In any weakly alkaline liquid, an iron cation may be expected to reach the O zone and displace two H^+ ions to react with further OH^- ions giving water as shown in Fig. 2.12. This is the reaction predicted by energy considerations – whether the current density is low or high; it will provide the same electric transfer as if the Fe^{2+} ion entered the liquid, and a greater drop in free-energy; it may be expected to occur, and to provide a protective film of oxide.

It requires only a very low concentration of sodium hydroxide to provide the same pH value as fairly concentrated sodium phosphate. This might suggest that NaOH should be a more effective inhibitor than Na_3PO_4 or Na_2HPO_4. The reverse, however, is found to be the case. If in dilute sodium hydroxide the OH^- locally becomes exhausted, corrosion may start; in the phosphate solution fresh OH^- will be formed by hydrolysis, the phosphate constituting in effect a 'reservoir' of OH^-. Moreover, the sparing solubility of the iron phosphates may be useful in the event of local exhaustion of OH^-. Mayne and Menter,[136] studying films stripped from iron which had stood two days in disodium phosphate (Na_2HPO_4) solution, found that they consisted mainly of cubic iron oxide (Fe_3O_4, γFe_2O_3 or an intermediate solid solution), but in places particles of $FePO_4.2H_2O$ were observed. When tri-sodium phosphate (Na_3PO_4) had been used (which would provide more OH^-), the film was entirely oxide.

In the case of sodium silicate there is another possibility; at any point where the alkali becomes exhausted, silicic acid or silica is likely to be deposited, and silica is indeed found on steel surfaces exposed to waters treated with silicate to regulate corrosion. It is generally believed that water brought to a given pH by means of silicate is less corrosive than if sodium hydroxide had been used to produce the same pH value.

Protection of steel reinforcing wires in concrete

A practical example of the use of alkali as inhibitor is provided by the fact that steel wires or rods can be used as reinforcements in concrete, and steel frames in buildings embedded in concrete in general construction work (1960 vol., pp. 302–8; 1968 vol., pp. 130–5; 1976 vol., pp. 185–90). The reactions involved in the setting of cement liberate alkali, and this, under favourable circumstances, prevents corrosion of embedded steel, provided that the concrete cover is sufficiently thick and compact; if channels exist through which CO_2 can penetrate, destroying the alkali, corrosion may start. It will also start if chlorides are present; NaCl or $CaCl_2$ is sometimes deliberately added to a concrete mix to accelerate the hardening, or to prevent freezing in cold weather; of course, near the sea, salt can penetrate if the concrete cover is porous or insufficiently thick.

The presence of chlorides is particularly to be avoided if reinforcement wires are to be held in tension, but even with unstressed reinforcements, chlorides are undesirable, they are especially dangerous if the wires, at the time of burial, carry small rust-spots; under such circumstances, the rust-covered points will receive no alkali, and the combination of small anodes with large cathodic area (the unrusted part which receives replenishment of alkali) can provide a dangerous situation.

A point to be recollected is that the provision of alkali is due to the setting reaction and will not continue for ever; on the other hand, if the concrete cover is porous, the supply of CO_2 will continue indefinitely. It would seem that a reservoir of alkali is needed. This is sometimes provided by application of a slurry of neat concrete to the reinforcing bars before surrounding them with concrete.

Page has provided electron micrographs showing the formation of a lime layer at the interface between concrete and steel. (C. L. Page, *Nature*, 1975, **258**, 514.) His discussion, based on the use of a Pourbaix diagram, leads him to recommend the plan suggested above of applying a cement slurry. Other methods of dealing with the menace include the use of zinc-coated wires as reinforcements; fears that the zinc coating will prevent a strong bond between reinforcement and concrete may have been exaggerated, although if impure zinc is used the evolution of hydrogen may have an undesirable effect. Addition of a soluble inhibitor to the concrete mix (a plan which has sometimes been advocated) is probably dangerous, especially if the steel carries rust-spots; the production of small anodes surrounded by a large cathodic area is then a definite risk. The relative merits of the various protective methods, have been compared in reports from the Battelle Institute (U.S.A.) and from the Building Research Station (U.K.) (1976 vol., pp. 185, 188, 190).

A research by Treadaway on the corrosion of steel reinforcements in concrete exposed to marine conditions deserves attention. (K. W. J. Treadaway, *Chem. Ind*., 1976, 348.) Long-period tests were carried out at exposure sites, and also laboratory tests under galvanostatic or potentiostatic conditions with cycling between -1400 and $+1000$ mV (relative to the saturated calomel electrode). Tests on the site showed that after two years there was generally little attack if the $CaCl_2$ content of the cement mixture had been kept below 1·5%, but serious corrosion of strong steel occurred at higher contents; galvanized steel suffered loss of zinc. There was more corrosion on pre-rusted steel than on shot-blasted steel. On stressed steel there was some breakdown of bond between reinforcement and concrete.

Condensed metaphosphates as inhibitors

Whilst the inhibitors mentioned above (including the orthophosphates) seem to work by the provision of OH^-, the group known as linear polyphosphates certainly act in a different way, since they are effective in waters which are distinctly acid. The inhibition proceeds best if plenty of Ca^{2+} and a trace of Fe^{2+} is present; an acid water can hold Ca^{2+} in solution without precipitation, and may produce the necessary Fe^{2+} by action on iron vessels or pipes.

The best known inhibitor of the class is generally known as **Calgon**. It was at one time called **sodium hexametaphosphate** (a misleading description) and was assigned the formula $(NaPO_3)_6$; it would be better to write $(NaPO_3)_n$, where n is a number exceeding 6. Since the best results require the presence of Ca^{2+}, a glassy polymer of sodium and calcium metaphosphate known as **Micromet** is convenient; this is only slowly soluble, and a corrosive water percolating through a mass of Micromet particles of suitable size may become relatively non-corrosive.* The combination of sodium and zinc metaphosphates is even more effective; commercial products containing Na and Zn are available, but are clearly not suitable for treating drinking water.

Work at Teddington by Butler (1968 vol., p. 76) has brought out the value of using these combinations of inhibitors. By the addition of metaphosphate and calcium to water containing 2500 ppm of NaCl, which is normally very corrosive, the attack can be reduced to a very low rate. One of the most intractable problems in water engineering is the suppression of corrosion by water containing chloride after attack has started. The addition of an inhibitor of the alkali-producing type may be worse than useless since the inhibitor will not reach the points already screened by rust, and there is risk of the combination of small anodes and large cathodic area. The metaphosphate combination seems to solve the problem.

Butler advised 50 ppm of P_2O_5 and 30 ppm of Ca^{2+}, but says that when once a restraining film has been formed, some laxity of the maintenance of the inhibitor concentration can be tolerated – an important point in situations where analytical control is not possible. The main disadvantage of this method is that it works best under conditions of flow; for stagnant conditions (and most water supply systems have some stagnant regions) a higher concentration of inhibitor is needed.

The reason for the effectiveness of such combinations has been

*A view widely held today is that polyphosphates in the presence of calcium ions function as cathodic inhibitors. The soluble calcium polyphosphate complex sheds its end members yielding calcium orthophosphate, which slowly settles on or migrates to a metal surface. This is an example of inhibition by solution precipitation.

brought out by Lamb and Eliassen[168] using a radioactive tracer technique. They showed that water containing calcium polyphosphate and also a trace of iron develops positively charged particles, probably hydrated ferric oxide containing metaphosphate, which are deposited on the cathodic points, thus interfering with the cathodic reactions which would otherwise maintain corrosion.

Corrosion and inhibition at different concentrations of the same salt

Brasher has suggested that practically all salts are corrosive at very low concentrations and inhibitive at very high ones. (D. M. Brasher, *Nature*, 1962, **193**, 863.) This appears to be generally true, but there may be more than one factor at work. Chapman showed that dilute fluoride solutions corrode iron about as quickly as chloride solutions, but above 0·8M the attack suddenly ceases; this may be due to the solubility of ferrous fluoride being depressed when excess of fluoride ion is present, so that the case becomes analogous to that of silver in chloride solution. (A. W. Chapman, *J. chem. Soc*., 1930, 1546.) Similarly, sodium carbonate or bicarbonate cause rusting on iron at low concentrations and keep it bright at high concentrations; at borderline concentrations, behaviour depends on other factors – such as prior surface treatment or the concentration of oxygen in the gas-phase.

Some of the cases studied by Brasher, notably the inhibition produced by concentrated nitrate solution, may be due to the more complete phalanx of oriented oxygen-containing anions set up at the high concentration. It is interesting to note that oxygen, which usually helps the wet corrosion of iron at atmospheric pressure, may prevent attack at high pressures; this may be partly due to direct oxidation at any places when a film develops a discontinuity; probably, however, the more rapid cathodic reduction of oxygen over the main film-covered area raises the potential into the passivation range.

Crack-heal in films

Borderline concentrations, where a given reagent behaves either as corrodent or inhibitor according to the character of the surface exposed to it, have been much used in measuring the probability of corrosion, by means of drop methods similar to those described on p. 90. They are also useful in studying the alternate cracking and healing of films. There have been many variants of the experiments performed, but they have all led to the same conclusions. A simple experiment will be described, with a diagram (Fig. 6.4) somewhat idealized in the interest of clarity; (only one scratch-line is shown, although in the actual researches, many scratch-lines were engraved.)

Rectangles of steel, abraded, degreased and tinted by heating in air for 40 minutes at 200°C, were engraved with scratch-lines and placed in a sloping position in a $NaHCO_3/Na_2CO_3$ buffer solution (each salt at 0·011M, which was the borderline composition for the steel used); the upper part of the rectangle, used as a handle, remained unwetted.

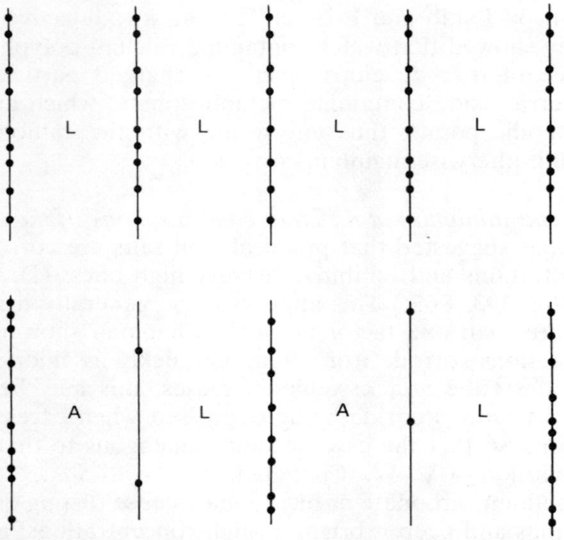

Fig. 6.4 Pattern of rust spots after exposures to liquid (L) or to air (A)

After 5 minutes the specimen was withdrawn into air in such a way as to come into contact, during emergence, with a filter paper rectangle soaked in the same solution, and was exposed horizontally to air for 5 minutes, with the wet filter paper clinging by capillarity. It was then replaced in the liquid for a further 5 minutes, the filter paper, which detached itself as the steel re-entered the liquid, being preserved to show the pattern of rust-spots produced at points on the scratches where local corrosion had started. After 5 minutes fresh immersion, the steel was again withdrawn, in contact with a second piece of filter paper, and in that way a series of papers was obtained, showing the pattern of breakdown points after each liquid immersion. On the whole *the pattern remained constant*; occasionally an old rust-spot would fail to reappear; occasionally a new one would arise, but these were exceptional events. Provided that there was no direct exposure to air (only exposure through the wet filter paper), a corroding point generally continued to corrode and very few new ones appeared.

If, however, between the immersions, the specimen was dried and exposed directly to air, the effect on the pattern was marked. After air exposure (A), all, or nearly all, the old spots disappeared, and only a few new ones arose; after the next liquid exposure (L) new spots arose, but in positions almost always different from those which had disappeared. Evidently air-exposure heals the faults in the film where corrosion has been occurring, and new faults occur spontaneously on fresh immersion in the liquid. (This cannot be attributed to some chemical action of the salts on the film-substance, since when their concentration is increased, the corrosion is prevented.) It is clear that the film is developing cracks spontaneously, but it seems that only in certain situations will corrosion start at the new cracks.

In the case of the first (LLLL) experiment, the number of rusting points does not steadily increase with time – as would be expected if every fresh crack led to attack. The reason clearly is that, as soon as a considerable number of corroding points have been established, they will provide protection to iron exposed at fresh cracks, by shifting the potential in a sense unfavourable to anodic attack. Corrosion at these points will therefore not develop. (The manner in which vigorous corrosion prevents attack in its neighbourhood is shown by the immune bands of Fig. 2.6, the protection being here provided by corrosion at the sheared edges.) Only after the original faults, where corrosion had been proceeding, have been healed by air exposure, will a situation arise in which attack can develop at fresh points.

Drop experiments

Early Cambridge measurements[134,917] of corrosion by drops placed on steel surfaces suggested that some inhibitors interfere with the anodic, and others with the cathodic reaction. Drops of sodium chloride, sulphate or nitrate solution (where both anodic and cathodic products would be soluble) were found to attack steel more rapidly than drops of the distilled water used for making the solution. Drops of zinc sulphate solution produced much less attack, and sodium phosphate practically none. The first would be expected to produce zinc hydroxide as cathodic product, and an adherent grey-buff deposit was observed, the colour being due to interaction between the anodically formed ferrous salt with the zinc hydroxide. The phosphate left the steel unchanged in appearance; it was then thought that the invisible film was an iron phosphate, but later Mayne's work (p. 184) suggested that it generally consisted of oxide.

Causes of inhibition

Is inhibition due to simple adsorption or compound formation?

For decades, there has been controversy between those who explain inhibition by the adsorption of molecules of the inhibitor unchanged at places where otherwise corrosion would start and those who attribute it to a film of some known compound formed by reaction between the metal and the inhibitor. That adsorption may be the first step in establishing inhibition is likely enough, but it can hardly be regarded as the *explanation* of inhibition, since it has also been put forward as the explanation of a failure to inhibit. We are told that chromates inhibit because CrO_4^{2-} is adsorbed, but also that chlorides prevent inhibition because Cl^- is adsorbed.

Some critics of the idea that a protective film of a compound is responsible for passivation have argued that such an explanation is unacceptable because the amount of inhibitor taken up by the metal is less than that needed even for a monomolecular layer of a compound. But there is no need to demand a complete monomolecular layer. If there is already an air-formed film of invisible oxide, it may be sufficient for the inhibitor to repair or reinforce that film at places where the film would keep cracking owing to internal stresses or is leaky owing to some structural defect. At such places, it may be necessary to establish a block of compound several molecules thick, before the internal stress is used up or the point of leakage sealed. Thus the thickness of the required layer, averaged over the entire surface, may work out to be either less or more than the thickness of a uniform film one molecule thick.

There is in fact evidence that the matter produced on the surface when inhibition has been established is indeed thickest in places most likely to suffer attack. Page and Mayne (1976 vol., p. 94), studying the distribution of the blocking substance produced when an azelate is used as an inhibitor for iron, have found the thickness of the layer to vary from point to point; it is significantly greater than the average at grain-boundaries where the looseness of the structure would be favourable for the escape of iron atoms as ions; the blocking substance is probably a ferric azelate (perhaps basic), but its composition does not affect the argument.

There is, indeed, sometimes no sharp difference between the two theories of inhibition. If the inhibitor is one capable of forming a compound with the metal, then at a place where the 'film' is one molecule thick it can be described either as a film of adsorbed inhibitor or a film of compound, according to the level which we regard as the interface between film and basis metal. (See 1968 vol., Fig. 7, p. 63.) But clearly, for protection against some highly corrosive liquid (e.g. an acid), a substance capable of forming a compound which can be built up as three-dimensional blocks capable of isolating sensitive spots may offer a better chance of success than something which is adsorbed unchanged. This expectation seems to be supported by the facts. Probably everyone would ascribe the resistance of Pb to H_2SO_4 or Ag to HCl to blockage by compounds known in the massive state ($PbSO_4$ or AgCl). Other cases, which were formerly in doubt have been brought into line by Russian work. Putilova and her colleagues found protection of iron by benzoate to occur above, but not below, pH 5·5; they found also that on mixing a ferrous salt solution and a benzoate solution, a precipitate occurs above, but not below, pH 5·5; apparently the precipitate is $[Fe_3(C_6H_5CO_2)_6](OH)_3$ – discovered by Weinland and Herz in 1912. (I. N. Putilova, S. A. Balezin and V. P. Barannak, *Metallic Corrosion Inhibitors* (translated by R. Byback), Pergamon Press, 1960.) The effectiveness of quinoline as an inhibitor for HCl is attributed to such compounds as (C_9H_7NH) $FeCl_4$; quinoline is relatively ineffective in H_2SO_4.

Nevertheless there are many cases of inhibition best explained by assuming simple adsorption. At least in some of the cases mentioned later, it is impossible to attribute inhibition to a compound capable of existing in the three-dimensional state.

Anodic and cathodic inhibitors

In cases where inhibition can be assigned to the formation of a definite compound, it is necessary to ask whether that substance is interfering with the anodic or the cathodic reaction. Interference with either reaction should slow down current but each method may introduce dangers. An anodic inhibitor, added in a quantity just insufficient to prevent attack altogether, may restrict it to small areas, setting up the dangerous combination of small anode and large cathode and producing intensified attack. On the other hand, a cathodic inhibitor (such as a thiourea derivative), used to prevent excessive attack on steel during the removal of scale by pickling in acid, may interfere with the discharge of H^+ at adjacent atoms, necessary if the hydrogen is to escape as H_2. This interference leads to entry of atomic H into the metal, with production of crack-formation or embrittlement (see p. 156).

An interesting possibility is that on some materials, an oxidizing agent which provides the possibility of a rapid cathodic reaction (its reduction) may raise the potential above the passivation level, so that current ceases to flow and corrosion is checked. At one time, this was being seriously considered as the normal basis of inhibition. Cartledge has performed a service in establishing a case of inhibition (by pertechnetates, which are extremely effective inhibitors) under conditions where there can be no reduction of the substance in question.

It is clearly important to decide whether an inhibitor is affecting the cathodic or anodic reaction. Chyzewski used a simple electrochemical cell to assign an inhibitor to its class. (E. Chyzewski and U. R. Evans, *Trans. electrochem. Soc.*, 1939, **76**, 215.) In the sections which follow cathodic and anodic inhibitors are considered separately. In one case (chromates) the situation is complicated by the fact that, although chromates are essentially anodic inhibitors, they can sometimes retard the cathodic reaction also (in strongly acidic solutions, however, they greatly *stimulate* the cathodic reaction); they will be treated as anodic inhibitors.

Cathodic inhibitors

Calcium and magnesium salts

In addition to zinc salts, mentioned on p. 162, the salts of calcium and magnesium can in certain circumstances act as cathodic inhibitors – a matter of some importance owing to their presence in natural waters.

It will be convenient first to consider (Fig. 6.5) the behaviour of a nearly vertical iron plate partially immersed in magnesium sulphate

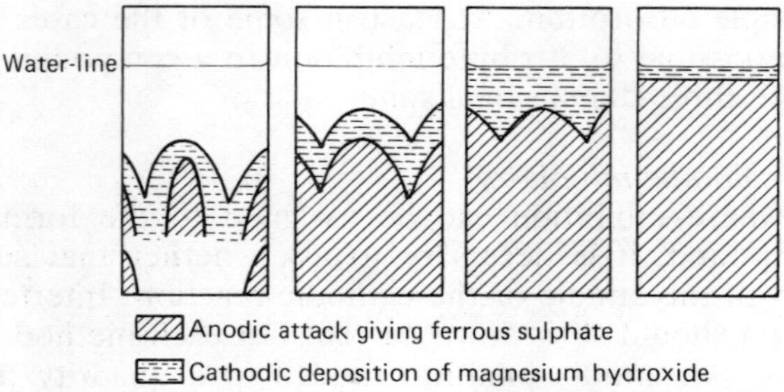

Fig. 6.5 Stages in attack of iron by magnesium sulphate solution (schematic)

solution. Corrosion starts at susceptible points, and spreads downwards and sideways, just as in sodium sulphate (or chloride) solution, producing dome-shaped areas of anodic attack. Whilst oxygen exists in the liquid, the cathodic reaction takes place just above and outside these dome-shaped areas, but – in contrast to that obtained from sodium salts – the cathodic product, magnesium hydroxide, is sparingly soluble, and coats the iron around the dome. The layer formed seems to interfere with the cathodic reaction perhaps by preventing oxygen in the liquid and electrons in the metal from coming close enough to each other to react. Anodic action then sets in at the coated area, and the ferrous salts formed interact *in situ* with the magnesium hydroxide, converting it first to a bright green ferroso-ferric hydroxide (probably with Mg^{2+} partially replacing Fe^{2+}) and finally to a pale brown ferric hydroxide, which, being formed from the originally adherent cathodic product, clings much more closely to the metal than ordinary secondary rust formed midway between anode and cathode. The brightness of the green and paleness of the brown (as compared with ordinary hydrated magnetic and ferric hydroxide) are evidently due to magnesium, the presence of which has been detected.

Whilst the original cathodic areas are becoming anodic, the cathodic reaction is being pushed steadily upwards; fresh areas become coated with cathodic product and then become anodic. Finally the cathodic reaction is confined to a strip about 1 to 2 mm broad along the meniscus; this strip becomes coated with a white layer, but, unlike the rest, it does not become green or brown. As the cathodic area is in the final stage narrow, and coated with a clinging substance which probably obstructs the cathodic reaction, the attack is much slower than in a sodium salt solution. In the author's experiments, magnesium sulphate of all concentrations between 0·01M and 0·50M gave practically the same amount of corrosion after twenty-six days; very dilute solutions (0·001M) produced more rapid attack.

Calcium salts behave in a similar manner. In a solution of calcium bicarbonate, the main constituent of many natural waters drawn from chalk or limestone, the OH^- formed by the cathodic reaction will interact with the calcium bicarbonate to give calcium carbonate, which is later replaced by hydrated magnetite and then by pale clinging rust possessing considerable protective properties.

It is today recognized that the physical character of the chalky scale produced by different waters greatly affects the protection afforded. M. E. D. Turner discovered that a compound present in some well waters favours the formation of an 'egg-shell' scale, which is much more effective than the nodular type formed when the compound is absent; this discovery is probably important both for ferrous and non-ferrous metals. (M. E. D. Turner, *Chem. Ind.*, 1963, 517.) The compound conferring this uncovenanted benefit was not identified, but was probably derived from sewage pollution.

Hard water flowing through a steel pipe

If a *hard* natural water containing calcium bicarbonate and oxygen is forced through a pipeline, the incipient corrosion process should coat the interior with a protective coat of rusty chalk or chalky rust; after the period needed for formation of this coating, the water delivered at the far end should be sufficiently free from dissolved or suspended iron compounds to be used for domestic purposes, including the washing of fabrics. This will not be true of the *soft* waters from igneous rock districts or acid waters from moorlands, unless they have first been treated; it will not even be true of hard waters, if they contain sufficient excess of carbonic acid to prevent the immediate precipitation of a calcium carbonate layer in physical contact with the steel. The latter type of water, however, may become less corrosive if it is cascaded through air, or if air is bubbled through it – processes which remove much of the excess carbonic acid, and also introduce oxygen. In film-forming water, oxygen usually exerts a favourable influence, ensuring that ferric rust is precipitated close to the metal; if oxygen is deficient, protection by films is often less effective, and the water drawn from the far end of the pipe may contain enough dissolved ferrous compound to stain fabrics or ceramics by precipitating ferric hydroxide on subsequent exposure to air.

The four types of carbon dioxide

A solution containing calcium bicarbonate ($Ca(HCO_3)_2$) and nothing else can only exist at very low concentrations. Much higher concentrations of calcium bicarbonate can, however, be obtained by shaking powdered calcium carbonate with water containing carbonic acid; the higher the concentration of carbonic acid, the larger the amount of calcium bicarbonate which can enter the water before equilibrium is set up with the solid calcium carbonate. Any water saturated with calcium carbonate and oxygen (e.g. one which has been shaken with marble dust and air) should, if brought into contact with iron, quickly deposit a chalky film which, although incapable of stopping corrosion entirely, can at least render it slow. If, on the other hand, the water contains more carbon dioxide than is needed for equilibrium, immediate precipitation of chalk cannot be expected, and a time interval must elapse before the retardation of corrosion can be brought about. Any carbonic acid in excess of the amount needed to keep the calcium carbonate in solution is known as **aggressive carbonic acid**.

It is possible to divide the carbon dioxide in a water into four parts:

(1) The carbon dioxide needed to form $CaCO_3$.
(2) The carbon dioxide needed to convert $CaCO_3$ to $Ca(HCO_3)_2$.
(3) The carbon dioxide needed to keep the $Ca(HCO_3)_2$ in solution (i.e. prevent it from being metastable in relation to solid $CaCO_3$).
(4) Any other carbon dioxide present. This will be the aggressive carbon dioxide or aggressive carbonic acid.

Clearly the amounts of (1) and (2) in a given water are equal; that of (3) depends on (1) or (2), the relation being shown in Table 6.1. The free carbonic acid, as determined by the water analyst, represents only (3) and (4) – provided that the titration is carried out rapidly; the figure obtained on slow titration may include part of (2).

To test for the presence of aggressive carbonic acid, the pH of a water should first be determined. A sample is then left overnight in a stoppered vessel containing marble dust; if next day the pH value has

Table 6.1 Relation between the different kinds of carbon dioxide*

Carbon dioxide of the first kind. mg dm^{-3}	Carbon dioxide of the third kind. mg dm^{-3}
20	0·5
40	1·75
60	4·8
80	11·5
100	25·0
120	47·0
140	76·4
160	112·5
180	154·5
200	199·5

Notes (1) The carbon dioxide of the second kind is equal in amount to that of the first kind.
(2) Any carbon dioxide present in excess of the first, second and third kinds, represents aggressive carbonic acid (carbon dioxide of the fourth kind).
(3) The relationship will be different if other ions are present.

*J. Tillmans and O. Heublein, *Gesundheitis-Ingeweieur*, 1912, **35**, 669.

risen, it is a sign that the original water contained aggressive carbonic acid; such a water is likely to give trouble on iron or steel. The criterion is of little value for zinc or lead, where protection depends more on the formation of a carbonate of the heavy metal than of calcium carbonate; to test a water for plumbosolvency (power to dissolve lead and thus become poisonous), it is better to leave a known quantity of the water in contact with a lead coil overnight, and estimate lead both in the liquid and in any sediment thrown down. It is commonly stated that any water containing *free* carbonic acid is capable of dissolving copper, but the facts are less simple.

Water treatment to avoid aggressive acid

Many acid waters from igneous rocks or moorland districts attack iron pipes, producing damage to the pipes or discoloration of the water as delivered. Apart from the danger of perforation and leakage, the carrying power of the pipes may be reduced by the formation of voluminous **tubercles** – elongated blisters full of magnetite. These troubles can often be reduced by passing the water slowly through beds of limestone, so as to ensure that the water issuing from the beds is in equilibrium with calcium carbonate, or by treating the water with enough lime, sodium carbonate or sodium hydroxide to neutralize the aggressive carbonic acid, or other acid present (moorland waters often contain quinic and other organic acids derived from plants). Such treatment often greatly improves the behaviour of the water towards iron or steel, although sometimes it merely replaces red-water complaints by tuberculation, which troubles the water consumer less but the water-undertaking more. The method is not well suited to water containing much sodium chloride.

It should be noted that the amount of sodium hydroxide or carbonate generally added merely serves to remove aggressive acid (carbonic or otherwise). Unless a much larger quantity of alkali is added, it will *not* prevent the water from producing rust on *freshly abraded* iron. After sufficient soda has been added to remove aggressive carbonic acid, any further additions will be used up in precipitating calcium carbonate or magnesium hydroxide, and this 'softening' action (however desirable from other standpoints) will not prevent rusting and may even increase it. Only when precipitation of hardness has ceased will further additions of alkali begin once more to raise the pH value, and at that stage the total corrosion produced by the water will decline. If chlorides and sulphates are not present in large amounts, the water might, with sufficient additions of soda, become non-rusting to abraded iron, but such highly alkaline water would be unsuited for many purposes; if chlorides or sulphates are present, an ill-judged addition of alkali may convert comparatively harmless general attack into dangerous localized corrosion. In general, alkali-treatment is not carried so far as to give a completely non-rusting water.

Anodic inhibitors

General

The alkaline inhibitors mentioned on p. 184 act by interfering with the anodic reaction. They are extremely effective if added in sufficient amount, but the localized attack sometimes produced requires special discussion. This will first be provided, and a brief account of chromate and nitrite inhibitors will follow.

Intensification of attack

If a drop of Na_2CO_3 solution of a concentration just insufficient to prevent attack completely is placed on steel, corrosion will start along the drop-boundary – the place which, with sodium chloride solution, would be immune. Similarly, a plate of steel partially immersed in sodium carbonate of borderline concentration will suffer attack along the water-line, which, in sodium chloride, would be free from attack in the early stages. If increasing amounts of sodium carbonate are added to a sodium chloride solution, the lower corroded areas steadily contract (Fig. 6.6(a), (b) and (c)), but, just as corrosion seems about to disappear altogether, a new zone of intense attack appears along the water-line (Fig. 6.6(d)), and still higher concentrations of inhibitor may prevent attack entirely (Fig. 6.6(e)). The corrosion produced at the water-line when both chloride and carbonate are present together occurs at a much higher total salt concentration than in chloride-free carbonate solution (more carbonate is needed to restrict the corrosion in presence of chlorides than in their absence), and both the total attack and intensity of attack is much higher. Thus a chloride–carbonate mixture of borderline composition is more destructive than a simple carbonate solution of borderline composition.

Dangerous inhibitors

Sodium carbonate is called a dangerous inhibitor, because, if added in quantity just insufficient to abolish corrosion completely, it localizes the attack and renders it more intense than if no inhibitor had been introduced. Sodium or potassium hydroxide, phosphate, silicate, chromate and nitrate are dangerous in the same sense. Early Cam-

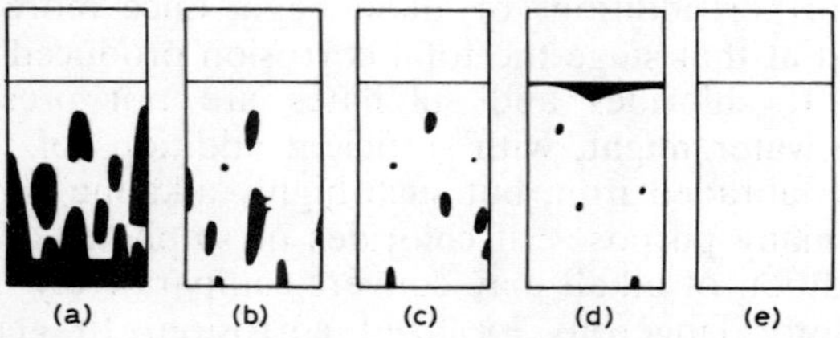

Fig. 6.6 Corrosion of steel by (a) sodium chloride and (b) to (e) sodium chloride with increasing amounts of sodium carbonate

bridge experiments showed, for instance, that the critical addition of one of these inhibitors to a water containing sodium sulphate and/or chloride quickly perforated a steel sheet, generally at the water-line; the same water, without inhibitor, produced a larger total of attack in the same time, but no perforation, the corrosion being better distributed. (U. R. Evans, *J. soc. chem. Ind.*, 1927, **46**, 347T.)

The intensification of attack is due to use of an anodic inhibitor for dealing with a case where the control is cathodic. The addition of the inhibitor *restricts the area* suffering corrosion, even when the amount added is insufficient to cause these areas to vanish completely. The cathodic area is not diminished, but, if anything, enlarged; thus the total corrosion will not, at first, be greatly altered, and the *intensity* of attack (corrosion per unit area of the part affected) will be *enhanced*. At first sight it might perhaps appear that the intensification would become infinite just before the anodic area vanished, but this alarming expectation is not realized, since, when the anodic area becomes sufficiently small, anodic control and approach resistance begin to restrict the current flowing. Nevertheless, the intensification caused by insufficient treatment with an anodic inhibitor under conditions of cathodic control may very seriously increase the local rate of penetration as compared with an uninhibited system.

The particular factors leading to this dangerous state of affairs are absent when the inhibitor is cathodic or when the control is anodic. Nevertheless, a cathodic inhibitor may sometimes produce intensification of attack. Concentrated zinc sulphate solution produces a film which has been known to break down at the water-line, where attack becomes more intense than in an uninhibited solution. Perhaps in the concentrated solution, water is abstracted from a hydroxide, producing a film of zinc oxide, an electronic conductor capable of acting as a large cathode.

Cause of water-line attack

It is not surprising that the special immunity met with at the water-line on steel partly immersed in sodium chloride, or at the drop-margin under conditions of drop corrosion, should disappear when an inhibitor is added. The immunity obtained in plain chloride solution is related to the fact that the product formed at the cathodic zone along the water-line or drop-margin is itself an inhibitor (sodium hydroxide). If the body of the solution contains an inhibitor, the probability of attack should become much the same everywhere. But it may still fairly be asked why the water-line zone (or drop-boundary) becomes *more* prone to attack than the rest, as shown in Fig. 6.7.

The explanation was suggested in 1938 by Schikorr in a private letter to the author. He considered that the tip of the meniscus constituted an inaccessible crevice, where replenishment of inhibitor would not readily occur; if the supply became insufficient, corrosion at the water-line would set in. This explanation was confirmed by Peers,[150]

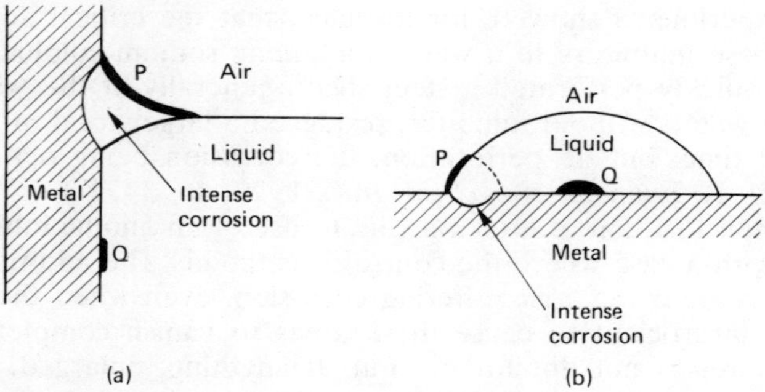

Fig. 6.7 Intense water-line corrosion of steel in liquid containing borderline concentration of inhibitor, (a) on partly immersed plate (b) below drop. At P, membraneous rust appears at the water/air interface and fails to protect, whereas at Q, solid matter is built up at an incipient discontinuity and prevents corrosion from developing

who made a detailed study of water-line attack, removing certain objections to Schikorr's theory which had been regarded as serious. He reached the conclusion, however, that the accumulation of chloride (or similar) ions which migrate towards any incipient anode would be more important in the meniscus crevice than elsewhere, and that this would enhance water-line attack in some cases.

Intense attack at contacts with non-conducting substances

Since crevices are formed when the surface of a metal is screened by contact with some foreign body (e.g. a glass rod, or a string tied round the metal specimen), intense attack is sometimes met with at the screened area; this will occur when the content of inhibitor is high enough to stifle attack over the main part of the surface, but not to prevent it in the crevice, either owing to lack of replenishment or by the corrosion product adhering to the wrong interface. Here again the dangerous combination of large cathode and small anode becomes operative. The danger is greatest if the liquid is in motion outside the crevice, although stagnant within it; this will ensure replenishment of inhibitor at all points outside, and the large cathodic areas will receive a steady supply of the cathodic stimulator (usually oxygen, unless the anodic inhibitor itself acts as cathodic stimulator); the total current will be strong and the attack intense.

Pitting caused by incorrect inhibitor treatment

Apart from intensification at a water-line and places shielded by foreign bodies (e.g. glass or string), exceptionally susceptible places (e.g. sites of sulphide inclusions in iron) may in a chloride–carbonate or chloride–chromate mixture of borderline composition suffer anodic attack, with the rest of the surface acting as large cathode. If the secondary corrosion product accumulates on the metal (as on the upper side of a horizontal specimen), the anodic area may gradually spread, because the product screens an ever-increasing area of metal from inhibitor; if so, the attack should not remain intense. On the lower side of a horizontal plate or disc supported some way above the bottom of the containing vessel, the spreading of the anodic area may not occur, since the corrosion product falls away. Under such conditions small deep corroded areas (pits) may be produced, as found by Homer using carbonate–chloride solutions of borderline composition.

Chromates

The prevention of attack on iron by a liquid containing sufficient CrO_4^{2-} is generally due to interference with the anodic reaction; the co-existence of Fe^{2+} and CrO_4^{2-} is impossible, just as is the co-existence of Pb^{2+} and SO_4^{2-}; on mixing solutions of K_2CrO_4 and $FeSO_4$, an immediate precipitate containing both Fe and Cr appears. The absence of attack on Fe in the presence of CrO_4^{2-} is completely analogous to the absence of attack on Pb in presence of SO_4^{2-}; the fact that the blocking substance is not ferrous chromate does not invalidate the argument that any anodic attack will lead to sparingly soluble matter in physical contact with the surface and likely to stifle further attack. In early work, Hoar stripped the film from iron which had been immersed in K_2CrO_4 and found it to contain Cr as well as Fe (T. P. Hoar and U. R. Evans, *J. chem. Soc*., 1932, 2476.) The presence of Cr in the film has been confirmed by several investigators using radioactive tracer methods.[152] Mayne and his colleagues (1976 vol., p. 98), have found a spinel containing both metals to be present. The early idea that the protective film is a mixture of iron and chromium hydroxides probably requires modification; the presence of a spinel is significant in view of Hoar's argument that salt-like oxides (and spinels can be so described) possess specially protective character.

Iron will remain uncorroded indefinitely in water containing chromate (provided that little or no chloride or similar salt is present); but if it is transferred to chromate-free water of a character which, in absence of chromate would attack iron, corrosion will recommence. However, Darrin[157], has shown that after long immersion in concentrated chromate, it is possible to transfer a specimen to dilute chromate (such as would normally allow attack) and yet corrosion does not start. Probably the internal stresses have become partly used up, e.g. by a crack–heal sequence in the concentrated solution (p. 187).

Nitrites

The use of a nitrite as a component of Vernon's successful mixture for treating automobile cooling waters is mentioned on p. 205. In general, however, the performance of a nitrite is somewhat unpredictable, and certain bacteria, if present, can cause its oxidation to nitrate. It has been stated that nitrites are more tolerant to Cl^- than are chromates, but that the danger of pitting and perforation is greater if the necessary amount is not added. An interesting suggestion of their action has been offered by Cohen, who considers that NO_2^- ions are adsorbed through the two O atoms, since the distance O—O in NO_2 is not very different from that of Fe—Fe in metallic iron. (M. Cohen, *J. electrochem. Soc.*, 1974, **121**, 191C.) There is also adsorption of OH^-, and the removal of NH_3 leaves a monolayer of FeO:

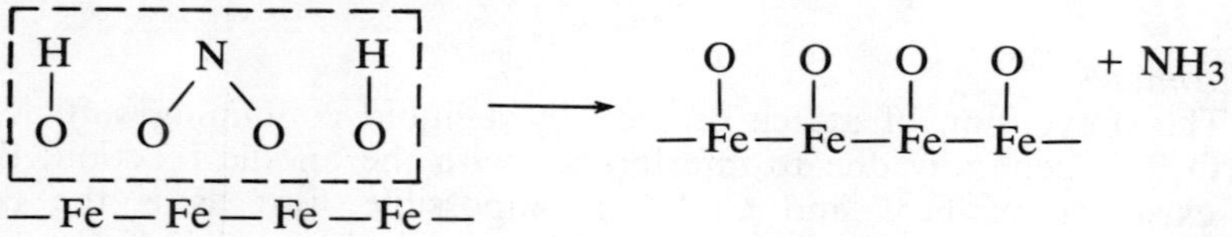

Organic inhibitors

Compounds containing sulphur

Considerable success has been achieved by Hackerman[180] in explaining inhibition in acid solution in terms of the donation of electrons from adsorbed molecules. He has emphasized the superiority of organic inhibitors containing sulphur over those containing nitrogen or oxygen alone. If, for instance, benzthiole 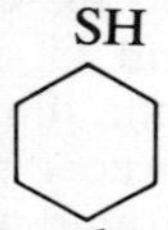 attaches itself at a point which would normally suffer anodic attack, through the (negative) sulphur atom, it will increase the electron atmosphere at such a point, and thus render more difficult the detachment of (positive) metallic ions. By attaching radicals at suitable points on the benzene ring, the negative character of the region can be enhanced and the inhibition made more effective; a positive group at the *m*-position, or a negative group at the *o*-position should do this. Similarly an aromatic body containing nitrogen, such as aniline, 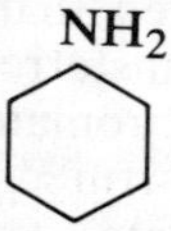 attached at a point on the metal which would otherwise favour the cathodic reaction should diminish the electron atmosphere and provide cathodic inhibition; this effect also can be enhanced by suitable substituent groups. The theory explains (1) why different substances can cut down attack by acid under circumstances where no oxide film could survive, (2) why some substances inhibit the anodic reaction, others the cathodic reaction, and others both (in a degree varying with concentration), (3) why, of two isomers, one may be far superior as an inhibitor to the other and (4) why some of the best pickling restrainers contain both sulphur and nitrogen.

It would seem that sometimes the inhibition obtained from an organic compound is really attributable to a secondary product (1976 vol., p. 93). Thus Horner has been successful in retarding the corrosion of iron, zinc and aluminium by dilute hydrochloric acid through the addition of dibenzyl sulphoxide, but attributes the main effect to dibenzyl sulphide. Italian work with a radioactive tracer technique confirms that the sulphide is a better inhibitor than the sulphoxide, being more strongly adsorbed; but the sulphoxides themselves possess inhibitive power in most cases. (Dimethyl sulphoxide, however, stimulates corrosion.)

Several other cases will be mentioned below where the decomposition product is a better inhibitor than the substance added. It may be asked why the more effective substance is not added in the first place; sometimes this may be difficult owing to unavailability or limited solubility.

Onium compounds (1976 vol., p. 94.)

Phosphonium and arsonium compounds are commonly grouped together under the name 'onium compounds'. Horner has found them to exert inhibitive action, which is, however, really attributable to secondary products; some retard the cathodic and others the anodic reaction; exceptionally good results are obtained if a cathodic and an anodic inhibitor are used together. In some cases, the secondary product responsible for inhibition has been identified. Triphenyl-benzyl-phosphonium chloride produces triphenyl-phosphine as a layer only one or two molecules thick which slows down the attack. An elegant nucleation method has been developed by Karagounis for testing an identification. A sample of the substance suspected of being the true inhibitor is melted and cooled down just below its true melting point to a temperature where no spontaneous crystallization would occur. Then iron powder, previously coated with the onium compound is added; if the suspicion is correct, crystallization will take place.

Amines and imines (1976 vol., p. 92.)

Hackerman has investigated inhibition by imines of the general formula $\overline{(CH_2)_n NH}$; where n, the number of the imine groups is 10 or more, inhibition is good; at 5 it is poor. The reason may be that the strain in the ring, which is relatively low when n is about 5, becomes greater when n increases; this strain may cause rupture of the ring with recombination to produce polymers capable of providing good protection. If that is correct, it affords another case of protection by a secondary product – but of a different type.

The amines, $CH_3-(CH_2)_n-NH_2$, have also been studied; here also the lower members seem to be the most promising inhibitors.

Unsaturated compounds

Putilova's work shows that the inhibition of the attack on steel by HCl often improves with rise in temperature; this suggests that the cause of inhibition cannot be due to physical adsorption which would decline with increasing temperature. Other Russian work (1968 vol., p. 68) shows that acetylene derivatives are effective inhibitors: probably they break up and the fragments recombine to form polymerization products as a hydrophobic surface film. If acetylene compounds are used along with inhibitors containing nitrogen, the effect of the combination is unexpectedly good; many other examples of this **synergistic** (mutually helpful) effect are known.

Relationship between structure and inhibitor (1976 vol., pp. 104–9.)
Most good organic inhibitors contain an S or N atom, as a member of a ring consisting otherwise of C atoms; attachment of the molecule to the metal appears to be through the S or N atom, and the efficiency of inhibition is altered greatly if one of the H atoms attached to a C atom in the ring is replaced by some substituent group. The effect depends on the nature of the substituent group, which may be **electrophilic** (tending to acquire an electric charge) like NH_3^+, or **nucleophilic** (tending to dispose of a charge) like COO^-; it also depends on the position where replacement occurs on the ring, which may be 1, 2 or 3 places from the point of attachment (the *ortho- meta-* and *para-* positions). It would be wrong to think that the substitution causes an entire electron move into or from the ring; the charge transfer will be less than one electron, and, whether it is a loss or a gain, it will be distributed *unequally* over the atoms of the ring; its effect in increasing or decreasing electron density in the metal will vary with the distance between the substitution point and the attachment point.

Now organic chemists, studying changes quite unconnected with corrosion, are studying the effect of substituent groups on the speed of reactions. Largely owing to the work of Hammett, an equation has been established between the rate constant of a reaction k_0 when no substituent is present and that measured, k, when it is present; this is $\log (k/k_0) = \rho\sigma$ where ρ is a constant depending on the type of reaction and independent of the substituent, whilst σ (called the **Hammett constant**) depends on the substituent but not on the reaction-type. Thus in hydrolysis, ρ will have a different value from that observed in bromination, but the same value whether the substituent is NH_3^+ or COO^-; σ will have a different value for NH_3^+ from that observed on COO^-, but each will keep its σ value the same whether the reaction is hydrolysis or bromination. (This only applies to the *meta-* and *para-*positions; the principle is not valid for *ortho-*compounds.) A straight line can be obtained when $\log k$ is plotted against σ for different substituents; different reactions will have different gradients since ρ will be different. An interesting point is that the gradient in the region representing an acceleration of a reaction (ρ being positive and $k > k_0$ is the same as in the region representing retardation (σ being negative and $k < k_0$). Hammett's book, pub-

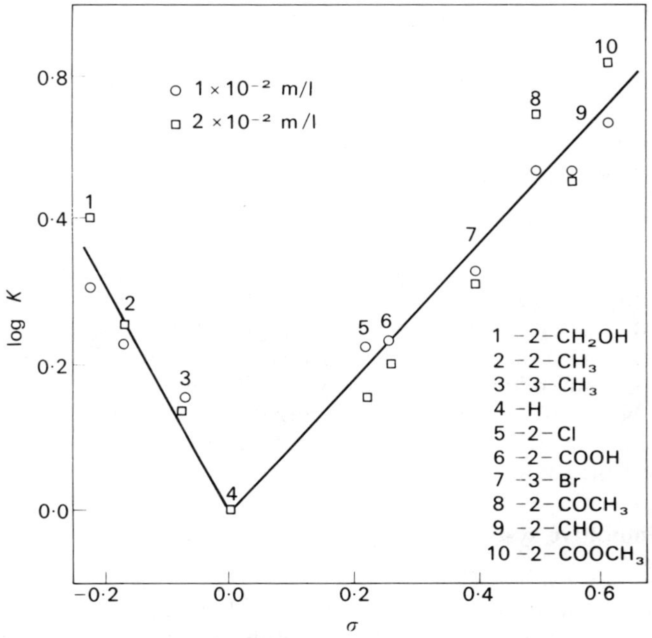

Fig. 6.8 Log K as a function of Hammett's constant (σ). K is i_0/i_R where i_0 is the corrosion rate in presence of thiophene (without substitute) and i_R is the rate in presence of a substituted thiophene (Z. Szklrske-Smialowska and M. Kaminski)

lished in 1940, contains values of σ for over 40 groups and of ρ for 40 reactions. If corrosion reactions are electrochemical and depend on a cathodic reaction which consumes electrons, and an anodic reaction which releases them, a change in the electron density brought about by the attachment of a molecule may be expected to influence the corrosion rate. Workers in Russia, Poland and Japan have used the Hammett principle to test this matter in different ways. In Fig. 6.8, log K is plotted against σ and the points fall close to two straight lines which intersect at the point for $\sigma = 0$ (meaning that there is no substitution). The result of the plotting is the more satisfactory in that the values of σ were mostly established by chemists completely disinterested in corrosion; the straightness of the lines passing between the points would seem to provide good evidence of the manner in which the electron density can affect the corrosion velocity.

However cases have arisen in chemistry where the Hammett equation has failed to explain the facts, and it has become evident that a substituent can do other things besides conferring or abstracting an electric charge. It can, for instance, produce a geometric effect on the manner in which the adsorbed particles can be packed and the effectiveness of the shielding of the metal. For equal masses of adsorbed

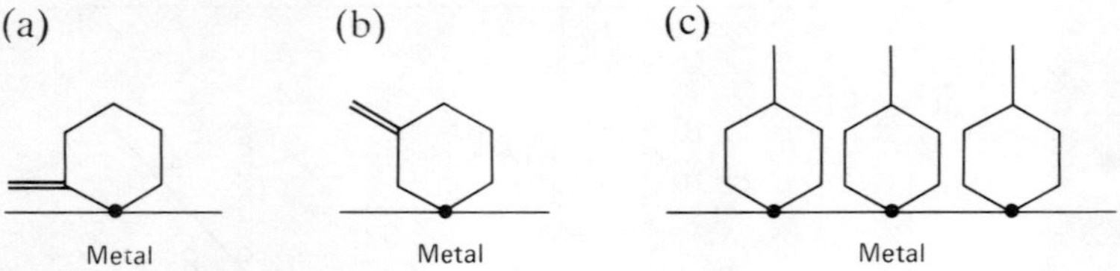

Fig. 6.9

material, a molecule having its side-chain in the position shown in Fig. 6.9(a) may produce better shielding than one with the side-chain as in (b). The situation shown in (c) suggests that a larger number of molecules could be packed in close array; more material per unit area would be demanded for inhibition, but if this is provided, inhibition should be more complete. It is perhaps too early to say whether arguments on these lines will serve to explain cases where the Hammett principle is not entirely successful.

Other inhibitive systems

Inhibition by colloids[179]

In his classical work, Friend showed that agar, egg albumen and gum tragacanth greatly retard the action of acid or water on zinc or iron; agar depresses the action of lead acetate and of copper sulphate on zinc. Gelatine diminishes the action of sulphuric acid on iron, and Morris showed that, in doing so, it is removed from the body of the liquid. There seem to be many unidentified inhibitors in food-products – perhaps stimulators also – and it is impossible to predict behaviour by studying a pure solution of the principal ingredient; lemonade and vinegar affect some materials differently from citric or acetic acid of equivalent concentrations. In many cases of inhibition by organic bodies, it is likely that they produce obstruction by adsorption on the surface.

Oil inhibitors

The 'soluble oils' used for inhibitive purposes are clear liquids containing oil, water, soap and an amphipathic substance such as a fatty acid or alcohol. Hoar and Schulman[169] regard them as water-in-oil emulsions with particles too small to produce milkiness. When a small quantity is poured into a larger volume of water, an oil-in-water emulsion is formed which may be milky or clear, according to the size of the oil particles (increase in soap/oil ratio diminishes the size); each particle carries a negative charge. These emulsions may be quite stable, unless large amounts of salts are present. The addition of the oil greatly reduces the corrosive power of the water. The emulsified oil may act partly as an adsorption inhibitor, and give a thin oil film over the whole surface. Probably it is also an anodic inhibitor, since

Fe^{2+} ions, each carrying two positive charges, can neutralize and coagulate the negatively charged oil particles, so that blobs of oil are preferentially formed over just those points where corrosion would normally start, and consequently prevent it from developing.

Inhibitors for motor cooling systems

The inhibitors added to the water used in motor cars, which, in winter at least, usually contains glycol to prevent freezing, probably control the anodic reaction, although their manner of action is still rather uncertain, and adsorption probably plays a part. They can be discussed only briefly. The problem is made difficult by the presence of many dissimilar metals (or alloys) in contact. Sodium benzoate and sodium nitrite is a good combination, developed by distinguished work at Teddington.[172] In absence of benzoate, nitrite protects cast iron, but increases attack on soldered joints, probably through the formation of complex anions; if benzoate is present, steel, cast iron and solder are protected, provided that there is a short initial period of heating. If aluminium alloys are present, there may be some attack if the pH rises too high. Another combination is triethanolamine phosphate (TEP) and sodium mercaptobenzthiazole (NaMBT). TEP alone gives protection if no copper or copper alloy is present, but stimulates attack on copper, producing copper compounds which then stimulate attack on aluminium. NaMBT prevents attack on copper, and the combination then becomes effective; it does not prevent attack on nickel, however, and is best adapted for systems where corrosion has not been allowed to start.

Weibull has reported the results of some tests in Sweden involving 51 vehicles; the median distance was 15 000 km, but one test extended over 100 000 km. The object was to compare the behaviour of various mixtures, and although reproducibility was poor, the trends are not in doubt. It is satisfactory to note that the procedure which proved most reliable was that involving sodium benzoate and sodium nitrite – the mixture originally evolved under difficult war conditions at Teddington by Vernon and his colleagues (1968 vol., p. 79).

Volatile inhibitors

Prevention of corrosion at metal surfaces which will not be immersed in a liquid but may become damp with condensed moisture, is best achieved by means of volatile inhibitors, which deliver into the gas phase molecules capable of dissolving in a moisture film. These may be basic substances, capable of rendering the moisture film alkaline, or at least preventing it from becoming acidic. Ammonia has been used for the purpose; it is often effective in preventing attack on steel, but actually stimulates attack on copper. Certain substituted ammonias are better in this respect.

Today there are two distinct types of volatile inhibitors used in *boiler-condensate* systems.

(1) **Film inhibitors**, practically **insoluble** in water, but added as **emulsions** to produce 10–15 ppm. They become adsorbed on steel surfaces where condensation occurs (e.g. in central heating systems based on steam) and produce a **hydrophobic** surface. The adsorptive and non-soluble properties required are found in **straight-chain primary or secondary amines** containing 10–18 carbon atoms.

(2) **Water-soluble, basic amines**, such as **morpholine** and **cyclohexylamine**. These are useful additions to boiler waters likely to contain carbon dioxide.

In **packaging**, volatile amine salts are much used. Here the degree of volatility is important, being a desirable property in a completely closed system, whereas it must be restricted if there is danger of the inhibitor escaping; the vapour pressures of some volatile inhibitors are recorded by Turnbull.[532] Of the numerous effective inhibitors, two may be mentioned as illustrating extreme cases. **Dicyclohexylamine nitrate**, with a low vapour pressure (about 0·02 mm Hg at 25°C), is suitable for protection in a restricted space, or for impregnating packing paper (not necessarily in a completely sealed package, since the low volatility prevents rapid loss). In contrast, **cyclohexylamine carbonate** is much more volatile (0·3 mm Hg at 25°C) and is suitable for protecting steel in large, *enclosed* spaces. Both substances may attack certain non-ferrous metals and non-metals. A discussion of the substances is provided by E. H. Evans and R. I. Baraclough. (*Chem. Ind*., 1961, 1980; 1962, 805.)

The volatile nitrites owe their inhibitive properties mainly to the NO_2^- anionic radical. Some authorities consider that the cationic group plays a part, but it is probably less important.

In some circumstances, inorganic substances which absorb moisture will serve to prevent the start of atmospheric corrosion. Silica gel has been used, and also lime; the latter serves to absorb acid fumes.

Wrapping paper impregnated with inhibitors is today available and plays a useful role in the prevention of corrosion; if the inhibitor used is a mildly volatile substance, corrosion should be avoided even at places where there is no contact between the wrapping and the metal article. Paper used for wrapping metallic articles should always have the lowest possible quantity of chloride; this requirement is difficult to fulfil – a point emphasized by van Muylder, who recommends impregnation of wrapping paper with the volatile inhibitor dicyclohexylammonium nitrite. He states also that impregnation with sodium benzoate has given good results in warm humid atmospheres, under some, but not all circumstances; it can, however, set up corrosion on aluminium, owing to the alkalinity. (J. van Muylder, Cebelcor, *Rapports techniques*, F377.)

Inhibitive pre-treatment before painting

General

Paint applied to an untreated metal surface often behaves badly. It may start to peel off; or it may fail to prevent corrosion. The two troubles are often interconnected. Clearly if, owing to poor physical adhesion, paint peels, corrosion will start. Conversely, voluminous corrosion products formed locally below the paint coat may push it up in little hemispheres, causing it to crack; what is often more serious, alkali accumulating as the cathodic product of quite trivial attack, may creep between metal and coating, loosening the latter, or softening the coating by interaction with either vehicle or pigment (p. 228). For both types of trouble, a pre-treatment that provides a layer into which the paint will key and which will inhibit incipient corrosion, will be beneficial. Several have been worked out.

Phosphating of steel and zinc[562]

Some of the earliest phosphating liquids consisted of phosphoric acid saturated with an iron phosphate. Just as calcium carbonate dissolves in carbonic acid solution as bicarbonate, so ferrous phosphate dissolves in phosphoric acid as ferrous hydrogen phosphate. When such a liquid acts on iron, the anodic product is likely to be a ferrous phosphate, whilst the pH rise at the cathode should cause deposition of ferrous phosphate; given crystalline continuity between anodic and cathodic products, a protective coating might be expected, but in practice the result is usually unsatisfactory; probably oxidation of ferrous to ferric phosphate by air, with subsequent reduction by the metallic iron on wet days, causes undesirable volume changes.

It is better to use a solution of zinc phosphate or manganese phosphate in phosphoric acid where alternate passage between two states of oxidation is unlikely; if, in such a case, there is a suitable crystallographic relationship existing between the anodically formed iron phosphate and the cathodically formed zinc or manganese phosphate, good adhesion should be obtained. Further study of crystallographic relationships along with the electrochemical reactions is desirable, but Holden[563] has described the layer of crystalline, insoluble phosphates as 'integral with, and tightly adherent to, the basic metal'. Much of the extensive research on phosphating processes has been empirical, but some theoretical aspects have been studied. The acceleration of the process has engaged much attention. So long as the cathodic reaction is evolution of hydrogen or reduction of oxygen, the formation of the phosphate layer is likely to be slow. Many easily reducible substances have been tried for speeding up the cathodic reaction. Copper salts were used at one time, but the presence of copper appears undesirable; today oxidizing agents like nitrates, chlorates and organic compounds such as nitroguanidine are used as accelerators. It is now possible to obtain good adhesion by a dip lasting only a few minutes, instead of perhaps an hour. Tests carried out at Woolwich[565] show that an accelerated manganese bath gives results as good as the corresponding non-accelerated bath, but that for zinc phosphate, acceleration is disadvantageous.

Phosphating is much used for steel, especially on motor-car, cycle and refrigerator parts – as well as on nuts, bolts and washers. Oxide-scale must first be removed by pickling or grit-blasting, and then the surface must be cleaned with an organic solvent, by vapour degreasing or a hot alkaline wash. The composition of the bath varies according as the phosphate layer is to afford a basis for paint, or merely to be covered with oil.

Pre-treatment of zinc and galvanized iron

It is well known that paint applied to new galvanized iron does not adhere well; this is sometimes attributed to surface grease and sometimes to attack on the zinc by formic or other acid produced during drying of an oil paint (in which case vehicles which dry by evaporation should be harmless). One method of overcoming the trouble depends on a phosphating treatment. An alternative treatment known as the **Cronak Process**[586] consists in a short dip (perhaps 5–10 seconds) in sodium dichromate solution containing a carefully adjusted amount of sulphuric acid. This produces a film containing both trivalent and hexavalent chromium; it probably acts as a mechanical barrier as well as a reservoir for inhibitive CrO_4^{2-} ions.

Pre-treatment of aluminium

Aluminium alloy surfaces are frequently anodized (p. 181) before painting, either in chromic acid (the solution used by the original inventors, Bengough and Stuart) or in a sulphuric acid bath, usually containing glycol and glycerine. Oxalic acid processes are also available, sometimes using a.c. as well as d.c. In general, anodized coatings should be sealed in boiling water or chromate; Bradshaw and Clarke[251] find that sealing in hot water changes tensional stress to compression – indicating a strong plugging action.

The use of an external e.m.f. is inconvenient, and processes which build up a protective layer by simple immersion are popular; here the liquid must supply the cathodic reaction, and most of the baths contain chromates, chromium being found in the layers produced. The M.B.V. process, of German origin, consists of a dip in a hot solution containing sodium carbonate and chromate; newer baths contain chromic and phosphoric acids, sometimes with fluoride as an accelerator. Such processes are used in the aircraft and motor industries.

Another method for obtaining good adhesion on aluminium depends on a 'wash primer', and the treatment has also proved useful on steel and other metals. **Wash primers** (also called **etch-primers** or **pre-treatment primers**[568]) constitute a large class of preparations.

Some are available in a form ready for application, and others as two-package systems, the contents of the two containers being mixed at the last moment; the one-package system is more convenient, but the other usually gives best results. A typical two-pack wash-primer consists of basic zinc chromate dispersed in a solution of polyvinyl butyral resin, to which, just before application, is added a thinner containing phosphoric acid; drying is complete in perhaps an hour at room temperature, or in 10 minutes on stoving at about 120°C. The coating produced is mainly organic, consisting of an oxidation product of the butyral; an inorganic layer containing trivalent chromium is present at the base and may contribute to the good adhesion; researches at Washington suggest that the stifling of the corrosion is largely due to zinc phosphate tetrahydrate, precipitated on the potentially active areas.

Pre-treatment of magnesium

The fact that magnesium can evolve hydrogen from salt water and that magnesium hydroxide, formed in pure water, is not completely insoluble and, being alkaline, might loosen certain paint coats, makes pre-treatment of magnesium alloys advisable. Many successful processes have been developed.[589] A bath with sodium dichromate and nitric acid is used for rough castings and sheet. Where dimensional tolerances must be strict, the 'R.A.E. half-hour bath' – a boiling solution containing potassium and ammonium dichromates, with ammonium sulphate and ammonia – is suitable, the constituents being adjusted to provide the pH level appropriate for the alloy. A bath with manganese and magnesium sulphates along with dichromate gives films in 3–10 minutes at the boiling point. Many anodizing processes have been put forward, mostly involving fluorides and relying on the low solubility of magnesium fluoride.

7

Protective Coverings

Introduction

Need to protect iron and steel

Ferrous alloys – generally known as steels – include some of the cheapest structural materials available to the human race, and on the whole the most mechanically suitable. The high strength, combined with convenient ductility or malleability, would make them preferred materials for most purposes even if there were no price advantage. Against this must be placed their density, which puts them at a disadvantage as compared with the alloys of aluminium and magnesium for aircraft purposes, and their relatively low resistance to corrosion. The latter disadvantage can be largely met by the application of protective coatings. Such coatings form the main subject of the present chapter, although there will be occasional references for coatings designed to protect non-ferrous metals.

Classes of coatings

The protective layer for steel may consist of a non-ferrous metal, which can be deposited in various ways. In some cases the iron or steel article can be dipped in a bath of molten metal, and emerges carrying a coating (**hot dipping**). Alternatively it can be made the cathode in a bath containing a compound of the metal to be deposited (**electroplating**). Again it can receive a shower of tiny molten droplets of metal delivered from a pistol, which contains the metal to be deposited either as powder or as wire (**spraying processes**). Metal can also be deposited from the vaporous state or by heating the article to be protected in a mass of powder; this generally produces an alloy layer (**cementation**).

A more recent development*, known as '**ion plating**', is carried out in a vacuum chamber containing a tungsten filament on which is placed the metal to be deposited. The chamber is exhausted to 0·001 33 N m^{-2} and then argon is introduced to give 1·33 or perhaps 6·5 N m^{-2}. A potential difference of 2–5 kV negative between the earthed parts of the chamber and the article to be deposited sets up a glow discharge. At first the tungsten is unheated and the article is

* D. G. Teer, *Trans. Inst. Met. Finishing*, 1976, **54**, 159.

subjected to positive ray bombardment which removes all contamination. After about 20 minutes, when cleaning is complete, the tungsten filament is heated by a strong current; ions of the metal to be deposited can then reach the clean surface, producing a strongly adherent layer. In a variant of the process, a high-power electron beam provides the vapour source.

Alternatively the coating can be non-metallic in character. If this is a transparent layer, it is known as a **varnish** or **lacquer**. If a solution of a resin in a volatile solvent is applied by brush or by spraying, and allowed to dry by evaporation of the solvent, a coating is easily obtained, but has limited protective power and often poor adhesion. Generally a drying oil is introduced, and the conversion into a solid layer in such a case will depend only partly on the evaporation of the solvent, and largely on the oxidation or polymerization of the non-volatile constituents. Sometimes the changes require treatment in an oven; the coatings thus attained (**stoving varnishes**) may confer superior resistance to corrosion. In general, however, protection is improved by introducing a powder or pigment into the mixture to be applied, which is then known as a **paint**; the drying of a paint can generally take place without heating. Usually several coats of paint are applied to the surface requiring protection; the lowest coat contains an inhibitive pigment, a substance having passivating properties similar to those of the inhibitors described in the previous chapter; but it must be only sparingly soluble (otherwise it would soon be washed away). The other coats are designed to protect the lowest coat from mechanical erosion; the outermost coat may be chosen for decorative purposes. A paint as applied may contain at least five constituents, the pigment, drying oil, resin, solvent and possibly a plasticizer which remains with the resin after the volatile solvent has evaporated. In the drying process, evaporation, oxidation and polymerization all play their parts.

In recent years, two-pack paints have come into favour. These contain two compounds, each of which will remain unchanged indefinitely in the pot; if they are mixed together, however, the mixture rapidly solidifies to give a hard coat, owing to interaction between the two compounds; there is little or no evaporation during the setting stage, and good adhesion can be obtained – provided that the metallic surface is clean before coating takes place. In one useful combination, the compounds are an epoxy resin and a polyamide respectively. All these matters will be further discussed later in this chapter.

Metallic coatings

Cathodic and anodic coatings

If the coating is a metal, or a paint richly pigmented with metallic powder, the situation will vary according as the coating is cathodic or anodic to the basis metal. On steel, coatings of copper or nickel will normally be cathodic, whereas coatings of zinc and (unalloyed) aluminium will usually be anodic. A cathodic coating will not prevent rusting if there are pin-holes or cracks in it. Indeed, if the coated steel is immersed in a strongly saline (and therefore highly conducting) water, the combination of large cathode and small anode may make the attack at the gap locally more intense than if the steel had been left uncoated. This intensification of attack at gaps is more probable with copper than with nickel coats. It becomes less likely as the conductivity of the water diminishes. In atmospheric attack, with only condensed water or raindrops wetting the coated steel, the attack at a gap is usually less intense than that which would take place if the steel were uncovered. In the days before chromium plating, cycle handlebars were commonly plated with nickel alone. Even when the coating contained pores, there was much less corrosion than would have occurred on bare steel.

In contrast, a coating of zinc or aluminium can protect steel exposed at gaps in the coat, provided that the current density upon the steel is sufficient. The coating is to some extent destroyed in conferring the protection, but, since the corrosion of the coating must spread sideways, and that of uncoated steel would at least partly extend downwards, the sacrifice of the zinc or aluminium may be justified (despite the fact that each is a more expensive material than iron). However, when the gap in the coat has been sufficiently enlarged by the corrosion process, the current density in the centre of the gap will cease to be adequate and attack on the steel will start. For that reason, it is important to choose a metal for the coating which will be corroded only just fast enough to confer protection. The choice will depend on the corrosive environment. In some experiments carried out by the author (1960 vol., p. 640), steel carrying discontinuous aluminium or zinc coatings was tested in salt solution and in a hard water rich in calcium bicarbonate; in such cases aluminium, used as anode, generally provided less current than zinc, owing to the oxide film on the former. In salt solution, both metals afforded cathodic protection to the exposed steel; but, since the zinc delivered more current than the aluminium, it was consumed more quickly, and rusting started first on the zinc-coated specimens. In the hard water (essentially calcium bicarbonate solution) which was less corrosive, the aluminium was attacked too slowly to provide sufficient current density for protection, and the steel suffered rusting from the outset. Thus the aluminium-coated specimens, which escaped rusting for a long time in the more corrosive liquid, developed rust quickly in the less corrosive one.

Protection by porous sprayed coatings of zinc or aluminium is probably sacrificial at the outset, but zinc or aluminium compounds fill up the pores and provide mechanical exclusion; if zinc or aluminium hydroxide is deposited on the steel at the bottom of each pore by the cathodic reaction, the excellent protection obtained can easily be understood. The zinc-rich paints, discussed later, protect in an analogous manner.

Coatings of noble metals

The chemical resistance of noble metals might seem to recommend them as coatings, but apart from high cost, they have other disadvantages. They are strongly cathodic to iron, so that at any discontinuity intensified attack may be expected. An early attempt made to put gold-plated steel-nibs on the market proved disastrous, since the soft gold was rubbed away at the tip, exposing the steel, which suffered rapid wastage. Where cost permits, thick coats of noble metals which are free from discontinuity can be obtained mechanically; **copper-clad** steel plates are made by rolling the two metals together.

Electrodeposition of noble metals is carried out for special purposes. Electroplated tableware (usually a nickel–copper–zinc alloy plated with silver) is well known, whilst silver deposits have been used to protect 'heavy-duty' bearings on aircraft engines from corrosion by oxidized oil. Palladium plating is applied to silver watches and cigarette cases to prevent tarnishing, whilst rhodium plating has been applied on reflectors for searchlight purposes or for beacons at airports.

Silver deposited from a silver nitrate solution consists of a limited number of distinct crystals, and would be useless for protection. The deposit obtained from a cyanide solution (such as $KAg(CN)_2$ or $NaAg(CN)_2$ fulfils requirements, being fine – almost structureless.* Silver anodes are employed, and, provided that the bath contains excess of free KCN or NaCN, the attack on the anode compensates for loss of silver at the cathode; if excess of cyanide is not present, deposits may be formed on the anode, obstructing the attack.

Complex cyanide solutions are also used in the deposition of gold, indium and alloys such as brass. In plating copper upon steel, deposition is often started from a cyanide bath; if a sulphate bath were used at the outset, incoherent copper would be thrown down by simple replacement. After a thin deposit of cyanide copper has been formed, deposition is usually continued from a copper sulphate bath, containing an addition agent such as gelatin, since from such a bath a higher current density (and hence more rapid deposition) is possible.

*A plating bath which contains the metal as complex ions (like $Ag(CN_2)^-$) will, on cathodic treatment, produce a deposit by a mechanism more complicated than a bath containing simple cations (like Ag^+). In the latter case, the reaction is probably loss of one or more electrons to form an atom of metal, and is likely to proceed most readily where a metallic crystal is already present, and more readily on parts of the cathode near to the anode than on remote areas. Thus a crystal, once started, will grow to a large size and the deposit will be rough or porous, whilst on articles of irregular shape the deposit will be thin or absent on remote parts. With a bath containing complex ions, these two undesirable features are avoided. A bath which gives a sufficient deposit in recesses or remote areas is said to have good **throwing power.**

The use of complex baths is not confined to noble metals. Zinc is deposited from a cyanide bath if the articles are irregular, on account of its good throwing power. For less complicated shapes, a bath of zinc sulphate containing organic additive agents may be used.

Coatings of nickel and chromium

Nickel deposits covered with a thinner chromium coat are much used for the protection of steel or brass. The nickel protects the steel or brass against corrosion, whilst the chromium protects the nickel from fogging (p. 98); the hardness of the chromium is a further asset. The combination, generally described as **chromium plating**, is familiar on cars, furniture and household fittings. Unlike many metals nickel can be deposited as a coherent coating from a simple sulphate solution, without organic additions; boric acid may be added to control the pH value, and either nickel or sodium chloride to prevent passivity of the nickel anode. The old Watts bath, as published in 1916, contained nickel sulphate, nickel chloride and boric acid. In modern practice, organic brighteners are commonly present. Chromium is generally deposited from a bath containing chromic acid and sulphuric acid. The $CrO_3:H_2SO_4$ ratio greatly affects the results. The hexavalent chromium is partly reduced to the trivalent condition, and the actual production of metal appears to take place within a thin film of chromium chromate, which may act rather in the same way as an organic addition agent, producing a bright, almost structureless, metallic deposit.

There are obvious attractions in using **electroless deposition** – which dispenses with an external e.m.f. Articles immersed in a bath containing a nickel salt and sodium hypophosphite become coated with a nickel–phosphorus alloy. On many surfaces, this proceeds well, but it fails on tin and lead, so that soldered joints remain uncovered.[614]

Bright electrodeposits of nickel are generally anodic relative to nickel deposited from a bath free from brightener, and the idea became prevalent some years ago that bright deposits provided less protection. The fact that bright nickel is anodic to semi-bright has, however, been utilized in **double-layer** coatings, where first a semi-bright layer is deposited and then a bright layer outside it. When atmospheric corrosion has locally penetrated through the outer coat, it turns sideways on reaching the inner (cathodic) layer, and does not burrow deeper, as would occur if the whole coating were of a single character. That this sideways diversion of attack really occurs is made evident by the photographs of Melbourne and Flint. (*Trans. Inst. Met. Finishing*, 1962, **39**, 185.) Edwards and Carter have reported promising results from an increase in the thickness of the chromium layer. (*Trans. Inst. Met. Finishing*, 1963, **40**, 48.) Normally this is kept thin (0.25 μm is conventional) – partly at least because thick chromium develops cracks (probably for reasons analogous to the cracking of thick oxide, as discussed on p. 22). Thick deposits reasonably free from cracks can be obtained by modifying the plating conditions. The superior corrosion resistance obtained by increasing the thickness has been demonstrated by outdoor exposure tests.

In some circumstances, however, it may be preferable to use a coating in which a network of fine cracks has intentionally been produced. An originally crack-free coating may develop cracks in service, which sometimes extend into the bright nickel immediately below the chromium. If the plating conditions are adjusted to provide a network of fine cracks, this is unlikely to occur. In the plating on a motor car, the choice will depend on the degree of rough service to be expected. Flint thinks that for normal service conditions double-layer nickel (or copper covered with bright nickel) covered with conventional chromium, should possess sufficient ductility to resist penetration by cracking. If exceptionally severe treatment is expected, however, a single coating of semi-bright nickel, buffed after deposition and then covered with micro-cracked chromium, is probably the most suitable combination (1968 vol., p. 239).

Another interesting possibility is the use of a bath containing trivalent, instead of hexavalent, chromium (1976 vol., p. 311); it is perhaps too early to state whether this will become a standard procedure, but reports are encouraging. In one process, aluminium is introduced into a highly agitated bath and becomes included in the coating*.

Tin coatings

Although used to some extent for covering copper cooking utensils, copper or lead water pipes and also brass condenser tubes, the main use of tin is in the canning industry. The **tin-plate** used by canners consists of dead-mild steel carrying a tin layer usually 2·5–5·0 μm thick; the number of pores per unit area decreases as the thickness increases, but, however pore-free the plate may be when manufactured, the tin layer is usually broken during the fabrication of the can. In fruit juice (probably the most corrosive substance which the canner must face), steel is generally cathodic to tin (which forms complex anions with the fruit acids), so that intense attack on the exposed steel does not occur, although the hydrogen evolved at the cathode may distend the can. Except with a few fruits (e.g. bilberries), failure of cans by hydrogen swelling is more frequent than by perforation. The composition of the steel affects behaviour. Further information is provided elsewhere (1960 vol., p. 649; 1968 vol., p. 241; 1976 vol., p. 307).

A factor to be considered is the lacquering of cans. Its primary purpose is to avoid discoloration of the fruit, whose anthocyanin constituents can act as hydrogen-acceptors at a cathode and thus change hue. Lacquering prevents this, but the lacquer coats as applied some

*C. A. Addison and E. C. Kadward, *Trans. Inst. Met. Finishing*, 1977, **55**, 41.

decades ago, instead of preventing the swelling of cans, rather increased that trouble. Today, improved stoving lacquers which are based on synthetic resins are available; by using two coats (the first applied to the flat sheet before the can is 'formed', and the second sprayed on to the finished can to repair any cracks produced in forming), the number of swollen cans can be greatly diminished.

Up to 1939, tin was usually applied by hot-dipping, but during the Second World War, electroplating, which gives a more uniform coat and thus economizes the valuable tin without increased risk of swelling or severe corrosion, came to be widely employed. The process involves the continuous passage of steel strip through cleaning baths and then through electroplating baths, in which tin is deposited at high current density. The deposit, on emergence, is matt white, but the strip, after passage through rinsing tanks, is carried through a furnace where heating by radiation, conduction or high-frequency current momentarily raises the tin just above its melting point – a procedure known as **flow-brightening**. The strip then passes through a chromic acid or chromate solution (with or without applied current) and finally receives an oil film. The appearance is now lustrous, and the material is suitable for soldering – a matter of importance for the production of cans.

The thin tin coats are apt to contain pores, exposing the steel; the number of pores decreases as the thickness increases. The places where the iron is exposed can be shown by placing the sheet in water of high purity, heated to 95–100°C for 3–6 hours, which produces adherent rust-spots at most of the holes; another method consists in applying paper soaked in a solution of potassium ferricyanide and sodium chloride, which shows up discontinuities as blue points. Neither method is accurate in revealing the exact number of pores present, and other methods are available for use where such information is important.

Problems are introduced by the presence of nitrates (1976 vol., p. 307). Nitrates have long been used in the curing of bacon, but carrots and green beans sometimes cause detinning of a can interior owing to their nitrate content. The action depends on the pH value, which may change on storage, so that behaviour is difficult to predict.

Lead coatings

Molten lead refuses to form a coat on iron unless alloyed with tin. Steel coated with lead containing 15–25% tin is called **terne plate**. It is used in roofing and also finds employment as drums for paints, and cans for petrol; in regions where the native population relies on the refuse dump as the source of its cooking utensils, outbreaks of lead poisoning have resulted. Lead can be deposited from a fluoborate or fluosilicate bath with colloid additions; (fluoborate baths are also used for tin and cadmium). In all these cases, organic substances added serve to prevent coarse crystalline growths, by adsorption on the growing crystal, in many cases producing a bright deposit – thus saving polishing costs.

Zinc coatings

A layer of zinc may be obtained by dipping the article to be coated (previously cleaned by pickling) in the molten metal, usually covered with a layer of some flux capable of removing the last trace of oxide from the article as it passes through the layer. In coating steel with zinc, ammonium chloride is often placed on the surface of the molten zinc to produce a flux layer which comes to contain $ZnCl_2.NH_3$ as well as free $ZnCl_2$ and NH_4Cl; or a layer of flux may be applied to the steel articles and dried by heat before the articles enter the molten zinc.

In most cases the layer produced on steel dipped in molten metal is complex; on the outside there may be relatively pure metal, with one or more alloy layers below. In the case of zinc-coated iron, the inner alloy layer is brittle; it can be made thin by shortening the time during which the iron is kept in the zinc, but this increases the risk that certain spots will receive inadequate covering.

Such a method (*hot galvanizing*) has long been used for protecting steel wire, sheets, pipes, tanks and cisterns. Being anodic to steel, zinc will generally provide protection at a gap (except in hot water), but it is destroyed in the process and thus relatively thick coatings are called for, if the protection is to continue. Thick coatings are apt to be brittle, and the exposure of the steel at a crack, even if it does not lead to immediate attack on the steel itself, may accelerate that on the zinc. On the other hand, thin coatings will have short lives. Coatings which are thin in some places and thick in others introduce disadvantages. To make the best use of the zinc applied, uniformity of thickness must be aimed at, and for articles free from re-entrant areas, electrodeposition may provide better uniformity than hot galvanizing; the coating is itself different, being free from the alloy layers present at the base of hot-galvanized coatings. On wire, electrodeposition is often preferred. In the **Tainton process**, plating is carried out in long narrow troughs, along which the wire passes continuously; the electrolyte is acidic zinc sulphate produced directly from the ore and suitably purified; the wire is made the cathode, whilst the anodes consist of lead containing 1% silver. The contacts delivering the current to the moving wire are made of copper, protected, except at the tips, with rubber; any zinc deposited on these tips grows out as a tree, which is knocked off at intervals. The wire emerging, plated, from the end of the trough is washed, dried and polished with rotating blades tipped with tungsten.

Iron intended for roofing, or for the walls of sheds, is coated by dipping in molten zinc and then corrugated; it is normally painted after erection. If the galvanized iron is not to be painted, a thick coat is essential, or rust will soon appear, even in country air. Statements have been made that the life of the coat (up to the appearance of rust) is proportional to the thickness. These are based on the behaviour of specially prepared samples having coats of very uniform thickness exposed under the simplified conditions of the test station, and do not necessarily apply to service conditions on houses or other structures. The lodging of drops at crevices, or drippings from high roofs on to restricted areas of low ones, are factors which may shorten life, but the main reason for the departure from the law of proportionality is lack of uniformity in thickness. Many years ago, two different structures of galvanized iron were erected at different times near a house in the Scilly Isles. One lasted over 15 times as long as the other, but it was certain that the *mean* thickness was not 15 times as great. Apparently both had been prepared by hot-dipping, and the worst one had probably received a thin non-uniform film which at certain places was very thin indeed.

Similarly, for use below water, a thickly zinc-coated steel should generally last longer than a thin one. A greater thickness is desirable for soft than for hard water and indeed it is mainly for waters rich in calcium bicarbonate that galvanized iron tanks and cisterns are useful. When fresh, the unalloyed (outer) coating can confer cathodic protection on any exposed steel against cold water; but the alloy coating gives no such cathodic protection, and even for unalloyed zinc the polarity is reversed at high temperatures, so that the steel exposed at cut edges of galvanized iron receives no protection in hot water – as shown by Kenworthy and Smith.[207]

The steel bars used as reinforcements in concrete are often galvanized by dipping in molten zinc. This has become important in connection with off-shore constructions in the North Sea oil fields. Cl^- passing through the concrete and acting on ungalvanized steel may produce a serious situation. Galvanizing provides protection and the bond strength between metal and concrete is believed to be improved by the zinc layer, probably because some roughness is produced by the attack of lime on the zinc, giving calcium zincate. However, high-strength steel may become embrittled through the entry of hydrogen produced in the reaction, and it is well to introduce chromate into the cement or to treat the zinc-coated steel in a chromate bath before it is embedded in the concrete.

Coatings of zinc or aluminium can also be obtained by **spraying**[599]; an engineering structure can thus be covered after assemblage. The surface is grit-blasted and the metal applied as a shower of tiny globules (often molten) which on striking the surface flatten to give a scaly porous coating. The spray of globules is obtained from a ‘pistol’ which is moved over the surface. In one form, the metal enters the pistol as wire which is fed forward into a special blow-flame, so that the tip is molten; a cloud of droplets is expelled from the nozzle. In another type of pistol, the metal is introduced as metallic powder.

Alloy coats can be obtained by **cementation** (1960 vol., pp. 64,

654). Steel articles packed in zinc dust mixed with zinc oxide and heated at 380–450°C (or in the case of springs as low as 250°C) become covered with a zinc–iron alloy layer, possessing protective properties; the process is called **sherardizing**. Similarly, if steel articles are heated in aluminium powder, alumina and ammonium chloride, a layer of iron–aluminium alloy is formed; this is the basis of the **calorizing** process.[64]

Sherardized coatings are frequently applied to small bolts and similar factory-made articles, and greatly add to the effective life, although, since the layer consists largely of an iron–zinc alloy layer, a yellowish corrosion deposit is formed at a fairly early stage; this does not mean that protection has ceased, and it should be distinguished from the true rust, which appears when corrosion has penetrated to the steel basis.

Ballard and Mansford have supplied information about the performance of steel sprayed with zinc (1968 vol., p. 226). Lock gates on a canal near Paris were free from rust after 16 years. Zinc-sprayed sluice gates on British rivers have remained rust-free for 11 years. Barges on the Thames and elsewhere, sprayed with zinc and then coated with bituminous paint, were in fairly good condition after 8–17 years – although a little retouching was needed.

The diving structure at the Cambridge University swimming sheds, sprayed with zinc from a powder pistol and then coated with zinc chromate and a pale blue paint (which probably added little to the protection) remained unchanged for 18 years; after 28 years there was a little rust at inaccessible corners.

Zinc-rich paints

Instead of introducing an inhibitive pigment into paint, metallic zinc can be used, and such paints can provide cathodic protection to steel exposed at places where the coating has been damaged. Zinc-rich paints were mentioned on p. 213, in connection with coatings of metallic zinc obtained by spraying; electrochemically the systems are closely related. Some further information may be provided at this point.

In Mayne's tests, plates of steel coated with zinc-rich paint, pierced by a scratch-line, and placed in sea water, remained two years without the formation of loose rust by attack on the exposed steel. Electrical tests indicated that in the early stages the zinc was in connection with the steel, which received cathodic protection. Later, however, the electrical connection ceased (probably because the lowest portions of the zinc had become destroyed through anodic attack); yet protection still continued, probably because, in the earlier stages a film of zinc chloride had been deposited on the exposed steel.

In a paint containing 92% of metallic zinc and 8% of vehicle, it is found that the film can act electrically as a continuous conducting mass and is capable of providing 'cathodic protection' to iron exposed at a gap – in much the same way as a coating of zinc obtained by hot-dipping. Each particle in a mass of dried paint has many neighbours, and if every particle is in good contact with each of its neighbours, at least some of these neighbours – it would seem – could consist of a non-conducting substance without interfering with the provision of conducting paths throughout the mass. It is reasonable to expect that, provided the vehicle content is kept sufficiently low, it may not be necessary to maintain the zinc-content as high as 92%; some of the zinc particles could surely be replaced by non-conducting particles. There is some prospect of obtaining paints lighter and cheaper than a 92% zinc paint. There are, however, limitations to the extent to which the zinc content can be reduced, and much work requires to be done before a satisfactory paint of relatively low zinc content becomes available. For the moment, the zinc-rich paint with 92% or more zinc seems to give the best protection.

Such a high zinc content cannot be obtained with all vehicles. It cannot be reached, for instance, with linseed oil, although a mixture of metallic zinc and zinc oxide in linseed oil provides a paint which has considerable merits; in Britton's outdoor tests, coatings of such a paint were found to provide some protection at a gap left in the coat. The discovery that exceptionally high pigmentation with metallic zinc could be achieved by using chlorinated rubber as vehicle was patented by Dyche-Teague in 1931. Later, Mayne found that, by using polystyrene, pigmentation up to 95% could easily be obtained.

The chlorinated rubber and polystyrene paints dry by evaporation of the solvent (which is xylene if plasticized polystyrene is used). Drying by evaporation may be expected to produce coatings which are not very adherent to the steel basis, because the solvent is first lost from the outermost layer of the film, and the paint below the nearly solid layer remains liquid or soft for a long time after the surface appears to be 'dry'; a little consideration will show that when at last the coat does dry, it may be imperfectly joined to the metal. If a mixture can be found which becomes solid through some interaction between two components, with little or no loss of solvent, good adhesion becomes theoretically possible; one plan is to supply two components in different packages; these two components, if kept apart, will remain liquid indefinitely, but if mixed and quickly spread out on the steel surface, a satisfactory solid coating should be obtained. Such a system provides a possible basis for metallic zinc paints; zinc-rich paint with an epoxy vehicle, 'cured' (hardened) by means of polyamide, provides a coating which is very adherent – provided that the surface to which it is applied is scrupulously clean.

Inorganic vehicles have also been used for metallic zinc. Paints have also been made by adding zinc dust to water-glass. One of these has given admirable protection in Australia when used for the coating of an important pipeline; in that case, there was curing by heat. Opinions of the relative merits of the inorganic and organic types of zinc paint vary considerably; one authority states that the inorganic type provides longer life in a marine atmosphere, but that for immersion in fresh water the organic type gives the better performance (1976 vol., p. 305).

Aluminium coatings

As stated above, protection has been obtained on steel by grit-blasting the surface, and then spraying with aluminium from a pistol, followed by painting. This process was used throughout the Second World War in the protection of the hydrogen cylinders used for balloon barrages; despite the rough usage and adverse climatic conditions to which these cylinders were exposed, it gave excellent protection. The method is now used extensively in the protection of structural steelwork. Exposure tests organized from Cambridge on sand-blasted steel, coated with aluminium from a wire pistol, showed admirable behaviour in urban, marine, mixed and country atmospheres. Even where no paint had been applied, specimens exposed 7 years in London remained free from rust, although the aluminium showed some deterioration; where paint had been applied, the aluminium coating was completely unchanged. In country air, paint keeps its colour better when applied on aluminium than when placed directly on steel.

More recent experience confirms the good protection afforded (1976 vol., p. 299). Scott states that the best results are obtained when the purity of the aluminium is 99·5%. Long-term protection – up to 15 years in a marine atmosphere – has been provided for steel by a thin Al-spray coating (0·076 mm). In the case of Al-sprayed steels, a slight stain of rust appearing in the early days of exposure does not mean that protection is failing. The rust acts as a wick, and moisture establishes the cell Al/Fe, which reduces the rust and provides cathodic protection to the iron, so that the disfiguring colouration disappears. This temporary rusting is more marked in pure atmospheres than in industrial atmospheres, where the acid in the moisture favours the action of the cell Al/Fe. Bailey states that coatings 0·076 mm thick, can give protection for over 12 years in severe marine and industrial atmospheres. If there is sealing with vinyl lacquer or paint, 20–30 years protection may be expected. Tests by Stanners at Shoreham lasting $8\frac{3}{4}$ years in a marine industrial atmosphere suggested that Al-spraying generally provides better protection than Zn-spraying, whilst spraying with a Zn–Al alloy gave promising results.

There has been considerable interest in the relative merits of zinc, aluminium and their alloys as protective coatings. Recent tests at Rome* indicate that aluminium is superior in marine atmospheres but that in the absence of oxygen both provide protection, aluminium

*R. Bruno and M. Memmi, *Brit. Corrosion J.*, 1976, **11**, 35.

being consumed at a higher rate, so that zinc is preferred. In a marine environment with sulphide pollution, unalloyed zinc can lead to hydrogen embrittlement and alloy coatings seem preferable.

Cadmium coatings

At one time it was thought that cadmium would largely displace zinc, owing to the deceptively favourable results of certain salt spray tests, which failed to represent service conditions. Cadmium coats do, however, appear to possess superiority over zinc coats for articles intended for the tropics, as they provide protection under storage conditions of high humidity and high temperature. Cadmium plating is used in aircraft to prevent frettage and fatigue (p. 172). The toxicity of cadmium limits its use in certain situations.

Non-metallic coatings

Basis of protection

Many people still believe that the function of a paint coat on steel is to exclude the water and oxygen needed for rusting. However, paint coats are more pervious to water and oxygen than is commonly supposed. Mayne[543] has collected figures which suggest that the passage of water and oxygen through an ordinary paint or varnish film could take place sufficiently fast to allow rusting at the rate observed on steel exposed, unpainted, to an industrial atmosphere. He concludes that such coats 'cannot inhibit corrosion by preventing water and oxygen from reaching the metal'. Protection, he thinks, is in some cases due to the slow movement of ions through the paint film, so that the electrochemical reactions which occur rapidly on an unpainted surface, when wetted, take place very slowly. In other cases, the paints contain inhibitive pigments, such as red lead or zinc chromate; in the latter compound, both anion and cation possess inhibitive properties. Paints containing certain metallic pigments such as zinc prevent attack even at gaps in the coating initially by cathodic protection (p. 117).

Although Mayne's statements have not been seriously challenged, it is likely that the practical man continues to regard a paint coat as something which shuts out corrosive influences. One cannot in a moment get rid of the thinking-habits of a lifetime. Perhaps by the time that the average person has come to realize that the idea of a paint coat providing mechanical exclusion to corrosive influences has been wrong, that idea will have become right. Already research is in progress, aimed at determining why paint films are permeable and abolishing the permeability. A succession of experiments in Mayne's laboratory on unpigmented lacquer films shows that different areas of the same film may possess different characters. The difference is shown by the manner in which the conductivity changes when the film is in contact with salt solutions of different concentrations. In one type of film, which Mayne calls the D or Direct films, the film conductivity increases as the concentration of the salt solution increases. In the other type, the I or Inverse films, conductivity in the film drops with increase of salt concentration (which involves a decrease in water activity). Different varnishes differ in the frequency of the two types. In a certain alkyd varnish only 15% of the samples taken proved to be of the D type, but in epoxy–polyamide varnish (regarded today as more satisfactory material) 76% were of the D type. It is believed that the D type arises from cross-linking, and that films with greatly improved protective value could be produced if the cross-linking were more homogeneous. Swedish workers, using Mayne's work as the basis of experimental assessment of paint-substances, have measured the ion-exchange capacity of pigmented paint film; they believe that the measurement of this capacity at time intervals during the exposure of a paint system may provide information about progressive deterioration.

Permeability to Cl^- is clearly of great importance. The passivating pigments present in a paint film must be sparingly soluble; otherwise they would be washed away. If Cl^- ions arrive at the metal–paint interface more quickly than molecules of the inhibitive substance, passivity is likely to break down. Murray (1976 vol., p. 280) has studied three coatings. In cellulose acetate, permeation by Cl^- occurred at a constant rate. In epoxy–polyamine mixtures with ratio 1:1, permeation increases with time, but when the ratio is 2:1, it falls off with time.

Components of paints and varnishes

Most paints which dry by evaporation contain resins – synthetic or natural. In addition to the volatile solvent, there is usually a non-volatile organic substance to act as **plasticizer**; in the absence of such a compound, the films would become brittle and porous when the volatile liquid had evaporated. A plasticizer should be non-volatile, chemically stable, insoluble in water and possess lubricating properties; its role seems to be to fill voids in the molecular lattice and provide planes of easy glide. Useful plasticizers include dibutyl, dioctyl and dinonyl phthalates, and tricresyl phosphate.

In most paints used today, the setting of a coat depends less on evaporation than on oxidation. The drying paints are among the oldest but most complicated types of covering. In addition to resins, they contain a drying oil, which will introduce a mixture of glycerides of unsaturated fatty acids. These can take up oxygen at the double bonds, and the oxidized molecules link up together yielding a three-dimensional network, so that the liquid becomes solid; the oxidation and polymerization is favoured by compounds of metals which exist in two stages of oxidation. Such compounds are known as **driers**, and they may consist of salts of acids (such as linoleic) present as glycerides in the drying oil. Naphthenates, derived from acids containing rings of five carbon atoms – by-products of the oil industry – and also 2-ethyl hexoates are used as driers. The different metals assist drying in different ways, and more than one may be introduced; indeed lead, manganese and cobalt may all be present in a single paint.

Usually, drying oils receive refining or heat-treatment before being used in paint. The refining aims at the removal of objectionable minor constituents, particularly the slimy bodies known as 'foots', but the heat-treatment modifies the glycerides themselves. The products may be 'boiled oil' (obtained by heating with driers), 'blown oil' (prepared by passing air through the hot oil) or 'stand oil' (made by prolonged heating with exclusion of air and without driers); 'stand oil' is particularly valuable since small quantities added to other oils yield tough paint coats.

The principle to be followed in drawing up a painting scheme should be to apply to the steel a 'priming coat' containing an inhibitive pigment, such as red lead, basic lead sulphate or zinc chromate, and to follow it by one or more outer coats chosen to resist abrasion and exclude water, as far as possible; paints of this class are never completely impervious.

Red lead is frequently diluted with red iron oxide or with graphite; one authority held the view that iron oxide with only 20% red lead provides a better, as well as cheaper, paint than undiluted red lead. Similarly graphite paint containing red lead is a good inhibitive primer, although a paint containing graphite alone is dangerous; graphite is cathodic to steel.

The toxicity of lead compounds has encouraged endeavours to find an alternative for red lead. Tests have indicated that as an inhibitor zinc chromate is not seriously inferior to red lead except for application over rust. As regards keeping power, paints containing it are much superior to those made with the more inhibitive types of red lead; in priming coats it is often mixed with a relatively large amount of iron oxide. Zinc oxide (alone or in admixture with iron oxide) possesses inhibitive properties, but it is liable to be dissolved away by acid rain.

Metallic lead pigments provide inhibitive priming paints which keep well. Oil paints containing metallic lead probably develop, after being spread out in contact with air, the inhibitive substances identified by Mayne and van Rooyen (p. 226). However, alkyd paints pigmented with metallic lead give good results, and the inhibition is here probably due to the formation of lead hydroxide, which has been found to protect iron from attack by water, probably acting as a source of OH^-[548].

In any paint system, the outer coat chosen should be as nearly waterproof as possible, so as to prevent the inhibitor of the inner coat being washed away by water or converted into other substances (red lead, exposed in a polluted atmosphere may change to lead sulphate).

Inhibitive pigments are less necessary in the outer coats than in the priming coat, although some authorities recommend that they should be present in both. The water-proof character of outer coats can be improved if pigments of flaky form are introduced; when brushed out the flakes lie parallel to the surface and to each other; the path which a molecule must follow before it can pass through a film of a given thickness will clearly be longer than if the paint contained roughly spherical pigment particles. Flaky pigments include metallic aluminium, and the form of iron oxide known as **micaceous iron ore** (also called **flaky haematite**).

The principle that the *lowest* coat should contain an **inhibitive** pigment (e.g. red lead) and the *outer* coats an inert pigment chosen for its **physical** properties (e.g. iron oxide) is borne out by tests conducted in all parts of the world.

Function of inhibitive pigments

As stated above, certain pigments used in paint confer protection on iron even where it is exposed at a discontinuity; they include red lead, litharge, metallic lead powder and zinc chromate. Some of these substances diminish the corrosion of iron even when no oil is present. Lewis found that specimens of unpainted iron shaken in a tube with air and water (or air and a salt solution) suffered far less loss of mass when one of these pigments was suspended in the liquid; being a loose powder, it would provide no mechanical shielding of the iron surface. He also found that when a scratch-line was made on iron coated with red lead paint and a drop of M/1000 sodium chloride was placed on it, the iron exposed at the scratch suffered no rusting, although parallel experiments using paint pigmented with iron oxide (a pigment lacking inhibitive properties) invariably led to rusting at the scratch-line. There are, of course, limits to the inhibition conferred by red lead. In experiments where concentrated sodium chloride was used, rusting was not prevented.

There is an essential difference between red lead paint, which prevents corrosion by keeping the iron passive, and iron oxide paint, which, as suggested by Mayne, probably slows down the attack by increasing the resistance of paths connecting anodic and cathodic areas. This difference was shown in outdoor corrosion tests carried out at Cambridge by Britton (p. 230). In one series, different amounts of thinner were added to the paints. The iron oxide paints were found to give less and less protection as the thinner content was increased; the red lead paints, on the other hand, continued to give satisfactory protection after large additions of thinner. The general causes of passivity have been discussed in Chapter 6. Zinc chromate doubtless behaves in the same manner as a potassium chromate solution under conditions of immersion; to introduce potassium chromate into paint would give short-lived protection, because so soluble a substance would quickly be leached out. On the other hand, a chromate of extremely low solubility might fail to provide the necessary concentration of $(CrO_4)^{2-}$ ions. An intermediate solubility must be chosen, and zinc chromate seems appropriate for steel; the zinc ion may play a part by inhibiting the cathodic reaction, as explained on p. 219. Strontium chromate has been used in paints for protecting aircraft alloys.

The inhibition conferred by lead oil paints was attributed by Mayne and van Rooyen to the degradation products of the soaps, such as lead linoleate, present in the coat. They found that a saturated solution of 'lead linoleate' (meaning thereby the lead soaps of the mixed acids derived from linseed oil), if prepared in absence of air, does not prevent corrosion; the solubility of the lead linoleate is in fact extremely low. If oxygen is present, more soluble compounds are formed from the lead linoleate, including the lead salts of formic acid (H.COOH) and of azelaic acid (COOH.$(CH_2)_7$.COOH). Although lead formate is corrosive, lead azelate inhibits corrosion at pH values down to 4·6. A synthetic mixture was made up, based on the result of analysing the degradation products and was found to possess inhibitive properties. This accounts for the good anti-corrosion properties of lead oil paints.

Stoving varnishes

Materials are known which are liquid at ordinary temperatures but become solid when heated, because the molecules can join up to form a three-dimensional network, thus losing their translatory motion. Many synthetic resins have this power of thermo-hardening, and form the basis of modern stoving varnishes, or (if they contain pigments) stoving paints. The compositions are often sprayed on to the articles, which then pass through a heated tunnel. The heating may conveniently be produced by radiation; the exterior layer is then preferentially heated and becomes covered with a tough, adherent coat. Stoving varnishes are largely used on motor cars, cycles, metal furniture, typewriters, telephones and scientific instruments.

Tar and bituminous paints

Many mixtures containing pitches, tars or bitumens are known which are solid at ordinary temperatures but melt on gentle heating. These can be brushed on when molten, solidifying when cool; thicker coats can be cast or pressed in the plastic state on to the steel. A pipe can be supported by a chain and lowered into the molten mixture, so that interior and exterior are coated simultaneously; for pipe interiors a centrifugal application gives a smooth, mirror-like surface which facilitates water flow.

There are on the market numerous black compositions made from natural bitumen (rich in aliphatic hydrocarbons) and/or from industrial coal-tars (rich in aromatic hydrocarbons). Most of these can be applied by brush or spray and dry by evaporation at the ordinary temperature. In general, they are more sensitive to light than oil paints, easily developing a network of cracks when exposed to the sun; moreover, many of the older compositions provided coatings which were only serviceable over a narrow temperature range; either the coat became brittle and cracked in cold weather, or it became soft and ran in hot weather. Modern research is tending to overcome these disadvantages, and where light and temperature fluctuations are not excessive, these black coatings may give better performance than most other types of paints. Their use on pipes is an example, whilst some tar paints (usually prepared with lime additions to neutralize tar acids) have shown excellent performance at chemical works where oil paints had failed. Another situation where this type of paint has long been used, is for the boot-topping of ships, that is coating the water-line zone alternately exposed to wind and sea; comparative tests carried out by the Institution of Civil Engineers have confirmed their value at that level. Black coatings of this type have remained adherent when applied to galvanized steel which has received no phosphating treatment. There is, however, some variation between products from different sources; in the days when coal-tar was widely used the product from horizontal retorts generally performed better than that from vertical retorts.

Paints of the tar–bitumen class are somewhat sensitive to the condition of the surface to which they are applied, but by emulsifying bitumen in water, products may be obtained capable of application to wet surfaces; when the water evaporates, continuous coats remain. Moreover, excellent compositions are known containing bitumen and/or tar along with oils, adapted both for air-drying and stoving. Some of the constituents of tar inhibit the oxidation changes responsible for the 'drying' of ordinary oil paints, but they allow polymerization processes at high temperatures. One of the earliest forms of stoving enamel was based on this principle. Pipes coated with mixtures of linseed oil and bitumen, and then heated, were found after 37 years' burial near Pittsburgh, to be in excellent condition.

Glassy coatings

Iron may be coated with **glassy enamel** by applying to the pickled surface powdered glass of suitable composition, often suspended in liquid (usually water); the articles, after being dried in warm air, are heated strongly so as to melt the glass, which, under suitable conditions, flows over the whole surface producing a coherent coat. In general, several coats are applied, the bottom one being chosen for adhesion and the outer ones with a view to resisting corrosion or abrasion – according to the conditions of service envisaged. The composition for general purposes is a borosilicate glass containing fluoride and often lead; it is generally made from borax, felspar, quartz and cryolite. For acid-resistance, silica may be increased and titania introduced. The lowest coat usually contains compounds of cobalt or nickel which improve the adhesion; it is possible that a cobalt and nickel tee is formed on the iron, by simple replacement, and provides key-ing. The coefficient of expansion of acid-proof enamels is considerably lower than that of iron, the discrepancy being greater for sheet than for cast iron; it is best to use enamel of high coefficient for the lowest coat, building up with successive coats of increasing acid-resistance and decreasing coefficient.

Thick linings of glass are sometimes used in steel vessels, but are naturally brittle. Cement or concrete layers can be cast in contact with the metal, and are applied internally and externally to pipes; thinner layers can be obtained with a paint or slurry of Portland cement and water, perhaps containing a little glue.

Plastic coatings

Thick coatings of plastic or rubbery material – whether natural or synthetic – are often used for lining metal tanks, to keep corrosive liquid from reaching the metal. Obviously the material chosen must itself resist the liquid; it must also adhere well to the metal.

Many ingenious methods for the continuous application of much thinner plastic coats to steel strip are being worked out, and some are in operation.

Behaviour at a gap in a coating

If steel carrying a discontinuous coating is placed in a natural water, without stirring, the rust, which requires oxygen for its formation, is usually thrown down outside the gaps in the coat. Being very voluminous, it may cover up the coating, but, until so much metal has been destroyed as to undermine the coating at places between the discontinuities, the coating will remain adherent in its original situation. With coatings of flaky metal, as obtained from a spraying pistol (or even paints pigmented with flaky metallic particles), the flakes will remain, clinging at certain points to the metal, even after much corrosion has taken place. If the water is in motion, so that there is ready replenishment of oxygen, the rust will be formed close to the metal, and is more likely to seal the discontinuities and stop attack. The presence of calcium salts in the water aids the formation of adherent rust (p. 193), and a hard water will sometimes seal pin-holes in a coat by means of rust blobs, where a sodium chloride solution would cause continuous attack, with the formation of loose rust outside the coat.

Alkaline softening and loosening[552]

Early experiments at Cambridge showed that the distribution of anodic and cathodic areas on painted steel is often the same as on unpainted steel – but corrosion develops much more slowly. A steel plate varnished and partly immersed in salt water displayed an immune (cathodic) area near the water-line and a corroded (anodic) area lower down. Drops of salt solution placed on painted steel often produced alkali round the edge of the drop, which in the case of alkali-sensitive paint causes the paint to become soft and loose, so that the gentlest rubbing produces peeling sometimes over an area larger than the area originally covered by the drop; this suggests that alkali has been creeping between paint coat and metal. Where the paint contains an inhibitive pigment, peeling is much reduced. As in the case of unpainted steel, the corrosion occurs in the centre of the drop, which is evidently anodic, the cathodic perimeter remaining immune from attack. In some cases, an island of paint on the centre of the drop may remain adherent, showing that the peeling is not due to the anodic destruction of the steel but to the cathodic formation of alkali. Only liquids which produce alkali by the cathodic reaction cause serious peeling. Thus drops of distilled water or of Cambridge tap water (containing calcium bicarbonate) produce little or no peeling, even when they cause rusting. Magnesium, calcium and zinc salts produce much less peeling than sodium salts.

This phenomenon is characteristic of alkali-sensitive paints, which are generally those based on saponifiable oil-vehicles; such paints should be avoided for marine conditions. Non-saponifiable polyvinyl chloride/acetate paints pigmented with lead sulphate, non-leafing aluminium barytes and Burnt island red oxide have given good service in raft tests. Additions to the vehicle which increase alakli-resistance (such as copal) have been found to reduce peeling. As shown later, paints containing pigments attackable by alkali, such as metallic zinc, can suffer loosening under some conditions. Coatings of certain tar paints can become loosened by alkali creeping between coating and metal, even though there is no perceptible softening.

Softening and loosening is greatly accelerated if scratch-lines have been traced through the coat, exposing the steel – especially if the scratch passes through the water-line of a partly immersed plate or through the margin of a drop. If no scratch-line or other defect exists in the coating (intentional or otherwise), peeling will only occur rapidly if the vehicle is one exerting poor resistance to ionic movement. Mayne considers that for the formation of alkali on the steel at the cathodic region, three conditions must be fulfilled: (1) the coating must be permeable to water and oxygen, (2) the vehicle must carry a negative charge, (3) the corrosion current must be carried through the paint coat by Fe^{2+} ions at the anodic region and by Na^+ ions at the cathodic region. Some paints which readily peel when a scratch exists resist peeling when the coating is continuous. A scratch-line allows the ready formation of alkali on the exposed steel; peeling can then develop at an alarming rate. Specimens covered with a coating of iron oxide paint in linseed oil, partly immersed in sea water for a few days and then gently rubbed, were found to have the coat already very loose in the neighbourhood of the water-line; later, the loosening extended over practically the whole immersed area, and also spread upwards on the originally dry area, probably by alkali creeping between paint and steel.

Fortunately, under service conditions, alkaline softening and peeling are only likely on painted steel under the rather exceptional circumstances where alkali accumulates at one place. It is brought about by partial immersion in stagnant salt water, or by the presence of a single drop of salt water lodging at a point where evaporation is slow. It would not be produced by a shower of salt droplets falling on a surface at random, since, at any given point, alkali may be formed at one moment, but ferrous chloride at the next, which will destroy the alkalinity. On magnesium, however, the corrosion product will be an alkaline substance (magnesium hydroxide) even in the absence of salt, and loosening is more likely. This is probably the main reason why inhibitive pretreatment of magnesium alloys (p. 209) is so necessary if paint is to adhere.

Effect of surface conditions on performance of coats

Before applying a metallic coating, it is absolutely necessary to remove all scale, rust and grease. The removal of scale by pickling has been mentioned on p. 175. Grease may be removed by an alkaline bath containing a silicate, phosphate or aluminate; subsequent washing to remove alkali traces is necessary. Alternatively, dipping in an organic solvent, such as naphtha, is often employed. 'Vapour-degreasing' is favoured for some purposes; the article is placed in the upper part of a vessel containing boiling trichloroethene or other solvent below, so that the liquid condensing on the surface drips down and carries off the grease.

Whether the surface should be smooth or rough depends on the nature of the coating contemplated. On a surface intended for plating, all roughening should be avoided, and pickling should be conducted with sufficient restrainer to ensure this, the restrainer being chosen to minimize organic contamination of the surface. On the other hand, sprayed coatings adhere badly to a smooth surface, and the grain used in the last stage of grit-blasting should be an angular grit chosen to provide dovetailing for the coating.

In the application of organic coats, it is possible to quote examples of mill-scale or foundry-scale being left in position – apparently without harm resulting. For instance, cast iron and cast steel are usually painted without removing the scale, which – if free from sulphur – actually increases the protection; similarly the *adherent* portions of the scale on wrought iron can safely be left in position. Even on rolled steel, there is usually no removal of mill-scale if a thick bituminous layer is to be applied, or if the steel is to be buried in cement.

In most other cases, it is advisable to remove mill-scale and rust before paint is applied. The advantage of this course has been proved in tests organized from Cambridge and carried out at stations representing a number of atmospheres, as well as in a more extensive series conducted on large specimens by J. C. Hudson (p. 104). It is also indicated from service inspections; a survey of painted steelwork in

Holland made about 1938 suggested that almost all cases of rapid breakdown were due to mill-scale left on the metal. Sometimes the presence of mill-scale may appear advantageous in the early stages; where a poor paint has been applied over scale, it may develop rust stains less quickly than the same paint applied to clean steel, the scale acting as an additional coat. But when the time comes for repainting, it will be found that coats applied over mill-scale are rising from the surface, owing to rust formed below them at breaks in the scale; in such cases, drastic cleaning will be needed to obtain a proper foundation for the new coat, whereas paint applied to clean steel, even after it has become rust-stained, may sometimes remain sufficiently adherent for the new coats to be applied immediately upon it.

There are two separate reasons why mill-scale below paint is dangerous. First, although the scale may appear adherent, it may crack off, carrying the paint with it, if the steel is bent repeatedly to and fro, as will, for instance, happen on ships' plates. Secondly, even without bending, there may be serious rising of paint at any places where the scale has become damaged. The dangerous combination of small anode (the exposed steel) and large cathode (the scale-covered areas) leads to intense attack at the breaks of the scale; the fresh rust there formed appears below the paint (under conditions of atmospheric exposure) and, occupying a far larger volume than that of the iron destroyed in producing it, pushes away the paint from the metal, and causes failure.

If the mill-scale has been removed completely – whether by weathering, grinding, chipping, pickling or shot-blasting – and if the surface has subsequently become covered with rust, the question arises whether paint can safely be applied on that rust. Since the main constituent of rust is ferric oxide, often used as a pigment in paint, an affirmative answer might be expected. Actually paint can be successfully applied over the rust produced where a drop of distilled water has been placed on steel (previously cleaned by grinding) and allowed to evaporate. But rust formed on exposure to urban atmospheres generally contains ferrous sulphate, usually unevenly distributed. Mayne states that where paint is applied upon the rust formed on steel exposed out-of-doors, breakdown seems to occur at places where ferrous salts exist.

Paints for different situations (see 1976 vol., p. 287.)

It may be helpful to finish this chapter with some recommendations made in recent years by recognized authorities, although, like all generalizations, they must not be taken as applicable for all places and all times. In the procedure adopted by the (British) Gas Council for the coating of *pipes*, the external surface is pre-coated at the mill with a flood-coat of coal-tar enamel incorporating an inner fibre-glass reinforcement and an outer thermoglass wrap. The internal surface is coated with epoxy paint. On arrival at the site, the pipe-lengths are welded together, and the weld area is coated by the flood method.

For the painting of steel *bridges*, Smith and Day consider that in the U.K. a bridge should last 120 years and require maintenance only 10 or 20 times during that period; they consider that red lead is still unrivalled for a steel surface carrying rust, and that in such cases brushing is still the best method of application. Johansen has described the 65-years history of the Victoria Falls bridge in Zimbabwe, which has had to withstand corrosive conditions in a tropical climate. Up to 1963, oil paints were used, but in that year an epoxy–tar combination was adopted (2 coats of plain epoxy–tar and one coat of epoxy–tar containing aluminium); after 5 years, inspection revealed no corrosion or mechanical damage.

For *ships* there is still interest in lead paints – despite the health hazard. The paint favoured by the (British) Royal Navy in 1969 contained basic lead sulphate, non-leafing aluminium, barytes and iron oxide in a vehicle consisting of stand oil and tung oil with a synthetic resin; it is moderately resistant to alkali, and can be used for a sacrificial anode system of cathodic protection during the fitting-out period. If an impressed current system is adopted, epoxy–tar paint should be used.

For *aircraft*, Scott emphasizes the need for a paint which is itself chemically resistant and flexible; an epoxy paint cured with polyamide is recommended. A priming coat pigmented with strontium chromate provides inhibition; that pigment is sufficiently soluble to furnish CrO_4^{2-} ions at the rate required without being so soluble as to be washed away quickly.

On *automobiles*, the bright 'trim' is usually plated with a layer of semi-matt nickel covered with bright nickel and then a thin layer of chromium; the procedure is discussed on p. 215. The main part of the car body is painted. There has been dissatisfaction at failures caused by impact of grit particles carrying salt which damage the coating and start rusting; the damage to the coat occurs in winter when salt is applied to the road surfaces to prevent freezing, but the rusting mainly develops during the following summer. Such figures as are available suggest that the greater number of corrosion failures arise largely from success in avoiding mechanical failures; in the early days, a car ended its useful life through mechanical deterioration, and the end came after a short time, before corrosion had had time to become serious. Today the 'mechanical life' is longer and deterioration due to corrosion has time to develop. The matter is discussed by Hundy (1976 vol., pp. 9–12) who also discusses methods for improving the resistance to corrosion – at a price. One method, favoured in some quarters, is to use galvanized steel. If there is a zinc coating between paint and steel, corrosion, at a point where the paint has been chipped away by a stone particle carrying salt, will turn sideways; but clearly this sacrificial protection will not last for ever. Galvanized steel presents problems in drawing and welding, and Hundy considers that improved electropainting is likely to be the prime defence against corrosion. He provides pictures showing improvements already achieved, but states that corrosion resistance still tends to vary between steel coming from different plants – or even from different places within the same plant.

Electropainting – much used in the car industry – is likely to become increasingly important (1976 vol., p. 283). The liquid used in electropainting may contain a resin, probably an epoxidized alkyd, brought into solution by addition of a base, such as monoethanolamine or triethylamine. The article to be coated is made the anode of a cell and the H^+ produced by the anodic reaction throws the resin out of solution, so that it is deposited on the metallic surface; probably Fe^{2+} and Fe^{3+} help to make it insoluble. Another form of the process uses a resin dissolved in acid; here the article to be coated must be made the cathode.

8

Kinetics and Chemical Thermodynamics

The laws governing film growth

General

The simpler growth-laws were stated, but not proved, in Chapter 1. In the present chapter, derivations are presented for more general laws, of which the simple laws constitute special cases. The theoretical arguments advanced are not universally accepted, but they appear to explain the facts known today; it is doubtful whether some of the alternative arguments do so.

Mixed parabolic equation

When a film-free surface is exposed to an atmosphere containing oxygen, sulphur, iodine or other non-metal, the latter will be adsorbed, and cations from the metal will probably arrange themselves in appropriate fashion to form the first layer of oxide, sulphide or other compound; then further adsorption will occur, and the process will repeat itself. So long as the film remains thin, the rate of thickening will probably be determined by something independent of the film thickness; this may be (1) the rate at which cations in the metal can detach themselves from their neighbours, (2) the rate at which anions can be formed from molecules of the non-metal, or (where the non-metal is present only at low concentration in the gas-phase), (3) its rate of arrival at the metallic surface; there are other possibilities, but, at least in this opening stage, the rate of film-thickening should be independent of film thickness. However, as the thickness increases, it will become increasingly difficult for cations to move outwards (or anions inwards) sufficiently quickly to maintain the pace which the surface reactions would set in the absence of a film. At this stage, there are two resistances to the movement*, one

* The simple derivation of the mixed parabola based on the resistance concept was put forward by E. Fischbeck (1933) and W. Jost (1937). It is preferred to the author's earlier derivation (1924) for use in this book because the Hoar–Price calculation of the parabolic rate-constant is also based on the resistance concept. Those who criticize the Hoar–Price derivation on the grounds that movement through the film is in fact largely diffusion, not ionic migration, and who prefer the various treatments due to C. Wagner (*Z. phys. Chem.* (B) 1933, **21**, 25; and *Atom Movements* 1951, p. 153 (Amer. Soc. Metals)), may perhaps also favour the 1924 derivation of the mixed parabola (U. R. Evans, *Trans. Amer. electrochem. Soc.* 1924, **46**, 269).

independent of film thickness, and the other (in the simplest case) proportional to it. We can, therefore, express the increase of thickness y with time t

$$\frac{dy}{dt} = \frac{k_1}{k_2 + k_3 y}$$

where k_1, k_2 and k_3 are constants. Throughout this chapter, all the k's and K's are constants independent of t and y, but dependent on temperature. In the present case, the dimensions of k_2 and k_3 are, of course, not the same.) Integration gives

$$\tfrac{1}{2}k_3 y^2 + k_2 y = k_1 t$$

or

$$y^2 + Ky = K't$$

where K and K' are constants. This is the **mixed parabolic equation**. It serves to express the situation where the non-metal is present in small amounts. Wagner and Grünewald found it to be obeyed when copper was exposed to oxygen at low pressure, and Fischbeck when iron was exposed to steam, which is equivalent to a low partial pressure of oxygen.

The mixed parabola approximates to a rectilinear equation $y = (K'/K)t$ when the thickness is very small, and to the simple parabolic equation $y^2 = K't$ when the thickness becomes sufficiently large. Probably most curves representing oxidation at high temperatures, if carefully measured, would be found to show a straight portion at the outset, and an approximation to $y^2 = Kt$ later, but the straight part often goes undetected. At one time, iron was believed to oxidize at high temperatures according to the simple parabolic law, but Bénard, using a balance due to Chévenard which enabled the early part to be traced accurately, found that at 1000°C the curve was straight at first and then became roughly parabolic. There are cases, however, where rectilinear thickening may persist, and others where obedience to the simple parabola starts early in the oxidation. Some examples are given in Chapter 1 and may be recalled here.

Rectilinear equation

A constant growth-rate would be expected where the film-substance remains pervious to one of the reactants. Calcium exposed to oxygen at 500°C under the conditions of Pilling and Bedworth's experiments gives the straight line shown in Fig. 1.8, but recent work at Exeter suggests that different results might have been obtained, had the oxygen been quite dry. Another example is provided by the zinc exposed to the atmosphere by Vernon (Fig. 1.9).

Simple parabolic equation

Many metals, exposed to oxygen at high temperatures, show fairly good obedience to the law $y^2 = Kt$ over a considerable period.

Copper at 800°C provides an example (Fig. 1.8), but almost any metal could be used to illustrate parabolic growth over a suitable temperature range. It should be noticed, however, that obedience is only to be expected if the film-resistance is essentially ohmic (doubling when the film thickness doubles). Actually, the movement through the film is partly a **migration** of charged ions under an **electric potential-gradient**, due to the fact that adsorbed oxygen captures electrons from the metal phase, to form oxygen ions and leave the metal positively charged, and partly to **diffusion** under a **concentration gradient** (strictly an activity gradient) due to the fact that the film will contain more metal (less oxygen) near its inner surface than near its outer surface. In both cases, however, the rate of movement will be inversely proportional to the thickness, so that we can write

$$\frac{dy}{dt} = \frac{K_1}{y} + \frac{K_2}{y}$$

where the two terms represent migration and diffusion respectively. Now migration and diffusion are closely connected, both being controlled by the fact that a particle will only pass from one site of low energy to a neighbouring site if it possesses sufficient energy to jump the intervening hurdle. There is indeed a simple relationship between diffusivity D and mobility B, due to Einstein. This is

$$D = \frac{RT}{N_L} B$$

where T is the absolute temperature, R the gas constant and N_L Loschmidt's number. It should be possible to convert K_2 into electrical units, and thus obtain an expression for the parabolic rate constant K in $dy/dt = K/y$ (where $K = K_1 + K_2$) which contains only electrical constants. Wagner[848] was the first to obtain such an expression, but a simpler proof, leading to the same expression, was later provided by Hoar and Price[847], by considering a model in which the movement was supposed to take place solely under an electrical potential gradient. Since – whether the movement is due to migration, diffusion or both – the ultimate driving force is derived from the affinity of metal for oxygen, the rate should always be the same, unless the precise manner of driving the particles somehow alters the frictional or viscous resistance to the motion; if, for instance, an electrical field were to alter the shape of the moving particles, the assumption made by Hoar and Price would become invalid; such an alteration of shape is only likely at very high fields and if it did occur any other theoretical treatment would have to take account of it.

Hoar and Price picture the system metal/oxide/oxygen as analogous to a primary battery of internal resistance R_i ohms, within which the ions move to anode and cathode respectively; the two terminals are joined by a wire providing an external resistance, R_e ohms, along which flows the electrons. The current I (in amperes) is given by $I = E/(R_i + R_e)$, where E is the e.m.f. in volts. Likewise, on the oxidizing metal, the transport is equivalent to a current $I = E/(R_i + R_e)$, the resistances being *added* since they are *in series*; notwithstanding that both ions and electrons are flowing through the same material, the ionic and electronic parts of the circuit are *not* alternative (parallel) paths; *both* are needed if movement through the film is to take place. For if the flow of *either* ions *or* electrons were prevented, all passage would cease, just as current would cease in the case of the primary battery if *either* the cell was emptied of electrolyte *or* if the wire was cut. Hence addition is appropriate.

The specific conductivity (κ) of the film substance is defined as the current which will flow across a cube of edge-length 1 cm when 1 volt is applied to opposite faces. Now, in the general case, the movement of electrons, cations and anions may all contribute to the current, in proportions represented by three transport numbers n_e, n_c and n_a, where $n_e + n_c + n_a = 1$. Thus the specific *electronic* conductivity of the material is $n_e\kappa$ and the specific *ionic* conductivity $(n_c + n_a)\,\kappa$. The electronic resistance (R_e ohms) of a film of area A cm², and thickness y cm will be y/An_ck, and the ionic resistance will be given by

$$R_i = \frac{y}{A(n_c + n_a)\kappa}$$

Thus

$$R_e + R_i = \frac{y}{A\kappa}\left(\frac{1}{n_e} + \frac{1}{n_c + n_a}\right) = \frac{y}{A\kappa n_e(n_c + n_a)}$$

since

$$n_e + n_c + n_a = 1$$

Now by Faraday's law

$$dy/dt = JI/AFD$$

where J is the equivalent weight of the film substance, D its density, and F Faraday's number. Again, by Ohm's law

$$I = \frac{E}{R_e + R_i} = \frac{EA\kappa n_e(n_c + n_a)}{y}$$

Hence

$$dy/dt = \frac{J\kappa n_e(n_c + n_a)E}{DF}\,\frac{1}{y}$$

Comparing this with

$$dy/dt = k/y$$

we have a means of obtaining k, (also K', as defined in Table 1.4) from the electrical properties of the film-substance, and the free energy of formation of the oxide, which is ZEF, where Z is the valency of the metal. The proof just given is essentially that developed by Hoar and Price, although the equation had been developed previously by Wagner without using the concept of the electrolytic cell.

It has been assumed above that the conductivity is not seriously influenced by the partial pressure of oxygen. If this is not the case, a more general relationship can be reached. It can be shown, from the law of mass action, that the conductivity at pressure P can be written $\kappa_P = \kappa_1 P^x$. Then, if the equilibrium oxygen pressure corresponding to the compositions at the outer and inner faces of the film be P_o and P_i respectively, it can be shown that

$$\frac{dy}{dt} = \frac{J\kappa_1 n_e(n_c + n_a)}{DF} \cdot \frac{RT}{xzF}(P_o^x - P_i^x)\frac{1}{y}$$

where R is the gas constant, T the absolute temperature, and z the 'valency factor', i.e. the valency multiplied by the number of atoms in the molecule (for oxygen, $z = 4$).

Unfortunately, there are comparatively few cases where the values of the conductivity and transport numbers are known. Three of these cases, comprising the formation of oxide, sulphide and iodide films respectively, are collected in Table 1.4, and it is seen that the calculated and experimental values of K' (which is equal to Dk/J) are always of the same order of magnitude, and the agreement as close as can be expected.

Influence of temperature on growth-rate

In all cases of movement through a film, the diffusing particles, whether atoms or ions, have to move in jumps from one position of low potential energy to the next. In general, a particle will only be able to jump the intervening hurdle if it attains the requisite energy. The number of particles which possess a critical energy equal to or greater than W is proportional to $e^{-W/RT}$. Thus in the case of a metal obeying the parabolic law

$$dy/dt = k/y$$

k will be proportional to $e^{-W/RT}$; consequently a straight line should be obtained on plotting log k against $1/T$. Such a relation has been found for several metals – e.g. for copper and brass by Vernon and Dunn (p. 26); on iron, Portevin and his colleagues[69] found two curves intersecting at about 930°C, apparently owing to the $\alpha \rightleftharpoons \gamma$ transformation.

As pointed out by Hoar, the exponential relation between oxidation velocity and temperature follows immediately from the electrical conception of film growth. The conductivity of the film depends on the number of lattice defects and their mobility, and since both these things vary exponentially with the inverse of the temperature, it follows that the velocity constant, which is proportional to the conductivity, will vary exponentially also.

Direct and mixed logarithmic equations

In the range where the parabolic law is obeyed, the oxidation rate falls off steadily as temperature is lowered; below a critical temperature there is a change to a different (logarithmic) law; in this region of temperature, the growth slackens with time much more quickly. The fact that the rate in the early stages of film-growth is not very different above and below the critical temperature suggests that the mechanism is not fundamentally different (a point perhaps overlooked by those who have proposed rather special interpretations for the logarithmic law). For films which grow by outward movement of cations, it seems likely that, at places where oxidation continues, the mechanism remains *exactly the same*; according to this view, the rapid slackening of oxidation is due to the fact that over a rapidly increasing area the film and metal cease to be in contact and oxidation is almost halted.

That the gaps are really formed is known as a result of direct microscopic observation. It will now be shown that the development of such gaps should lead to logarithmic relations.

Let ϕ be the fraction of the area which remains free from gaps sufficiently broad to obstruct outward movement, and m be the gain in mass for a specimen of unit area (i.e. the mass of oxygen taken up). Whilst the whole surface is available, dm/dt is proportional to the uniform thickening rate

$$dm/dt = k_1 dy/dt,$$

where k_1 is a constant. When thickening becomes restricted to a fraction ϕ of the whole surface, this must be changed to

$$dm/dt = k_1 \phi dy/dt.$$

Two cases call for attention:
(1) where the gap-formation occurs at the early stage where oxidation (at points where there are no gaps) would still be **rectilinear**; (2). where it takes place more slowly, mainly during the later period where oxidation (in absence of gaps) would have become **parabolic**. They will be considered separately.

Case (1)

In the **rectilinear** stage of thickening, dy/dt is constant with time, so that

$$dm/dt = k_2 \phi$$

where k_2 is a new constant. J. N. Agar points out that k_2 is related to K and K' of p. 235, being equal to $K'/K \times$ (density of oxide) $\times$ (fraction by mass of oxygen in the oxide). Metallic cations pass from metal to oxide at a rate equivalent to dm/dt, but only a certain fraction of them will leave vacancies behind, and of this fraction the majority will diffuse into the metal and end up at dislocations; it is well known that dislocations act as sinks for vacancies. A small number will end up at the original surface of the metal, but those which end up at points where a gap already exists between metal and oxide will not increase the obstructed area; only those which end up at points on the original surface where there is no gap will contribute to fresh obstruction, and the chance of this happening is proportional to the unobstructed fraction. Thus the rate of increase of the obstructed area is proportional both to dm/dt and to ϕ, giving

$$-\frac{d\phi}{dt} = \frac{k_3\phi\, dm}{dt}$$

where k_3 is a new constant or

$$-d\phi = k_3\phi dm$$

so that

$$\ln \phi = -k_3m + k_4$$

where k_4 is an integration constant, giving

$$\phi = k_5e^{-k_3m}$$

where k_5 represents e^{k_4}, so that

$$dm/dt = k_2k_5e^{-k_3m}$$

or

$$e^{+k_3m}\, dm = k_2k_5\, dt$$

giving

$$e^{k_3m} = k_6t + 1$$

where k_6 is a new constant, assuming the relation to hold from $t = 0, m = 0$. Hence

$$k_3m = \ln(k_6t + 1)$$

which was written by Vernon and his colleagues as

$$m = k \log_{10}(at + 1)$$

This is the **direct logarithmic equation**.

Case (2)

At the stage where, but for cavities, the film would be thickening **parabolically**, the thickening rate on the unobstructed areas (dy/dt) is no longer independent of t and y, but is proportional to $1/y$, that is to

$t^{-1/2}$, so that

$$dm/dt = k_7\phi t^{-1/2}$$

where k_7 is another constant and

$$-\frac{d\phi}{dt} = k_3\phi\frac{dm}{dt} = k_3k_7\phi^2t^{-1/2}$$

Thus

$$d\left(\frac{1}{\phi}\right) = 2k_3k_7\,d(t^{+1/2})$$

giving

$$\frac{1}{\phi} = 2k_3k_7t^{1/2} + k_8$$

where k_8 is an integration constant, which must be equal to 1 if ϕ is to be 1 when $t = 0$, so that

$$\frac{1}{\phi} = 2k_3k_7t^{1/2} + 1$$

or

$$dm = \frac{k_7t^{-1/2}}{2k_3k_7t^{+1/2} + 1}\,dt = \frac{2k_7}{2k_3k_7t^{+1/2} + 1}\,d(t^{1/2})$$

giving

$$dm = \frac{2k_7}{2k_3k_7}\ln(2k_3k_7t^{1/2} + 1)$$
$$= k_9\log_{10}(at^{1/2} + 1)$$

which we may call the **mixed logarithmic equation**; it reduces to the direct logarithmic equation as one limiting case and the simple parabolic equation as another.

The direct logarithmic equation has, however, been found in cases where movement of non-metal is considered to proceed inwards into the metal, e.g. on abraded surfaces where there are paths of loosened structure. Here the explanation must be different. It may here be due to pressure closing up pores along which oxygen would otherwise pass inwards.[833]

The direct logarithmic equation was established by Vernon, Akeroyd and Stroud for zinc; behaviour depended on the purity of the zinc, but in general there was conformity throughout the range 250–400°C; Moore and Lee, using a different zinc, found logarithmic growth below 350°C and parabolic growth above 370°C.

On abraded iron, Vernon, Calnan, Clews and Nurse[836] found logarithmic growth up to 200°C, above which the growth-law became parabolic; this change-over from logarithmic to parabolic was confirmed by Eurof Davies, Evans and Agar on hydrogen-cleaned iron, but it occurred at a higher temperature slightly above 300°C.

When in 1955 the author arrived at the *mixed* logarithmic equation on theoretical grounds, no experimental case had been established. The following year, Mills,[837] studying copper in the author's laboratory, found a range of conditions where the measurements were inconsistent with the direct law but obeyed the mixed law. It may be mentioned that the mixed *parabolic* law had also been arrived at on theoretical grounds in 1924, although the first experimental demonstration was carried out by Fischbeck in 1933. The fact that theoretical arguments have *twice* predicted results *before* experiments were carried out is somewhat reassuring, as a sign that reasoning has on the whole been sound. It also suggests that the common belief that in metallurgy progress has always been made by accident or by empirical methods, the reasons being only arrived at long after the facts were known, is unduly pessimistic. On the other hand, all theoretical argument is speculative, and often the same equation can be arrived at in more than one way, which to some extent weakens arguments based on prediction.

General equation, with parabolic and inverse logarithmic equations as special cases

In discussing the parabolic law, it has been assumed that movement of ions under the field provided by the cell metal/oxygen has obeyed Ohm's law. There is really no justification for such an assumption, and indeed cases are known when the resistance is undoubtedly non-Ohmic. If, however, we start from the accepted fact that a cation will only pass from one point of stability on the lattice to a neighbouring point of stability if it possesses sufficient energy to surmount the intervening energy barrier, we arrive at a general equation with two limiting cases. This shows in a very simple manner why the parabolic law prevails at high temperature, with effective Ohmic resistance, whereas at low temperatures another equation (the inverse logarithmic) will produce a rapid slowing down of oxidation rate at very small thicknesses.

Consider passage across a given plane within a film parallel to the metal surface. In absence of a gradient, the number passing in either direction will be the same, being proportional to $e^{-W/kT}$, where W is the energy needed to jump a hurdle separating two adjacent sites of low energy, and k is Boltzmann's constant. If now, through the capture of electrons from the metal by adsorbed oxygen (giving oxygen ions) an electric potential difference, V, is established across the film, the energy needed for a jump in one direction is reduced to $(W - \psi)$, and for a jump in the other direction increased to $(W + \psi)$, where ψ depends on that part of V which falls over the inter-site distance at the plane under consideration, and is proportional to $1/y$.

The net movement, therefore, becomes proportional to

$$e^{(-W + \psi)/kT} - e^{(-W - \psi)/kT}$$

or to

$$K(e^{+X} - e^{-X})$$

where K represents $e^{-W/kT}$ and X represents ψ/kT.

The thickening equation is of the sinh form

$$dy/dt = K'(e^{X} - e^{-X}).$$

Expansion gives

$$dy/dt = K'[(1 + X + \tfrac{1}{2}X^2 + \tfrac{1}{6}X^3 + \ldots) - (1 - X + \tfrac{1}{2}X^2 - \tfrac{1}{6}X^3 + \ldots)].$$

If it is permissable to neglect the third and later terms, this becomes

$$dy/dt = 2K'X = K''/y$$

since X is proportional to ψ, and therefore to $1/y$.

Integration gives $y^2 = K''t + K'''$.

Apparently a parabolic equation should be obeyed under circumstances where the third and later terms can be neglected. This is permissible at high temperatures since $X = \psi/kT$, but not at low temperatures. It now becomes clear why parabolic growth is obtained at reasonably high temperatures, but not at low temperatures.

If the temperature is sufficiently low, the second exponential can be neglected, as being small in comparison with the first; the second exponential represents movement *against* the field set up by the cell metal/oxygen, and at the extremely small film-thickness which will persist at a low temperature, the field, which is proportionate to $1/y$, will be extremely strong; movement can then only occur in the direction where the field helps it and not in the opposite direction where the field opposes it. Thus rate of movement is proportional to $K'e^{x}$; it is shown elsewhere[824] that by an acceptable approximation, the growth-equation becomes

$$1/y - 1/y_0 = K \ln [a(t - t_0) + 1]$$

where y_0 is the thickness prevailing at time t_0; a will, of course, have the dimension $(\text{time})^{-1}$. This is known as the **inverse logarithmic equation**.

Thus theory predicts obedience to the parabolic equation at high temperatures and to the inverse logarithmic equation at low ones; the experimental facts agree with these predictions.

Hurlen (1976 vol., p. 41) has shown that over a broad range of conditions the general equation can approximate to a cubic equation $y^3 = kt$; conformity to this cubic equation has been reported by experimenters on several occasions. It is, for instance, obeyed at least approximately, during the oxidation of zirconium. It may appear surprising that obedience to the cubic law has not been observed consistently for all metals, but the fact is that the conditions where obedience might be expected are the conditions favourable to the formation of gaps between film and basis metal, which, as already shown, will lead to the direct logarithmic or the mixed logarithmic equation.

Obedience to the inverse logarithmic law at still lower temperature has been frequently observed. On iron, Gilroy's measurements (1976 vol., p. 18), continued for a year at ambient temperature, are consistent with the inverse law but not with the direct law. Hart (1960 vol., pp. 825, 831) found that the films which appear on aluminium at low temperatures obey the inverse logarithmic law, although in *humid* oxygen there is a period at the outset during which the direct logarithmic law seems to govern the situation. Godard has confirmed Hart's findings on aluminium exposed to air of between 52% and 100% r.h.; here again the early results are consistent with the direct logarithmic law, but later there is serious departure from it; between 1 day and 1000 days there is good agreement with the inverse logarithmic law. (H. P. Godard, *J. electrochem. Soc.*, 1967, **114**, 554.)

Berwick's work (p. 76) on the resistance of stainless steel to dilute sulphuric acid containing dissolved oxygen provides evidence that the protective oxide film thickens according to a logarithmic law. The same law was indicated by the results of Stern, using a different method. (M. Stern, *J. electrochem. Soc.*, 1959, **106**, 376.)

Lateral spreading of films

Studies of the early stages of oxidation suggest that, at least at the low pressures used in electron microscopy, oxide crystals first appear at isolated points and then spread laterally to meet one another. It is probable that a two-dimensional oxide (chemisorbed oxygen attached to the outermost layer of metallic atoms) is formed over the whole surface, but the formation of a small three-dimensional crystal requires nucleation energy; where once a crystallite has been formed, it will be easier to add to it than to start a new one. Beautiful electron micrographs of iron in the early stages of oxidation, produced by Bardolle and Bénard – also by Gulbransen, McMillan and Andrew – show embryo oxide crystals related to the crystal structure of the metal below, so that orientation varies from grain to grain.[837] The frequency of nucleation increases with oxygen pressure, so that at pressures higher than those used in the electron microscope, it is possible that reasonably continuous films of relatively uniform thickness are formed at an early stage – justifying the mathematical treatments generally applied. Curious outgrowths from the films, taking the form of whiskers or blades, may complicate the situation, but probably they add less to the mass of oxide than might be expected from their appearance.

The laws governing lateral growth of films of uniform thickness spreading in circles from nuclei are most easily obtained by the

author's statistical method[939], in several cases, alternative methods based on conventional mathematics are available; these are longer but lead to the same results. Since in fact the film does not spread in circles, the results have limited application to oxidation theory, and it is sufficient here to state the formulae obtained. There are two cases:
(1) If all the nuclei appear, distributed sporadically, at $t = o$, no nuclei being formed subsequently; the fraction of the surface remaining uncovered at time t is $e^{-\pi\omega\nu^2 t^2}$, where ν is the radial velocity of spreading (assumed constant), and ω is the **nucleation density** (defined by stating that the probability of a nucleus appearing in area element δA is $\omega\delta A$).
(2) If fresh nuclei continue to appear on the ever-shrinking uncovered area, distributed sporadically in space and time; here the fraction remaining uncovered is $e^{+\pi\Omega\nu^2 t^3/3}$ where Ω is the **nucleation rate** (defined by stating that the probability of a nucleus appearing within time element dt on area element δA is $\Omega\delta A$ dt).

The two expressions, of forms e^{-kt^2} and e^{-kt^3} respectively, both lead to sigmoid curves when the coverage is plotted against time; the rate of growth increases as the perimeter of the circles increase and falls off as the circles start to meet one another. This sigmoid form is to be expected in the more realistic case when the covered areas are not circular in shape. If the oxidation at this stage consists wholly in the spreading of a film of uniform thickness (thickening of the part already covered being neglected), the curve connecting oxygen uptake and time should be sigmoid also, and such sigmoid curves have been reported by several investigators.

Since oxidation does not, in practice, spread out as expanding circles giving a layer of uniform thickness, the equations based on such assumptions will rarely be obeyed. However, there chances to be one case when the theory can be tested (1976 vol., p. 33). Gray and Pryor, studying the oxidation of aluminium, found that around 525°C, two oxide layers are formed; the lower one connects cylinders of crystalline $\gamma-Al_2O_3$ spreading laterally from their points of origin and obeying the laws for expanding circles indicated above; apparently they arise from the inward movement of oxygen. It is covered with a film of amorphous Al_2O_3, thickening according to the parabolic law as a result of the outward diffusion of Al^{3+} ions.

Electrochemical measurement of wet corrosion today

Today, corrosion velocity is being measured by electrical methods based on a simple mathematic basis, which possesses certain analogies to the arguments just presented for the study of dry oxidation. The matter is discussed elsewhere[1023] but a brief review may be provided here. Simple electrochemical principles lead to a general expression consisting of the difference of two exponentials, representing movement in two opposite directions – helped or hindered by the electrical field. There are two limiting cases, one acceptable in situations where the later terms of the expansion of the exponentials can be neglected, and the other where the second exponential itself can be neglected. Application to the electrochemical study of wet corrosion shows that the relation between the current i and the polarization ΔV, should be such that at low values of i and ΔV a straight line should be obtained on plotting i against V, whilst at a later stage a straight line should be obtained on plotting log i against V – in accordance with Tafel's equation (p. 290). This leads to two different methods of obtaining a corrosion rate by electrical measurement.

It has sometimes been thought that the two methods require experiments on different specimens – which would be unfortunate, since two specimens, theoretically identical, often, in fact, behave differently. It is always a good principle, when comparing the results of different measuring methods, to carry them out on the same specimen; for instance, in Bannister's early measurement of silver iodide films (p. 2), the thickness was measured by three methods (gravimetric, electrometric and nephelometric) on the same specimen and good agreement was obtained. Mansfeld (1976 vol., p. 85) has shown that, in the present case, a careful choice of conditions allows both measurements to be made on a single specimen.

Heitz (1976 vol., pp. 82–4) has applied both methods to the attack of four organic acids upon iron. This was chosen because he considered it to be a case where there was a reasonable chance of finding a purely 'chemical' reaction (not divisible into anodic and cathodic components); if the attack was one which followed the pure chemical course, any electrical measurements would have given results bearing no relationship to the actual corrosion rate. In fact, however, his results, reproduced in Table 8.1, show clearly that the mechanism is electrochemical. The two methods produce figures agreeing well with one another and also with those obtained by chemical analysis. The table provides unusually clear evidence in support of the electrochemical mechanism of corrosion – an idea which was sometimes strongly denounced around 1925.

Table 8.1

	Calculated values of iron corrosion (mA/cm^2) obtained by		
Acid	Tafel method at high current value	Resistance polarization method at low current value	Observed value obtained by chemical analysis
Formic	0·31	0·34	0·35
Acetic	0·063	0·078	0·067
Propanoic	0·028	0·033	0·024
Butanoic	0·0008	0·0009	0·0005

Chemical thermodynamics

Pourbaix diagrams

A graphical method, due to Pourbaix,[900] has proved valuable in providing much pertinent information in concise form. Diagrams, with pH and potential as co-ordinates, present curves representing various chemical and electrochemical equilibria which should exist between metal and liquid, and such curves serve to delimit the conditions under which **immunity, corrosion** or **passivation** may reasonably be expected.

Each line represents a balanced reaction. A horizontal line represents an equilibrium involving electrons but not H^+ or OH^- ions; a vertical line represents one involving H^+ or OH^- but not electrons; a sloping line represents one involving H^+ or OH^- and *also* electrons.

For instance, in the diagram for iron (Fig. 8.1) the family of horizontals marked 19 show the potentials of the electrode equilibria

$$Fe \rightleftharpoons Fe^{2+} + 2e^-$$

at ferrous ion 'activities' equal to 10^0, 10^{-2}, 10^{-4}, 10^{-6} times normal according as the number attached is 0, -2, -4 or -6; they are spaced at intervals of 58 mV, owing to the operation of the principle

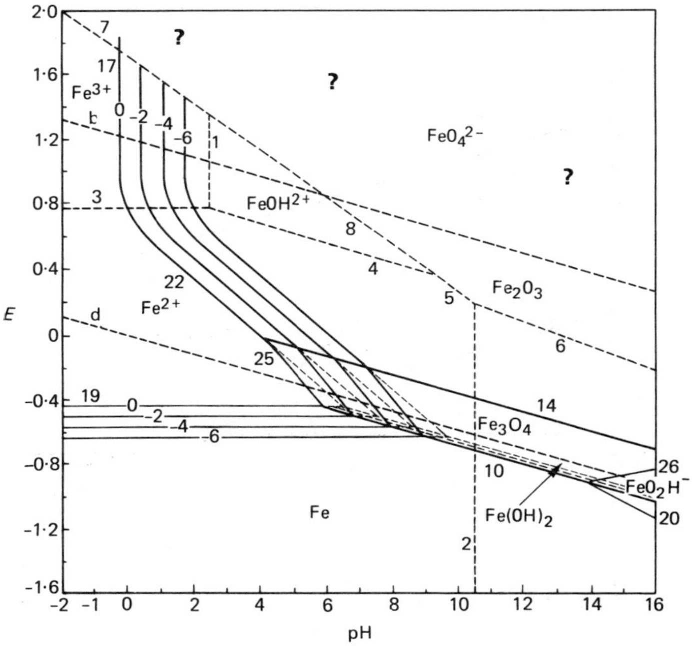

Fig. 8.1 Pourbaix diagram for iron at 25°C

discussed on p. 283. The verticals marked 17 represent the hydrolysis

$$Fe^{3+} + H_2O \rightleftharpoons Fe(OH)^{2+} + H^+,$$

whilst the family of sloping curves marked 22 represent the equilibria deciding the possibility of 'reductive dissolution' of solid ferric oxide to ferrous ions in the liquid – a matter discussed on p. 55.

$$2Fe^{2+} + 3H_2O \rightleftharpoons Fe_2O_3 + 6H^+ + 2e^-$$

Below one of the horizontals 19, corrosion is impossible when once the liquid has come to contain Fe^{2+} ions at the concentration appropriate to the line selected; if a higher concentration of Fe^{2+} existed metallic iron would be deposited; the area below the line can be regarded as the region of **immunity**. Conversely, at potentials raised above the line (whether produced by natural conditions or by application of an external e.m.f.), the change

$$Fe \rightarrow Fe^{2+} + 2e^-$$

will take place, and this area is the region of **corrosion.**

Above the sloping lines 22 the formation of a solid corrosion product becomes possible, but the formation of Fe^{2+} ions in the liquid also remains possible, so far as energy considerations are concerned.

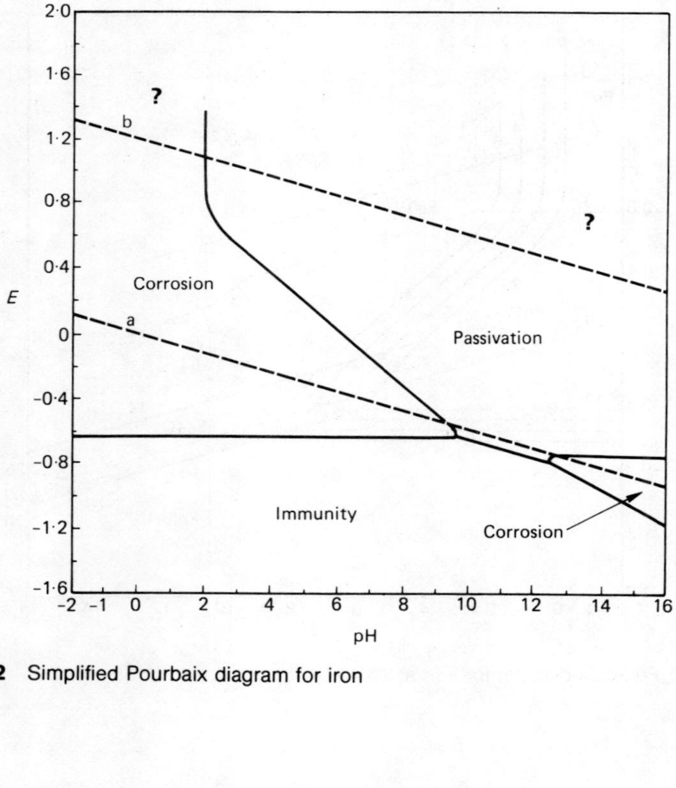

Fig. 8.2 Simplified Pourbaix diagram for iron

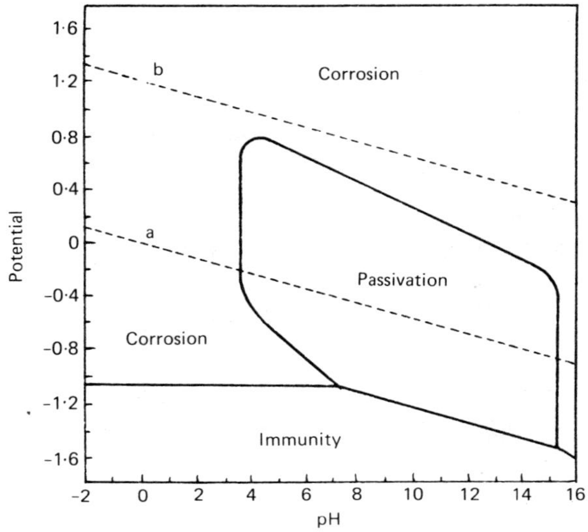

Fig. 8.3 Simplified Pourbaix diagram for chromium

However, as soon as a solid film has been produced, the entry of iron in solution is likely to be obstructed, and we may therefore consider this area the region of **passivation**.

A simplified form of diagram showing the regions of immunity, corrosion and passivation, without further details, is supplied in Fig. 8.2; the small 'corrosion' triangle on the right represents the formation of ferroates in a strongly alkaline solution. Similar simplified diagrams for chromium, zinc and copper are supplied in Figs. 8.3, 8.4 and 8.5. The area between lines marked b and a on Figs. 8.3, 8.4 and 8.5 represents the conditions under which water is stable. Above b it will be decomposed giving oxygen and below a it will give hydrogen.

Passivation and immunity

'Passivation' represents a type of protection essentially different from that established in the 'immunity' region, where corrosion is *impossible* on grounds of *energy*; in the passivation region corrosion is in general *unimportant* on grounds of *geometrical obstruction*; provided that the film-substance is physically suitable for protection, the corrosion rate will soon become negligibly slow. It would be possible to make the distinction clearer by replacing the labels 'immunity', 'corrosion' and 'passivation' by the words 'can't', 'does' and 'doesn't' respectively.

It should be emphasized that, whereas in the immunity region, the assurance against corrosion, being based on principles of thermodynamics, is absolute and independent of size, this is not true of

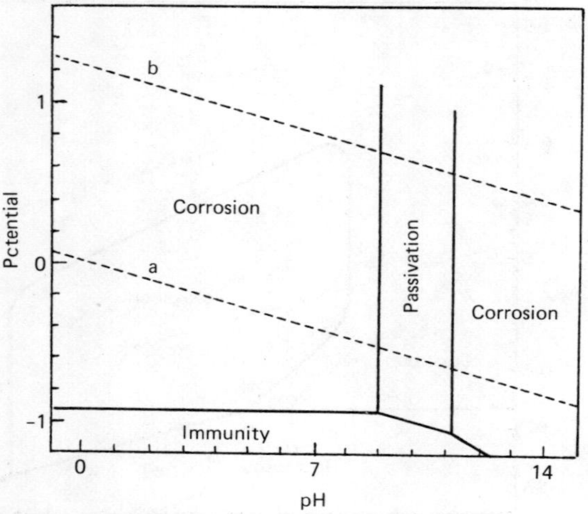

Fig. 8.4 Simplified Pourbaix diagram for zinc

the passivation region, where there is always a risk of some structural defect in a protective film, or a local reservoir of internal stress which will cause continued cracking. This risk may be small on a small area, but becomes greater as the area to be protected increases. Thus for protective schemes based on passivation, the size of the area under consideration must be taken into account; laboratory experiments on small specimens may fail to provide correct predictions of behaviour

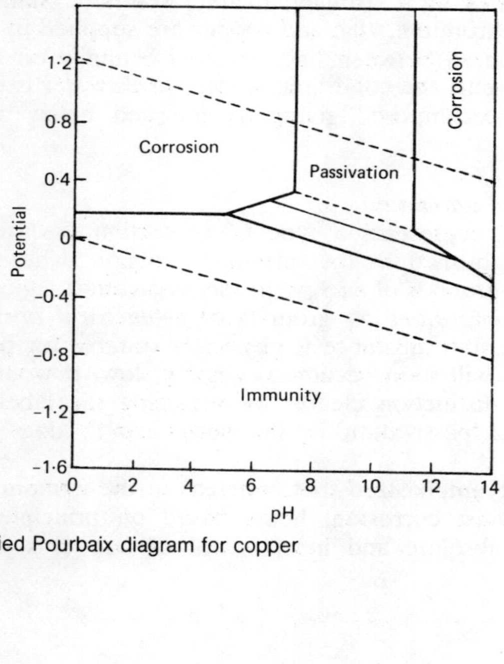

Fig. 8.5 Simplified Pourbaix diagram for copper

on the large areas involved in engineering practice. It becomes important to carry out probability experiments, and to consider the factors discussed in the short section of the following chapter devoted to the 'size effect'.

For a complete understanding of corrosion phenomena, the information provided by a Pourbaix diagram should be combined with kinetic and crystallographic data. The morphology of the various corrosion products is important in deciding their protective character, and the investigation of this subject in Feitknecht's laboratory deserves study. (W. Feitknecht, *Chem. Ind.*, 1959, 1102).

It is right to add the warning that exaggerated ideas of the possibilities of the graphical method have existed in some quarters – ideas for which Pourbaix himself is in no way responsible. Used in connection with other information, Pourbaix diagrams possess a value which can hardly be exaggerated. An atlas which includes the Pourbaix diagrams for all important systems has been published by Cebelcor (Centre Belge d'Étude de la Corrosion): M. Pourbaix, *Atlas of Electrochemical Equilibria in Aqueous Solutions*, translated into English by J. A. Franklin, Pergamon Press, 1966. This should be made available at every scientific library since each diagram represents a vast amount of information in a small space, which will be useful in many branches of physical chemistry and metallurgy. For every important metal, a diagram is provided, analogous to those of Figs. 8.1–8.5, showing the ranges under which corrosion, immunity and passivation may be expected. But in many cases there are other diagrams giving additional information such as the influence of pH on the solubility of various oxides and hydroxides, or on the composition of the solutions which will be in equilibrium with the metal at different values of pH and potential.

The diagrams in the atlas refer to a temperature of 25°C, but in some cases diagrams for higher temperatures are becoming available (1976 vol., p. 229).

9

Statistics and Size Effect

Foreword

Probability theory and statistics are today not universally taught at schools and colleges – which is to be regretted on general grounds; there is no branch of mathematics which has a closer bearing on the problems of everyday life. In the study of corrosion, some familiarity with these subjects is essential for understanding the irreproducibility of certain types of experiments, and the manner in which laboratory measurements may fail to predict large-scale behaviour. The distrust felt by engineers and plant chemists about laboratory results is on the whole justified, and the discrepancies arise from *two distinct causes*, connected with **statistical analysis** and **dimensional analysis** respectively; they are indicated – all too briefly – in this chapter. The treatment provided will not make the reader a statistical or dimensional expert, but it may perhaps help him in preparing for the discussion of problems with the expert who is today usually to be found in a large industrial concern.

Reproducibility and scatter

Effect of borderline conditions

It is well known that frequently two duplicate corrosion experiments do not lead to the same result. Discrepancy between duplicates is particularly marked when corrosive solutions have been treated with just enough *deterrent* to bring them to the borderline of passivity. Near the critical content of deterrent, it is found that some of the specimens remain quite unchanged, whereas others suffer intense local attack. A little thought will indicate that this lack of reproducibility is inevitable. When no deterrent is present, corrosion will start from a large number of centres, but as the deterrent concentration is gradually increased the average number of starting points per cm^2 will steadily diminish. Suppose that the specimens each measure 1 cm^2, and suppose that the amount of deterrent added is adjusted so as to make the *average* frequency of starting points one in 2 cm^2. It is clear that 'perfect reproducibility' would then demand half a starting point per specimen; this is meaningless, and the nearest approach to reproducibility would be obtained with one starting point on half the specimens and none on the rest; in actual practice a few specimens would be found to contain more than one starting point, but, whatever the distribution, it is certain that the specimens *cannot* all behave alike.

Bernoulli's principle

If the average number of starting points be N per unit area, and the area of a specimen be A, the average number per specimen is AN; this is known as the **expectation**. Even on a big specimen, a count would rarely show exactly AN; some specimens would carry slightly more, others rather fewer. The bigger the specimen, the larger will be the *absolute* deviation from AN commonly met with, but the deviations expressed as *fractions* of the mean value (AN) will become steadily smaller as AN increases, so that the proportionate disparity between duplicate experiments will tend to lessen as the numbers involved become larger.

This is simply an example of a general principle developed towards the end of his life by Daniel Bernoulli (1700–82) – a member of the famous Swiss family of mathematicians. Bernoulli's principle explains why some changes are reproducible and others not. In ordinary chemical reactions, we are dealing with vast numbers of molecules, and if the same experiment is carried out a number of times, the same result – within the limits of observation – will always be obtained. Corrosion changes in an ordinary salt solution, which begin at a large number of points per specimen, and soon spread out until a large part of the surface is suffering fairly uniform attack, are reproducible for the same reason; excellent agreement between duplicate experiments is obtained, provided that the experiments are carried out carefully, and under uniform conditions. It is only when a deterrent is added in amount sufficient to reduce the number of starting points per specimen to a small number that reproducibility must become poor.

Distributions

A large number of factors may cause the quantitative results of a number of duplicated experiments to be *distributed* about some mean value. One cause of variability is found in the fact that generally metallic materials are not uniform; two specimens cut from the same sheet differ physically or chemically from one another, especially where segregation has occurred in the ingot used in the rolling of that sheet. But even if the material were entirely uniform, error of measurement would still cause a certain 'scatter' among the values recorded.

In considering the type of distribution, it is necessary to ask whether continuous or discontinuous variation can take place. If all values between $-\infty$ and $+\infty$ are theoretically possible, including non-integral values, the **normal distribution** (see p. 256) is commonly met with; even where negative values would be either impossible or meaningless, the normal law may still express with sufficient accuracy the scatter of observational values immediately around the mean value. If only positive, integral values are possible (e.g. in counting the number of pits on a specimen, where 'half-pits' and 'minus-pits' would be meaningless), **Poisson's distribution** (see p. 259) is frequently encountered, provided that events are mutually independent; here the lower limit is zero, but there is no finite upper limit to the values obtainable. A third type, the **binomial distribution**, which defines the frequency of occurrence of integral values between zero and a finite upper limit, possesses historical interest to corrosion students, since Tammann based on it a statistical treatment of parting limits (p. 126). It is accorded prominence in books on statistics, perhaps mainly because the normal and Poisson distributions can be derived from it as limiting cases.*

Binomial distribution

If we carry out the kind of experiment which leads to one of two results, which we may call 'success' and 'failure' respectively, p being the chance of success and q of failure (so that $p + q = 1$), and if the experiment is repeated m times, the chance of obtaining exactly n failures will be given by the $(n + 1)$th term of the binomial expansion of $(p + q)^m$. For instance, if the experiment is repeated three times, the expansion becomes

$$p^3 + 3p^2q + 3pq^2 + q^3$$

so that the chance of no failures is p^3, that of exactly one failure is $3p^2q$, of two failures $3pq^2$ and of three failures q^3. If m be made very large, so that practically continuous variation becomes possible, and if $p = q = \frac{1}{2}$, a binomial distribution approximates in the central part of the series to normal distribution. If m is made very large, but p is very small (so that the 'expected' number of successes, mp, is still finite), it approximates to the Poisson distribution.

Distribution curves

Often, a knowledge of the mean value of a large number of experiments showing a wide scatter is of much smaller practical importance than a distribution curve showing the relative frequency of different values, since such a curve makes it possible to ascertain the probability that a given value will be exceeded. For instance, the average value for the depth of the deepest pit found on each of 1000 pipe sections is of little interest; what is of vital importance is to

*Actually, in a famous treatise discussing the probability of a miscarriage of justice, published in 1837, Poisson arrived at his distribution as an approximation to the binomial distribution under special conditions; but it is unfortunate that statistical writers should so often have adopted the historical derivation since, in the words of T. C. Fry 'there is another point of view which is of much greater practical importance'; Fry's book, '*Probability and its engineering uses*' (Macmillan) deserves special attention.

know on what proportion of the sections the pitting has exceeded a depth equal to the thickness of the pipe. A distribution curve can show this. Again, it may be desired to know the proportion of articles subjected to corrosion fatigue which will have lives shorter than a certain duration; a distribution curve can provide that sort of information.

The Histogram method of representing distribution often adopted in statistical textbooks is of limited value in practical corrosion problems. In that method the articles would be assigned to arbitrary classes according to their lives. Thus all those with lives between 0 and 1 year would form the first class (with mean life 0·5 year), all those with lives between 1 and 2 years the second class (with mean life 1·5 year), and so on; the number in each class would then be plotted against the mean life.† Now it is obvious that this method does not make full use of the data, since it is arbitrarily assumed (for instance) that *all* the articles having lives between 1 and 2 years survive for exactly 1·5 year, whereas really their lives are distributed. Furthermore, where the total number of articles is fairly small (less than, say, 100) it is necessary either to make the number of classes small, or to be content with a small number falling within each class; clearly the points cannot be expected to fall well on a curve, and where the total number of articles is still smaller (say, 20), the method becomes unusable.

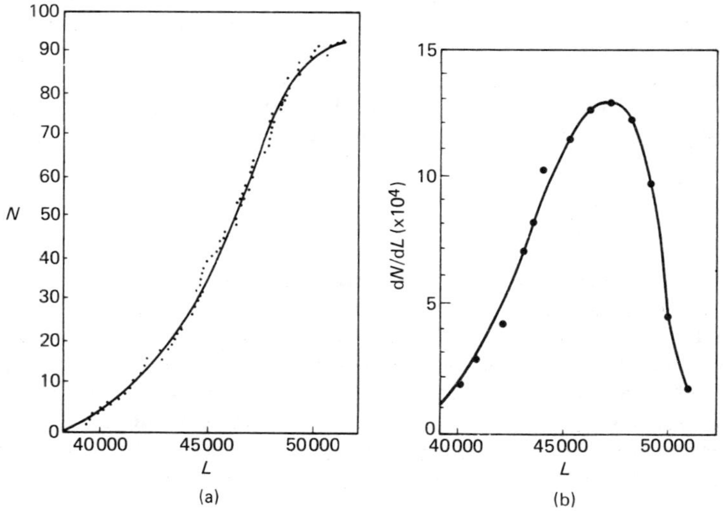

Fig. 9.1 Improved method of obtaining frequency curve

†It is customary to erect rectangular blocks of height representing the number in each class and spreading laterally over the period involved.

In the author's opinion it is far better to plot, against the life L, the number of articles N with lives shorter than L. The number of points available for the curve is then equal to the total number of articles, whereas in the histogram method it is only equal to the number of classes. Although the points may not fall *exactly* on a curve, the general shape of the graph will not be in doubt even if the number of articles is comparatively small; it will usually be a curve of the form shown in Fig. 9.1 (a) known as an **ogive**. If it is desired to obtain a curve showing the relative frequency of different lives, this is obtained by differentiation; the plotting of $\mathrm{d}N/\mathrm{d}L$ against L, using the data of Fig. 9.1 (a), gives the frequency curve shown in Fig. 9.1 (b).

Experience has shown that this method could provide information in an industrial problem where the histogram method had proved useless. Nevertheless there are occasions when someone who wishes to publish his experimental data may be wise to adopt the histogram form of expression, since it has found favour in the office and is understood also by the man in the street.

Normal distribution

If we imagine a vast number of small animals deposited suddenly at a spot on an infinitely long north–south road, and allowed to roam to and fro as they please (keeping to the road), it being supposed that at any moment a given animal is equally likely to move north or south, they will distribute themselves around their starting-point according to the normal law. Similarly, in a scientific research, errors of observation will usually cause the observations to group themselves symmetrically, roughly according to the same (normal) law, around the true value as mean – provided that there is no systematic error. This distribution is commonly defined by the statement that the chance of obtaining a deviation from the mean lying between δ and $\delta + \mathrm{d}\delta$ is

$$\frac{\mathrm{e}^{-\delta^2/2\sigma^2}}{\sigma\sqrt{(2\pi)}}\,\mathrm{d}\delta$$

where σ^2 represents the **variance**, defined as the *mean* value of δ^2 for numerous observations. By converting the law to standardized form, that is by expressing the actual deviation of a particular observation (δ) as a fraction of the **standard deviation** (σ), which is the square root of the variance, a *single* equation can be made to represent *all* normal distributions – irrespective of the absolute value of σ. Thus if τ is written for δ/σ, the distribution can be concisely defined by stating that the chance of obtaining a value between τ and $\tau + \mathrm{d}\tau$ is

$$\frac{\mathrm{e}^{-\tau^2/2}}{\sqrt{(2\pi)}}\,\mathrm{d}\tau$$

The probability of obtaining a deviation in a given range expressed in this standardized manner is always the same. Thus the chance of the

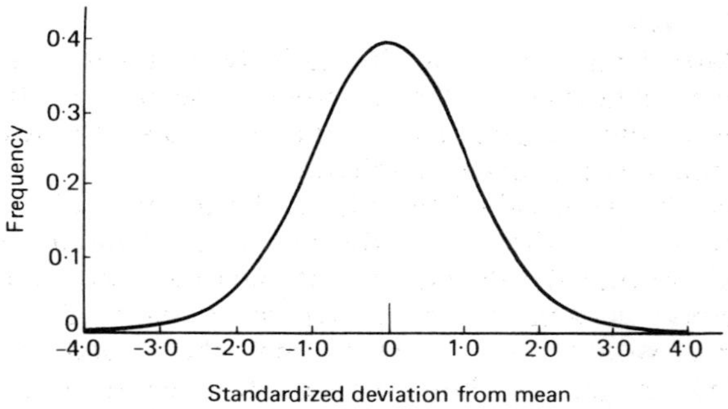

Fig. 9.2 Normal distribution curve

deviation exceeding σ is 1 in 3, that of exceeding 2σ is 1 in 20, whilst that of exceeding 3σ is only 1 in 370; there is a roughly equal chance of obtaining deviations greater or less than $\frac{2}{3}\sigma$.

The normal distribution, expressed as a frequency curve, is shown in Fig. 9.2 whilst the same data, expressed as a cumulative or ogive, is presented in Fig. 9.3. Here the ordinate represents

$$\frac{1}{\sqrt{(2\pi)}}\int_{-\infty}^{\tau_s} e^{-\tau^2/2}\, d\tau$$

giving the proportion of observations which show deviations (algebraically) lower than τ_s. More precise values of this integral, which provides information regarding the chance of exceeding any stated value of δ/σ, are recorded in the tables found in most books on statistics. Some notes on the use of such tables are provided on p. 259.

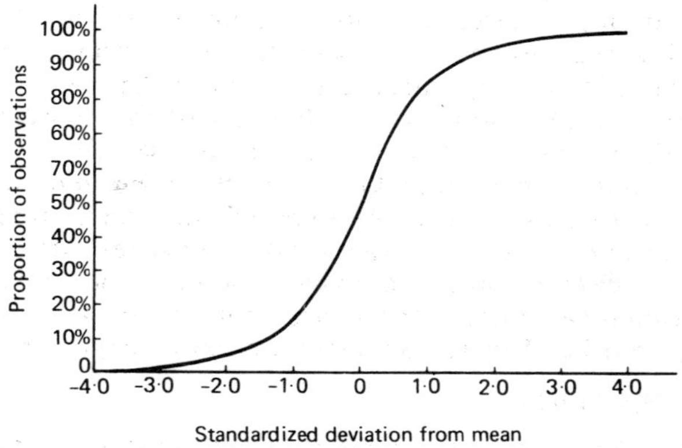

Fig. 9.3 Normal distribution expressed as a cumulative or ogive

Measures of scatter

The standard deviation, σ, or alternatively its square, the *variance*, is a convenient indication of the wideness of scatter of observations; the variance is the simpler conception and has the advantage of being, in certain circumstances, additive.

If we knew the 'true value' (which, in the absence of 'bias' or systematic error, may be defined as the mean of an infinite number of observations), so that the deviation δ of each measurement from that true value could be written down, then the variance derived from n measurements would be $(1/n)\Sigma\delta^2$ and the standard deviation $\sigma = \sqrt{(\Sigma\delta^2/n)}$. In general, however, the true value of the mean is unknown, and the mean of the limited number of measurements performed must be taken as the best available estimate of it; in such a case, the *best estimate* of the standard deviation can be shown to be $\sqrt{[\Sigma\epsilon^2/(n-1)]}$ where ϵ is measured from the mean; this estimated standard deviation is usually written s.

It is common knowledge that, although the mean of a limited number of experiments generally differs from the true value, as defined above, the error involved tends to diminish as the number of experiments performed becomes larger. If a large number of sets, each consisting of n experiments, were performed, the means of the sets would themselves be spread out around a 'master mean'; but they will be clustered more closely than the results within any one set. The standard deviation of the means of the sets (from the master mean) is known as the **standard error** (S_n); it can be shown that the standard error is only $1/\sqrt{n}$ of the standard deviation of the individual measurements, so that

$$S_n = s/\sqrt{n} = \sqrt{[\Sigma\delta^2/n(n-1)]}$$

which for large values of n is roughly equal to $\sqrt{[(\Sigma\delta^2)/n]}$.

Clearly the *absolute* error (Δ) of the mean of a set of measurements tells us little; it will look small if small quantities are being measured, and large if large quantities are under review, although it does not follow that less care has been taken to ensure accuracy in the second case. It is more satisfactory to write the error in a standardized form, by expressing it as a fraction of the standard error. The *standardized error* (as it may conveniently be called) is usually denoted by t and is sometimes known as the ***t*-function**. It is non-dimensional, so that its value is the same whether measurements are made in pounds and feet, or in grams and centimetres. Moreover, for a given type of distribution and a given value of n, the probability of reaching a certain value of t is the same, whatever the *absolute* scatter of the results may be. This valuable statistic can be expressed

$$t = \Delta/S_n = (\Delta\sqrt{n})/s$$

where Δ is the deviation of the mean of a particular set of n from the true value. G. F. Peaker adds the useful comment: 'Unless the number of observations is very small (4 or fewer) the standardized error will rarely exceed 2, whilst standardized errors exceeding 3 are very rare indeed. If an observed mean differs from the value of a mean expected on some hypothesis by a large standardized error, this constitutes a sound reason for suspecting the hypothesis, or alternatively for believing that there are systematic, as opposed to random, errors affecting the observation.' Values of the integral representing the chance of exceeding any stated value of t are recorded in tables found in books on statistics. For large values of n the t-distribution becomes normal, and tables of the normal integral would serve, but this will rarely be the case in corrosion problems. Different books present the information in various ways.

Poisson distribution

On p. 252 it was pointed out that on a number of specimens subjected to a corrosive liquid containing a deterrent, the number of starting points of attack may vary slightly from specimen to specimen. If we define the 'expectation' (E) as the average number per specimen obtained after counts on an infinite number of specimens, then under ideal conditions the probability of obtaining n points of attack on any specimen would be given by *Poisson's formula*

$$e^{-E}E^n/n!$$

The conditions for validity demand that the occurrence of one point of attack shall not influence the chance of the inception of attack at points around it. Often these conditions are not fulfilled. In some cases, the attack at one point may provide cathodic protection to potentially sensitive points in the surrounding region, diminishing the probability that attack will start. In other cases, corrosion product from the corroding points may infect points which otherwise would be just insufficiently sensitive to develop attack, by screening them from deterrent and thus causing them to start corroding. Either **protection** or **infection** will cause a departure from the predictions of Poisson's expression. It is not surprising that Mears and Brown[935] have found that the distribution of pits on aluminium is not that predicted by the Poisson formula.

Use of tables of probability integrals

Tables are available providing the numerical value of integrals such as are discussed on p. 257. For some problems, their practical use is obvious. If we are interested in pit-depths in the walls of a pipe, the important question is the risk of exceeding a certain depth, equal to the thickness of the wall, since this represents the risk of perforation; the integrals provide a measure of this risk. However, the tables are often used for significance problems, and here an imaginary example may help to clarify the issue.

A firm has for a long time been purchasing material for some industrial purpose and has been measuring the corrosion velocity of a number of samples taken from each batch as it arrives; the measurements recorded have enabled a distribution curve for the velocity values to be plotted – which may be approximately normal. On one occasion, the samples from a newly arrived batch show abnormally high corrosion velocities – surpassing anything met with in the past. The question arises: does this indicate that the material is below the standard hitherto supplied, so that a letter of complaint must be written? This is not immediately obvious, since the distribution curve hanging on the office wall (constructed on the basis of previous measurements) will probably have been drawn asymptotic to the horizontal axis (which theoretically it is supposed to touch only at $+\infty$); such a form of curve recognizes that, *on rare occasions*, the random samples may show abnormally high values, even though they are drawn from a batch equal in quality to that supplied before.

The table of normal integrals will show the probability (P) of *exceeding* the figure obtained in the tests on the samples. (Where the information available in the firm's records is limited, it may be necessary to use the t-integral.) Supposing that probability to be 0·01, then the argument put forward would be: 'If the material was up to the standard hitherto maintained, the odds are 99 to 1 against us getting these figures in the laboratory. It looks as though the material was below standard. Let us write that letter of complaint.'

This argument appears logical, but the question is at once asked: what is the value of P above which a letter of complaint ceases to be justified? Here the instructions of the books appear somewhat arbitrary. Weatherburn states, 'If P is less than 0·05 we regard our value of t as significant. If it is less than 0·01 we regard it as highly significant,' (C. E. Weatherburn. *A First Course in Mathematical Statistics,* Cambridge University Press, 1946.)

In practice, the decision is never entirely objective. The cautious man will demand a lower value of P before writing his letter than the firebrand. But in any case the choice will rightly be influenced by considerations outside those provided by the measurements – the reputation of the suppliers of the materials, for instance. Clearly some information about the reliability of the supplying firm is inherent in the statistics already accumulated, but there will usually be much from outside sources which cannot be expressed in figures, and yet which it would be folly to disregard. Statistics give no dispensation from the use of common sense. The value of statistical treatment is that it may help a decision in cases where common sense by itself would be powerless to decide.

Size effect

Statistical causes of discrepancy between results of laboratory tests and large-scale experience

The Pourbaix diagrams discussed on pp. 247–250 show two regions representing conditions where corrosion will usually not occur. In the **immunity region** corrosion cannot occur because it would represent an increase in the energy of the system; this is something which is as true for a large surface as for a small one; there is no size effect here (unless we are considering attack on very small particles, where the high curvature introduces factors connected with interfacial energy). In the **passivation region** a corrosion reaction will not cause an increase in energy, and may actually start, but will stifle itself by building a protective film. Here there may be a small chance that at an exceptional point, some structural defect may exist, causing the protection to fail; this might, for instance, be a place where a hard particle has settled on the metallic surface and has been rolled in, producing local stress capable of causing a protective film to break as quickly as it is formed. Whatever the cause of the failure of protection at certain spots, it is clear that the chance of a sensitive spot existing will increase with the area under consideration. Here, clearly, a size-effect exists, and small-scale experiments may provide a totally false prediction of results likely to be met with in engineering or industrial practice.

Consider a case where the true number of sensitive spots is one per dm^2, and a laboratory test is carried out on a specimen of area only 1 cm^2; the odds are 99 to 1 against a sensitive spot occurring on a given specimen; even if the tests are carried out, say, in triplicate, the result will probably indicate that corrosion need not be feared. Yet, in a plant exposing an area of 1 m^2, it is practically certain that a sensitive spot will be present somewhere, and corrosion must be expected. *Thus laboratory predictions will be completely misleading*.

The situation is most threatening where an inhibitor of the dangerous class (p. 196) is being added to prevent attack. Supposing that a water containing a small amount of chloride is being used for some industrial purpose. It may be decided to add chromate to prevent attack; if sufficient is added, there will be no corrosion; if the addition is insufficient, corrosion, concentrated on a few sensitive points, may actually produce thinning locally more severe than if no addition at all had been made.

Clearly it is necessary to measure the number of points per unit area where protection will fail under any particular liquid. This measurement cannot be made by counting the number of points where attack starts on a large specimen (even if work on full-size specimens was possible in the laboratory), since corrosion starting at one point diminishes the probability of it starting at neighbouring points; (the effect, sometimes styled 'cathodic protection', is better described as 'mutual protection' since the absence of corrosion at a point does not require a cathodic reaction but merely the avoidance of an anodic reaction). The mutual protective effect is well illustrated by Britton's observations on a steel which in fairly concentrated NaCl solution suffered attack only at the cut edges, where the internal stresses left by the shearing would keep cracking any protective film[114]. It might be thought that, if the edges were protected with varnish, corrosion would be prevented altogether; but when this was tried, it was found that corrosion now started at numerous points on the face which, on an unvarnished specimen, would have remained unattacked, since the anodic attack on the edges would shift the potential on the face in a direction unfavourable to the inception of attack.

What is needed is to perform a large set of experiments on specimens so small that the chance of corrosion starting on a single specimen is not great, and the chance of two starting-points on a single specimen is negligibly small. It is possible to conduct all these experiments on a single large specimen by the 'square drop method', (1960 vol., p. 936) and yet maintain independence; a metal specimen is painted with two sets of wax lines at right angles, dividing the surface into squares, on which drops can be placed; clearly, attack in one drop will not affect the prospect of attack starting in its neighbour. However, this procedure is open to criticism on the grounds that conditions over the area covered by any one drop are not uniform; in a salt solution, there will be anodic attack in the centre, whilst the edges, being cathodic, will usually escape attack. In any case, the experimental areas are too small for the problem now under consideration.

Since a very large number of experiments at each chromate concentration are needed, a simple type of containing vessel, obtainable in multiple, must be chosen. Possibly corked boiling tubes would serve; if completely filled with liquid, conditions should be reasonably uniform, and each tube could contain a specimen of area about $12{\cdot}0 \times 1{\cdot}8$ cm. If 10 values of the CrO_4^{2-} concentration are to be tested, and 100 experiments are to be performed at each of these, 1000 tubes will be needed. In the set of 100 tubes, the number developing rust would be recorded. Only when this number falls in or close to the range between 10 and 20 will the number possess accurate meaning; it will, therefore, be needed to perform rough, exploratory experiments to decide the range of CrO_4^{2-} concentration likely to produce results within that range, and then to perform accurate experiments on ten chosen CrO_4^{2-} concentrations. The reason for inaccuracy in observations below 10 is of course that measurements can only be integers; no reading between 9·0 and 10·0 is possible, so that at 10 a 10% error must be expected. Above about 20 a similar error will arise due to the fact that some of the specimens 'counted' as having one starting-point will really contain two starting-points (or would do so if the mutual protection effect did not influence results).

In some such way, it might be possible to establish an empirical equation defining the relationship between the CrO_4^{2-} concentration and the number of starting-points per unit area. The practical problem concerns a very much larger area than is represented by the specimens in the boiling tube, and the safety requirements will demand a very much lower probability of attack than that represented by the development of corrosion in 10 tubes out of 100. If, however, it can be assumed that the equation established by the boiling tube experiments would remain valid for the practical conditions, it will be easy to calculate the CrO_4^{2-} concentration needed to obtain whatever degree of security may be demanded. Clearly this involves wild extrapolation – which is always unsatisfactory. But it would seem that the order of magnitude of the concentration needed could be arrived at, and by allowing an ample factor of safety reasonable security should be possible. Without experiments carried out on a statistical basis, any decision about the concentration required must be regarded as guess-work.

Unfortunately, in recent years, very little study has been made of corrosion probability. Work, rather on the lines suggested above, would seem to be urgently required. Quite likely, the opening experiments will show the need for a considerable departure from the method outlined, but that makes it all the more important that such work should be started without delay. Possibly a reader of this book – someone without much previous knowledge of corrosion science – may have ideas about a sensible procedure; if, at the outset, he knows little of the subject, that will at least mean that he is free from the fixed ideas held by the corrosion specialist; there are occasions when initial ignorance may be a positive advantage.

Dimensional causes of discrepancy

Hoar and Agar[817], from the point of view of dimensional analysis, have discussed anomalous results likely to arise if experiments on small-scale models are used to predict the behaviour of large-scale electrochemical systems. It may be sufficient here to present a single case which can be solved – so far as solution is possible at all – by inspection. (An excellent account of dimensional analysis is provided by P. W. Bridgman in *Dimensional Analysis*, Yale University Press, 1931.)

Consider a large rectangular electrochemical cell with electrodes forming two opposite vertical walls; current passes from one to the other through a liquid, producing attack upon the anode. Can we use a similarly shaped cell with all dimensions made α times that of the large cell ($\alpha < 1$) to predict behaviour? Obviously the intensity of corrosion (corrosion per unit area in unit time) will only be the same for the large and small cell if the current density is the same; indeed, if the current density is not kept the same, behaviour may be radically different – one cell developing anodic passivity and not the other. Thus for faithful reproduction, the current density must be kept the same.

If the height and length were reduced α times, but the width (the distance between the electrodes) was not reduced, the current density would be preserved unchanged; this, however, would alter the shape of the cell, which may be undesirable for other reasons. If the width is reduced α times, the resistance per unit area of cross-section falls, and the current density is no longer preserved; experimental results are then likely to be completely misleading.

The answer to the problem suggested by Hoar and Agar is to reduce the conductivity of the liquid α times. This result can be obtained by reducing the concentration – generally to *about* α times the original. Provided that the only effect of reducing the concentration is to reduce the conductivity, and that the size reduction has no other influence than that considered, this represents the perfect answer. Unfortunately, the alteration in concentration may affect oxygen-solubility, which in turn may affect polarization at the cathode. Quite apart from changes in the liquid, the mere alteration in size may affect the rate of supply of oxygen to the cathode by convection or diffusion, and also the rate of removal of corrosion products from the anode. Thus, in practice, the problem is not solved completely by reducing the conductivity, but certainly the model becomes a more faithful representation of the full-scale plant if the conductivity is reduced than if it is left unchanged.

Final remarks

The object of these short paragraphs suggesting the possible misleading character of small-scale experiments (due to two distinct causes – statistical and dimensional), is not to provide a complete answer; in some cases, probably, no complete answer exists, whilst in others a long discussion would be needed to arrive at it. Rather, an endeavour has been made to indicate the character of the problem, and the sort of lines on which, in any particular case, conditions may be sought which at least would minimize the discrepancy. Unless the factors making for discrepancy are kept in mind during the planning of small-scale experiments, such experiments may be not only unhelpful but positively dangerous.

Historical Note

The chronological sequence of scientific discovery is rarely the logical one. To arrange the facts of metallic corrosion historically would conceal the true interconnection existing between them, and thus deprive them of significance. Nevertheless, in view of the prevailing interest in the history of science, many readers may welcome a short narrative showing how knowledge of the subject discussed in this book has grown. The note which follows should serve to indicate some names and dates associated with the advance of understanding, but it must be remembered that the credit for any particular discovery cannot be assigned to a single year or to a single person. If a recent investigator is cited as the discoverer, objection may fairly be raised by the quotation from older papers of passages which seem to contain the germ of the idea; yet to allot the entire credit to early investigators may be unjust to later ones, who have established as facts what had previously been mere suggestions.

At the dawn of history, the first metals to be used were those which were either found native, or could easily be reduced to the elementary state; such metals do not readily pass into the combined state, and their corrosion can have raised no serious problems. But with the introduction of iron, the problem of its corrosion must have presented itself, although it is an undoubted fact that some of the iron produced in antiquity is today more free from corrosion than much of that manufactured in later years. This may have been due partly to the fact that iron reduced with charcoal contained less sulphur than modern steel, but it may also be connected with the absence of sulphur compounds from the air in the days before coal was adopted as a fuel; for it is often the conditions of early exposure which determine the life of metalwork. Whatever the cause, ancient iron has in some cases remained in surprisingly good condition for many centuries; the Delhi Pillar is the example which has excited most interest, but others could be quoted.

For many centuries there seems to have been little curiosity regarding the causes of corrosion, although a few significant observations were made. As early as 1788, Austin noticed that water, originally neutral, tends to become alkaline when it acts on iron. He attributed the alkalinity to the compound now called ammonia; this was probably an error, since the alkaline reaction produced by most saline waters is due to sodium hydroxide, the cathodic product of the electrochemical corrosion process.

The belief that corrosion is an electrochemical phenomenon was expressed in a paper published in 1819 by an anonymous French writer, thought to be Thénard, and in 1830 his compatriot, de la Rive, attributed the fact that acid attacks impure zinc more rapidly than the relatively pure varieties to an electric effect set up between zinc and the impurities present. Faraday's researches, especially those conducted between 1834 and 1840, afforded evidence of the essential connection between chemical action and the generation of electric currents; indeed (since his laws of electrochemical action apply as much to anodic as to cathodic processes) he provided a quantitative basis for the observations of later investigators. One of Faraday's many interests was 'passivity' – the subject of a famous correspondence in 1836 with his Swiss friend, Schönbein. In 1790, Keir had discovered that iron, when placed in concentrated nitric acid, became 'altered' in properties; it now resisted relatively dilute nitric acid, although the dilute acid vigorously attacked iron which had not been treated in concentrated acid. It was this alteration in properties, persisting after the treatment was over, which intrigued Faraday and Schönbein; it seemed to merit a special name, and Schönbein suggested that iron in this strange unreactive condition should be called 'passive iron'. Unfortunately, the word 'passivity' came to be used later in other senses; one authority has described metal as 'passive' when dissolving freely, but in an abnormally high valency condition. The use of the same word in several different senses seems at times to have led to apparent disagreement between people whose ideas were not essentially very different.

Later, interest in the electrochemical mechanism of corrosion seems to have waned. Attention may have been diverted by alternative suggestions ascribing corrosion to the presence or formation of certain substances. Some of these suggestions are now seen to possess a modicum of truth, but the underlying ideas are themselves consistent with an electrochemical mechanism.

Between 1888 and 1908 the view was frequently advanced that acids were the agents mainly responsible for corrosion; particularly it was held that the rusting of iron would only take place if carbonic acid was present. It was shown, however, by Dunstan, Jowett and Goulding in 1905, by Tilden in 1908, and by Heyn and Bauer in the same year, that iron exposed to water and oxygen, with exclusion of carbon dioxide, underwent rusting. Acid is not needed for the corrosion of iron, and indeed some of the most intense forms of attack take place in salt solutions which have been made slightly alkaline. There is, however, one important type of corrosion – that associated with the liberation of hydrogen – which occurs much more rapidly in acid than in neutral water; in atmospheric corrosion also, acid impurities in the air greatly stimulate attack upon iron, although they are not necessary for it. In such circumstances, therefore, the view that acid is an arch-promoter of corrosion may be held to be justified, but in atmospheric attack – as shown by Vernon – it is not carbonic but sulphurous acid which is mainly responsible; indeed Vernon (1935) demonstrated that under certain circumstances carbon dioxide can retard the attack upon iron.

Another theory of corrosion arose about 1900 as a result of the detection of hydrogen peroxide during the corrosion of many metals; this gave rise to the idea that it acted as intermediary in corrosion processes. It is now known that hydrogen peroxide may be formed in electrolytic cells where oxygen has access to the cathode. Consequently, if the electrochemical mechanism of corrosion is correct, the presence of traces of hydrogen peroxide among the corrosion products requires no further explanation.

Whilst British chemists were disputing the claims of carbonic acid and hydrogen peroxide to be regarded as the agents responsible for corrosion the electrochemical standpoint was being developed in Sweden and elsewhere, especially in Arrhenius' laboratory by Ericson-Aurén and Palmaer, working separately and in collaboration from 1901 onwards. Their work dealt mainly with corrosion by those acids which act on metals eliminating hydrogen, and the results provided strong evidence that this attack was connected with the formation of microscopic cells, which they called 'local elements'. Somewhat similar views were expressed by the American investigators Whitney (1903) and Cushman (1907) in their discussion of corrosion by neutral liquids.

The electrochemical views put forward by these writers overlooked, to some extent, the part played by oxygen, and broader theories of corrosion by neutral liquids, which recognized the function of oxygen as cathodic stimulator, were provided by the Americans Walker, Cederholm and Bent in 1907, and the Englishman Tilden in 1908. Still later research, however, rather represented a return to earlier views, by indicating that oxygen is not absolutely necessary for the attack of water on iron. Experiments by the American investigators, Corey and Finnegan (1939) and de Kay Thompson (1940), suggested that iron is slowly attacked by oxygen-free water, with liberation of hydrogen, although in this case the final product is not yellow rust (ferric hydroxide) but black granular magnetite or sometimes ferrous hydroxide. The fact that at the boiling point pure water, or even dilute alkali, could attack iron with liberation of hydrogen, had been clearly demonstrated by the Germans, Thiel and Luckmann, in 1928, and was probably known at an earlier date.

Extensive researches into the facts of corrosion were carried out in Germany about 1908–10 by Heyn and Bauer, whose writings contain the germ of many ideas developed later. They were probably the first to carry out really comprehensive measurements of corrosion velocity in numerous liquids, embracing iron and steel in different conditions both alone and in contact with other metals. Amongst other things they quantitatively established the fact that attack upon iron was stimulated by contact with a nobler metal, whilst contact with a baser metal often conferred partial or complete protection (qualitatively this had long been believed to be the case, and indeed as early as 1824 Sir Humphrey Davy had proposed connection with iron or zinc as a means of protecting copper against sea water). The behaviour of iron in contact with other metals was further studied at a later date. In 1924, Whitman and Russell demonstrated that when three-quarters of the surface of a steel specimen is coated with copper, the total corrosion by running water may remain the same as that of a similar specimen completely bare, but is now concentrated on the bare quarter. This experiment provided an illustration of the manner in which corrosion may be intensified when a small anode is connected to a large cathode, a matter also brought out by the author's experiments in 1928, and by the extensive work of Akimow, with Clark and others, published in Russia about 1935.

Despite the importance of the so-called 'galvanic currents' set up by dissimilar metals, the notion – prevalent up to about 1923 – that corrosion currents were generally due to dissimilar metals in contact, e.g. the dominant metal and the impurities present in it, led to some erroneous conclusions. It spread the belief (definitely held by Palmaer) that perfectly pure and uniform metal – if it could be obtained – would be uncorrodible. Now in the hydrogen evolution type of attack (which Palmaer studied in his laboratory) impure metal really is attacked much more quickly than the purer varieties, and the varying influence of different impurities was afterwards brought out by the researches of the Czech, Vondràček, and the Latvian, Straumanis. But this is not true of all types of attack; as pointed out in 1931 by the German scientist, Tammann, impurities do not increase the rate of corrosion of zinc by a persulphate solution. Actually, electric currents can arise in other ways than by the contact of dissimilar metals; they can, for instance, be generated by differences in the liquid wetting different parts of a metallic surface, notably by differences in the oxygen content.

The fact that variations in oxygen concentration can set up electric currents should have been clear from the experiments of the Italian scientist, Marianini, performed as early as 1830, or from those of the Scot, Adie (1845), the German, Warburg (1889), and the Russian, Kistiakowsky (1908); but the significance of these researches was largely overlooked. In 1916, however, the American, Aston, laid emphasis on the part played by local differences in oxygen concentration in promoting the rusting of iron, and in 1922 his fellow-countryman, McKay, showed that currents could also be set up on a single metal by variations of metal ion concentration in the liquid.

Experiments carried out in the author's laboratory at Cambridge from 1923 onwards indicated that 'differential aeration currents' (set up by differences in oxygen concentration) played an important part in the corrosion of many metals, although, as emphasized by Bengough, oxygen concentration is only one of many factors determining the potential existing at different points on a metallic surface. Considerable discussion as to the relative importance of different factors took place between the two groups of investigators associated with Bengough and the author respectively, and an agreed statement was published in 1938.

In the years 1931–9 the electric currents flowing over the surface of metal corroding in salt solutions were detected and measured by several of the author's collaborators at Cambridge, including Bannister, Hoar, Thornhill and Agar; also, after his return to America, by Mears, in collaboration with Brown. The currents were found to be strong enough to account for the whole of the corrosion actually measured, in the sense of Faraday's Law, and thus the electrochemical mechanism of corrosion by salt solutions may be said to rest upon a quantitative experimental basis.

Nevertheless, the increasing evidence for the electrochemical nature of corrosion by salts and acids led in some quarters to the exaggerated view that the mechanism of every corrosion process is necessarily electrochemical. No doubt any chemical change involves the displacement of electrons, but if we reserve the term 'electrochemical corrosion' for those processes where current flows between anodic and cathodic areas situated at different parts of the metallic surface (with electrons flowing over paths long compared to the interatomic distance), then it is clear that many true corrosion processes are not electrochemical, since they occur in solutions where electrical conductivity is almost absent. The work at Cambridge has suggested that the reason why electrochemical action is so destructive is that it often leads to the formation of soluble compounds as the immediate corrosion products, whereas direct combination with oxygen would produce a sparingly soluble body in physical contact with the metal thus stifling further attack. Frequently the electrochemical mechanism leads in the end to a sparingly soluble body, such as rust; but if this is formed as a secondary product through interaction between the anodic and cathodic products at a perceptible distance from the point of attack, there is no stifling and attack continues.

Actually in those cases where electrochemical action would produce a sparingly soluble body either at the cathode or at the anode, it stifles itself just like any other form of attack. Consequently it is possible to classify substances which stop corrosion by coating the anodic or cathodic parts with layers of sparingly soluble compounds as cathodic and anodic inhibitors. The Polish investigator, Chyżewski, working in the author's laboratory in 1938, made an experimental classification of inhibitors into these two groups. The anodic inhibitors are the most efficient, but are apt to be dangerous, since when added in insufficient amount they often reduce the area undergoing corrosion more rapidly than they diminish the total amount of corrosion, so that the intensity of the local attack is actually increased by the addition. The author had shown in 1936 that this intensification was to be expected on theoretical grounds. Much work has been devoted to the search for an inhibitive system which is both safe and efficient; the work at Teddington on benzoates may be quoted as an example.

Many of the substances which form useful pigments in anti-corrosive paints probably act in rather the same way as the soluble inhibitors just discussed. This was indicated by the early work of Cushman and Gardner published in America about 1910. The distinction between inhibitive paints, which will prevent rusting even if the coating is not impervious to corrosive substances and the mechanically-excluding paints, which only give protection in so far as they prevent access of moisture to the metal, became increasingly clear in a series of field and laboratory tests, organized from Cambridge and carried out from 1930 onwards, by Britton, Lewis, Thornhill and particularly Mayne. Another type of anti-corrosive paints which will give protection even if gaps exist in the coats are those richly pigmented with zinc dust; here the exposed iron escapes attack largely because it is the cathode of the cell zinc/iron.

Although it might be thought that salt-free water would produce oxides or hydroxides which would stifle attack, this stifling does not always occur. Water containing oxygen may be responsible for rusting, if the conditions are such as to produce at the metallic surface the appreciably soluble ferrous hydroxide, and to permit of its conversion to the less soluble ferric hydroxide (yellow rust) at a slight distance from the point of attack, so that the rust is non-protective. Forrest, Roetheli and Brown (1930–1), in a series of instructive researches carried out at the Massachusetts Institute of Technology, showed that rust coats varied in protective character according to the rate of supply of oxygen to the surface, and the same principle was found to operate in statistical experiments with oxygen–nitrogen mixtures carried out by Mears (1935) during his work in the author's laboratory. These indicated that oxygen can depress the probability of corrosion starting within a given area, although, where once attack has set in, oxygen accelerates the corrosion rate; this distinction between corrosion velocity and corrosion probability has served to clear up several of the apparent contradictions which puzzled earlier investigators.

The view has been advanced by several investigators that oxides of iron can in effect act as oxygen-carriers. In 1921 Friend, who had already carried out an extensive series of researches on the action of numerous salt solutions on iron, suggested that a colloidal solution of ferric hydroxide behaves as a carrier for oxygen, passing alternately between the ferrous and ferric conditions. Later (1936–8) Herzog attributed a rather similar rôle to 'solid' rust, but combined this with an electrochemical mechanism. After long immersion in stagnant water, iron frequently becomes covered with an inner layer of magnetite, overlaid with an outer layer of ferric hydroxide. The magnetite is supposed to act as cathode towards the iron as anode, and the ferric hydroxide just above suffers cathodic reduction to hydrated magnetite. This may either lose water, reinforcing the magnetite, or may absorb oxygen from the air, returning to the state of ferric hydroxide. It seems very likely that under certain circumstances some such mechanism does operate. The measurements of Taylor and the author provide evidence that the atmospheric rusting of iron is connected with an electrochemical cycle of reactions; ferric hydroxide suffers cathodic reduction to magnetite, which is re-oxidized to the ferric state by air.

A study of certain cases of corrosion where a direct chemical mechanism might reasonably be expected, published by Heitz in 1968 (1976 vol., p. 83) has produced evidence that even here there are anodic and cathodic reactions, probably proceeding at the same places.

The electrochemical study of passivity was the subject of a long series of experiments started about 1927 in W. J. Müller's laboratory at Vienna, which placed the mechanism of anodic passivation on a mathematical basis; he had numerous collaborators, of whom Konopicky and Machu deserve special mention. A summary will be found in the 1976 vol., pp. 165–70.

Most of the laboratory experiments mentioned above have been of short duration, and have been concerned mainly with the opening stages of corrosion. In this respect, they were not entirely representative of natural corrosion processes, which continue for long periods. A feature of the extensive and very accurate work of Bengough, started about 1927, first in collaboration with Stuart, then with Lee and later with Wormwell, was the performance of experiments under strictly controlled conditions, extending over many years.

Whilst the mechanism of low-temperature corrosion was being investigated, high-temperature oxidation was also under study and here it was found possible to establish simple laws of growth. The scientific study of the oxidation process may be said to have begun with the classical work of the Americans Pilling and Bedworth published in 1923. The careful observations of Pfeil (1929–31), based on refined analytical work, showed that sometimes oxidation consists in the movement of metal outwards rather than the penetration of oxygen inwards through the scale. Further valuable information regarding the oxidation of iron and its alloys was obtained by the detailed metallurgical observations of the French investigators, Portevin, Prétet and Jolivet, published in 1934. About the same time, the German physical chemist, Wagner, was producing a number of papers, practical and theoretical, which showed that high-temperature oxidation is connected with the passage of ions and electrons through the growing scale; low-temperature tarnishing, due to sulphur compounds in the air, follows a similar mechanism. A mathematical relationship was established between the oxidation-rate and the electrical properties of the film substance; this shows that good resistance to oxidation may generally be expected where the electrical resistance of the oxide formed is high. A new and instructive derivation of Wagner's equation was published by Hoar and Price in 1938, whilst in the same year Price and Thomas, working at Cambridge, applied these new theoretical ideas to the development of a process known as selective oxidation, which produces enhanced resistance of copper alloys to high-temperature oxidation, and similarly improves the behaviour of silver alloys towards low-temperature tarnishing. Shortly before this, Miley, also working in the author's laboratory, developed an electrometric method for measuring the thickness of oxide films, depending on the quantity of electricity needed for their reduction, and was thus able to obtain curves showing the early stages of the rapid growth of invisible films on iron and copper exposed to air at ordinary temperatures. A very complete study of the oxidation of zinc over a range of temperature by Vernon, Akeroyd and Stroud was published in 1939.

Modern ideas of oxidation owe much to the theoretical speculations of Mott, published in 1939. Familiarity with the concept of film growth controlled by ions jumping from site to site over intervening energy barriers has been largely due to him – although the germ of the idea is found in the papers of earlier workers. Another of Mott's suggestions – that the growth of thin films is controlled by tunnelling of electrons – was withdrawn by its originator, but revived by the Germans, Hauffe and Ilschner. In contrast, the theoretical treatment of the early stages of oxidation published in 1948 by Cabrera and Mott, and widely accepted in Great Britain, received criticism from Hauffe and Ilschner. The practical and theoretical researches of the Hauffe group on the oxidation of alloys deserve special mention; in the author's opinion, the principles emerging may have application in other branches of corrosion science, but this work also has not escaped criticism.

Knowledge of oxidation has been greatly aided by the examination of the oxide films. Here X-rays and, more particularly, electron diffraction methods, have been most helpful; the work of Finch and Quarrell (1933–4) on electron diffraction methods deserves special mention. Less widely known, but holding great promise, are the optical methods depending on the changes in the ellipticity of polarized light reflected at a metallic surface – changes which are themselves modified when a film is present on the metal. This method, which in suitable cases provides information both of the thickness and refractive index of the film, was first used to study oxide films in Freundlich's laboratory in Berlin (1927), and it was elaborated by the Norwegian physicist, Tronstad (1929–39), and has been further improved by Winterbottom; in a different form it was used by Leberknight and Lustman in Mehl's laboratory at Pittsburgh about 1939. Other optical methods for obtaining the thickness of films depend on interference. The simplest of these methods was used by Tammann in his classical studies of the laws of film growth in the years 1920–6, whilst later a more accurate spectroscopic method was employed at Cambridge by Constable (1927–9).

Further information has been obtained by the author alone (1927), and with J. Stockdale (1929), by stripping the films from metal, or, as in his later process (1938), by their transfer to transparent plastic; films which produce interference tints on the metal usually display complementary colours after transfer. A form of film-stripping suitable for the chemical estimation of the constituents was developed by Vernon, Wormwell and Nurse in 1939, who established the fact that the polishing of the surface of stainless steel can lead to an enrichment of chromium, the protective element, in the surface film; evidence of analogous cases of enrichment had been obtained in 1938 by the Polish investigator, Dobinski, using the electron diffraction method.

Another field of research which has been developed concurrently is that of atmospheric corrosion. Vernon's extensive and accurate studies of the behaviour of metals exposed to the atmosphere, published at intervals from 1923 onwards, established simple laws connecting corrosion and time. One of the important results attributable to Vernon, Hudson and Patterson was the principle of critical humidity; it was found that frequently corrosion only became rapid in air when the humidity exceeded a certain value.

Very extensive atmospheric corrosion tests have been carried out for various Committees on both sides of the Atlantic. The American Society for Testing Materials were pioneers in this field, and have accumulated a quantity of valuable data. Particularly impressive are the results obtained by J. C. Hudson from 1929 onwards; he worked first on non-ferrous metals for the British Non-Ferrous Metals Research Association, and then on ferrous metals for the (British) Iron and Steel Institute. Hudson's extensive researches have established the relative resistance of different materials in numerous different atmospheric conditions, and in the case of iron and steel have indicated how best protection can be achieved by coatings of paint or non-ferrous metal.

Up to about the time of the war of 1914–18, most of the experimental work on corrosion had been conducted by pure scientists. In later years, a serious increase in the frequency of breakdowns due to corrosion led to the inception of a number of technical researches on industrial problems. In Great Britain some were started by individual firms, but most by research organizations receiving support both from the Government and the interests concerned. A good example is the work on the corrosion of condenser tubes sponsored first by the Corrosion Committee of the Institute of Metals, whose first report appeared in 1911, and later by the British Non-Ferrous Metals Research Association. The experimental work was carried out in the early years by Bengough and several colleagues, one of whom, May, took over the leadership when Bengough passed on to other branches of corrosion research. Condenser trouble had been so serious in the war of 1914–18 that at one time it was said to be causing the British Admiralty almost as much anxiety as the strategic situation. Largely as a result of the work of Bengough, May, and their colleagues, along with the industrial development of new corrosion-resisting alloys, the position so much improved that in the war of 1939–45, condenser trouble, although experienced, was not a major problem.

A cause of corrosion, once unsuspected, but much studied in recent years, is the part played by bacteria in soils. This was first recognized by the Dutch bacteriologist, von Wolzogen Kühr, about 1922, and the subject was further developed at Teddington and Manchester.

Another important development of the subject has been conjoint action, whereby chemical and mechanical influences, working together, produce far more destruction than either could cause working alone. Such conjoint action may arise where certain chemically aggressive agents are present at a surface where stresses operate; these may be internal stresses left after cold-forming or welding, applied unidirectional stress (as in the **stress corrosion** of aircraft alloys or stainless steel), and alternating stresses, as in **corrosion fatigue**. Other examples of conjoint action include **frettage**, where relative movement between two surfaces which rub to and fro against one another over microscopic distances causes disastrous conversion to oxide, **cavitation** connected with collapse of vacuum cavities in a moving stream of water and the **impingement attack** at places where large bubbles of low-pressure air come into contact with a metallic surface – notably at the inlet ends of condenser tubes. These cases are discussed in Chapters 4 and 5.

Likewise there has been in recent years, much work on protective measures, largely based on paints or coatings of metals more resistant to corrosion than the one to be protected, but also on the treatment of corrosive waters either by the addition of an inhibitor or by the removal of objectionable constituents (e.g. the de-aeration of boiler feed-waters). Developments are mentioned in Chapters 6 and 7.

Apart from technical influences, there has been in recent years a change in the type of person who studies corrosion. The average chemist has ceased to be interested in the reactions of metals, becoming increasingly absorbed in the compounds of carbon and of those inorganic elements (such as silicon, phosphorus and fluorine) which build atomic chains or networks. In contrast, the physicist has become concerned with oxidation and film-forming reactions generally, since these largely depend on lattice defects – a subject which is greatly interesting physicists. On the subject of film growth, however, physical chemists, in some countries at least, have played an important part – as exemplified by Wagner's work on oxidation mentioned on p. 236. Understanding of the conditions which will determine the occurrence or avoidance of corrosion has been greatly aided by the development of chemical thermodynamics. One of the most impressive advances has come from the Belgian, Pourbaix; today Pourbaix diagrams, in which pH and potential are the two co-ordinates, provide a vast amount of information in concise form, and indicate at a glance the range of conditions within which immunity, corrosion and passivation may be expected.

Another development has been the growing interest in corrosion as an economic problem. In 1922, Sir Robert Hadfield, after quoting an estimate of corrosion damage put forward in 1920 by the American, D. M. Buck, made a fresh investigation into the cost of corrosion. He arrived at an estimate of well over £700 million for 'the annual cost of wastage due to rusting of the world's iron and steel'; that included the cost of preventive measures, metal losses and replacements arising directly from corrosion. With unintentional humour, the editor of the journal publishing this astronomical figure placed on the very same page the announcement of an award of £100 to Dr Newton Friend to enable him to carry out tests on the corrodibility of iron and steel, adding optimistically, 'It is hoped that the results will be of service'. Other estimates of corrosion loss have followed, the most important being due to Hudson, Uhlig and Vernon; the latter in 1956 reached the conclusion that the total expenditure in the United Kingdom alone, including the cost of preventive measures and metal losses, must be of the order of £600 million. In 1971, a government committee under the chairmanship of T. P. Hoar found that the annual cost to the nation was £1365 million of which £250 million was connected with building and construction, £280 million with marine problems and £350 million with transport. (Report of the Committee on Corrosion and Protection: *A Survey of Corrosion and Protection in the U.K.*, chairman T. P. Hoar, H.M.S.O., London, 1971. See also T. P. Hoar, *Proc. roy. Soc. (A)* 1976, **348**, 1.) Possible savings from the prevention of corrosion were estimated at £310 million. The report of the Committee emphasizes the need for greater attention to corrosion in the training of engineers, designers and architects.

Even before this report, a gradually extending recognition of the economic importance of corrosion problems had led to a considerable expansion of facilities for education in the subject. When, about 1924, lectures on corrosion were started at Cambridge University, it was largely because the subject, as a branch of pure science, was felt worthy on intellectual grounds of attention by students. The same may have been true of Sheffield University where, at the suggestion of Professor Wynne (head of the Chemistry Department), A. W. Chapman – an organic chemist – included corrosion in a set of lectures on 'Chemistry for Engineers' which he was starting in 1924, although the industrial importance of the subject was doubtless kept in mind. The example of these two universities was for a long time not followed by others, and if, in the last few years, a large number of corrosion courses have been started at universities and technological establishments, this is probably due to the growing demand from industry for students trained in a subject which is too complicated to be acquired easily at the post-graduate stage. That does not mean that the courses available today neglect the scientific basis of the subject; it would seem that science and application are generally well blended, and the places where such instruction is given are starting to develop schools of research; special mention should be made of the efforts of Shreir in London and Ross at Manchester. In recent years, an important centre of Corrosion Research has been established at UMIST (University of Manchester Institute of Science and Technology) under Graham Wood – the first Professor of Corrosion Science in the country. A feature of the centre is its close linkage with industry, and an enlightened application of scientific principles to industrial needs.

Nearly every country where scientific research is seriously carried out now possesses its corrosion organization, and there are numerous specialist journals in which corrosion papers can be published. There have been International conferences at suitable intervals; such conferences were held in London (1961), New York (1963), Moscow (1966), Amsterdam (1969), Tokyo (1972) and Sydney (1975). There have been conferences devoted to special features of the subject, such as passivity and localized corrosion, whilst the admirable meetings at Ferrara devoted to inhibitors have a character of their own. Mention should be made of the congresses of the European Corrosion Federation which include papers of high standard. The (American) National Association of Corrosion Engineers has played a rather special part in the organization of international conferences and in the printing of the volumes containing the papers delivered. Nor is international cooperation confined to the holding of conferences. Pourbaix, using the organization of Cebelcor (Centre Étude Belge de la Corrosion) which, in effect, has become an international body, enlisted many years ago the services of scientists in different countries to collect information and carry out the calculations needed to construct a Pourbaix diagram for practically every metal; the large Atlas mentioned on p. 251 represents the result of this fruitful collaboration.

Electrochemical Appendix

General

Importance of electrochemistry

So many corrosion processes are electrochemical phenomena that a knowledge of scientific electrochemistry is essential for the understanding of the subject. This appendix may help those who find difficulty in reading the body of the book, where a certain familiarity of electrochemical principles has been assumed, but all who can afford the time are strongly advised to study a standard electrochemical textbook, or at least the excellent elementary treatment provided by Potter. (E. C. Potter, *Electrochemistry: principles and practice*, Cleaver-Hume, 1946.)

A piece of metal placed in corrosive liquid generally develops 'anodic' and 'cathodic' reactions, which balance one another; the first leaves electrons in the metal; the second uses them up. The two reactions usually occur on different areas, but sometimes (as in Wagner and Traud's work, mentioned on p. 68) on the same area. It is helpful to study them separately, and this is best done in electrochemical cells, often subjected to an e.m.f. from an external battery. That is the reason why discussions of *single potentials*, which some readers at first feel to be unrealistic, can be helpful.

Faraday's Law

If an electric current of I amperes is made to pass through an electrolytic cell, consisting of two metal electrodes immersed in a solution of an electrolyte, then, in the absence of secondary effects, the velocity of the corrosion suffered by the anode will be I/F gram-equivalents per second, where F is Faraday's Number (96,487 coulombs per gram-equivalent). This means that 1 ampere hour will dissolve 1·04 grams of pure iron (as ferrous salt), 1·22 grams of zinc or 3·86 grams of lead.

Ohm's Law

If the effective e.m.f. driving the electrolytic cell is E volts, the current in amperes (I) will be E/R, where R is the resistance of the cell in ohms; the corrosion velocity will be E/RF. However, in most practical cases, the e.m.f. causing corrosion is not applied from an outside source, but is furnished by the metal itself; owing to 'polarization' the value of E tends to diminish as the current density (current per unit area) increases, and the state of affairs is that represented by Fig. 3.7.

Generation of electric currents

It is known – largely from photoelectric research – that the energy needed to remove an electron from a metal varies greatly in different cases; in general, the removal requires more energy in the case of a noble than a base metal. If two metals, say copper and zinc, are placed in contact, the copper will (in absence of complicating factors) tend to absorb electrons from the zinc until the energy levels of the electrons in the two metals have adjusted themselves to oppose further passage of electrons across the interface. It is, of course, impossible to obtain a continuous electric current from a circuit consisting half of zinc and half of copper (Fig. A.1(a)), since the potential difference (p.d.) set up at one contact (P_1) will exactly balance that set up at the other (P_2) – provided that the two contacts are kept at the same temperature. If, however, the zinc and copper are joined through a solution of, say, potassium chloride (Fig. A.1 (b)), a current can be generated, the energy being provided by chemical changes. The passage of electricity across the plane XY will consist in the movement of *negatively* charged *anions* (each bearing an extra electron) towards the zinc, which is the *anode* of the cell (the pole towards which the anions move), and the movement of *positively* charged *cations* (atoms deficient of an electron) towards the copper, which is the *cathode*.* The actual reactions at the boundaries between the liquid and metals will vary in different solutions, but in general they will provide for a continuous consumption of electrons at the copper surface, and a continuous supply of electrons at the zinc surface, so that the flow of electrons from the zinc to the copper, at the junction, P, will not result in an accumulation on one side or an exhaustion on the other; thus a flow of electricity through the metallic conductors can continue.

Equilibrium conditions

Reversible cells

The cell described above, made by zinc and copper dipping into the same salt solution (potassium chloride), is not reversible; the reactions which occur when current passes through it in one direction are quite different from those which occur when current passes in the

*This definition of the anode and cathode as the poles towards which anions and cations respectively move applies both to current-producing and electrolytic (current-consuming) cells. Some confusion is often caused by the fact that the anode of a current-producing cell is the *negative* pole (i.e. the one which should be joined to the negative terminal of an ammeter), whereas, in an electrolytic cell, the anode is the *positive* electrode (i.e. the one which should be joined to the positive terminal of the dynamo or accumulator supplying the current).

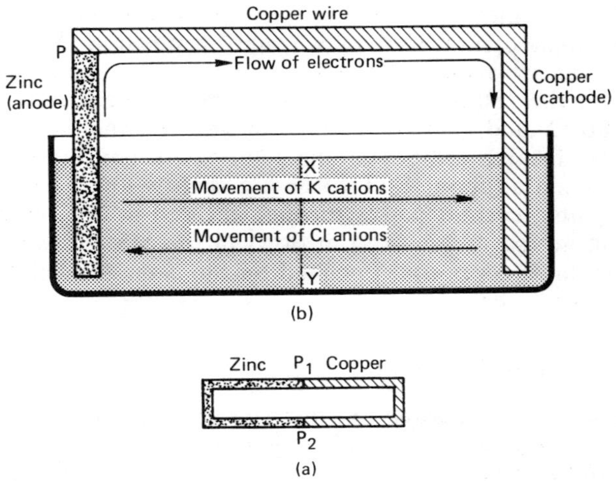

Fig. A.1 Movement of ions and electrons in a circuit

other direction. Certain cells containing two liquids are 'reversible' in the sense that a small departure from equilibrium conditions in either direction causes a flow of current in one sense or the other, the same reactions taking place, although in opposite directions. The Daniell cell, for instance, in which the copper is surrounded by copper sulphate solution and the zinc by zinc sulphate solution, with a porous partition separating the two liquids, is approximately reversible in that sense. Its e.m.f. may be obtained by balance on a potentiometer, a primitive form of which is shown in Fig. A.2; a variable potential difference is applied to the cell by means of the sliding contact, S,

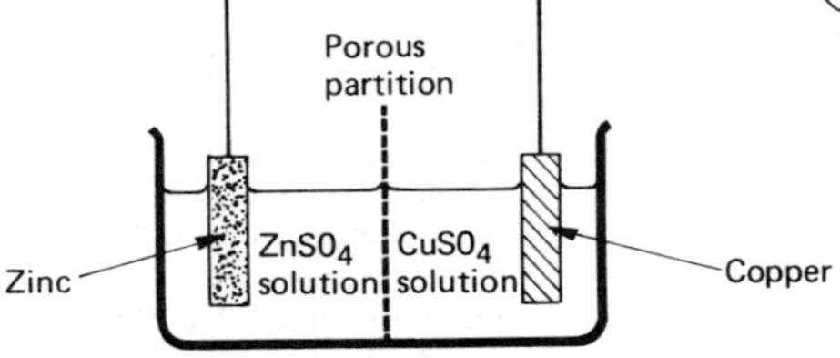

g. A.2 Daniell cell and simple potentiometer

and the slide is moved along the wire PQ (over which a potential gradient is applied from the cell C), until a position (X) is found at which the null galvanometer G shows no reading – which means that the e.m.f. of the cell is exactly equal to the potential difference between X and Q, and may be read off at once on the voltmeter V. In the case of a Daniell cell containing solutions of unit activity, balance will be obtained at about 1·11 volts; if a slightly lower p.d. is applied (say 1·10 volts), current passes through the cell in one direction, and if slightly more (say 1·12 volts), it flows in the other direction. In the first case metallic copper is deposited on the cathode, whilst zinc passes into the ionic state

$$Cu^{2+} + 2e^- \rightarrow Cu$$
$$Zn \rightarrow Zn^{2+} + 2e^-$$

In the second case, the same reactions occur in the opposite direction, metallic zinc being deposited and copper dissolved.

Instead of using copper and zinc, a cell of the 'Daniell type' may be made by immersing another pair of metals respectively in solutions of their salts. Any Daniell type cell may be written in the general form

M_1 | M_1 salt solution | M_2 salt solution | M_2

Concept of a half cell

Cells of the Daniell type are, in general, approximately reversible, although the e.m.f.s which they can balance vary greatly. Any such cell can be regarded as made up of two half cells

M_1 | M_1 salt solution

and

M_2 | M_2 salt solution

each of which contributes a characteristic p.d. The value of the p.d. of any half cell remains *unchanged*, irrespective of the nature of the other half cell with which it is combined. The total e.m.f. of the combination is equal to $E_1 - E_2$, where E_1 and E_2 are the p.d.s of the two half cells.*

This fact was at one time taken to indicate that the e.m.f. of a galvanic cell is associated solely with p.d.s occurring between the metal and the solution. Such a conclusion is now known to be incorrect. Frequently much of the p.d. associated with a half cell is located at the junction between the electrode metal and the wire (usually copper) used to join the half cell to the potentiometer.† Nevertheless, if

*This expression neglects the p.d. at the junction of the two salt solutions – which may be made low if they are connected through a concentrated solution of a salt such as potassium chloride or ammonium nitrate, where both ions have almost the same mobility.

†The p.d. between copper wire and, say, the zinc plate of a Daniell cell will be the same whether the copper is in direct contact with the zinc, or whether a brass terminal is interposed.

we include in our definition of the half cell a short piece of copper wire attached to metal M_1 or M_2 as the case may be, and consider the p.d. between the wire and the salt solution, the old conception of the e.m.f. of a Daniell type cell as the difference between the p.d.s of two half cells may usefully be retained.

Standard electrode potentials

If we choose a certain half cell as the arbitrary zero of a potential scale, we can assign a consistent value to the p.d. of any other half cell. The **standard hydrogen electrode**, consisting of platinum covered with platinum black, and dipping into acid of normal hydrogen-ion activity (1·2 molar hydrochloric acid) saturated with hydrogen at 1 atmosphere pressure, has been adopted as arbitrary zero. The potential of the half cell formed by any metal immersed in a solution of its salt of normal 'cation activity' is called the **standard electrode potential** of the metal in question.‡ Tables showing standard electrode potentials on the **hydrogen scale** are given in the endpaper table. The order in which elements fall in this table is called the potential series.

It is important to note that the standard electrode potential of a metal on the hydrogen scale really represents the e.m.f. needed to balance the cell formed by combining the standard electrode of the metal in question with the standard hydrogen electrode. The cell formed by joining a hydrogen electrode with a silver electrode is shown schematically in Fig. A.3. If the p.d. between the acid surrounding the platinum and a point P on the copper wire joining the hydrogen electrode to the potentiometer is *adopted* as the zero of our arbitrary scale, then the p.d. between the liquid surrounding the silver electrode and a point Q on the copper wire joining it to the potentiometer will be given, *on the hydrogen scale*, by the potentiometer reading when balance has been obtained.

As explained above, much of the potential drop is located at the junction between the metal and the copper wire, the rest being of course located between the metal and the solution. It would not affect the numbers given in the endpaper table if nickel wires were used in the place of copper wires; although this would shift the absolute value of the p.d. at the junction between wire and metal electrode, there would be a similar shift where the wire joined the platinum of the hydrogen electrode, and the measurement would remain unaltered. Similarly, any p.d. at the junction between the connecting wires and the alloy wire used for the windings of the potentiometer will not affect the numbers, since the p.d. at the entrance to the potentiometer will balance the p.d. at the exit.

‡The 'activity' of a dissolved substance is the 'effective concentration' which (if adopted in the place of the actual concentration) would cause the Law of Mass Action to be obeyed. For dilute solutions it often differs little from the actual concentration. The exact evaluation of the activities of single ions presents difficulties.

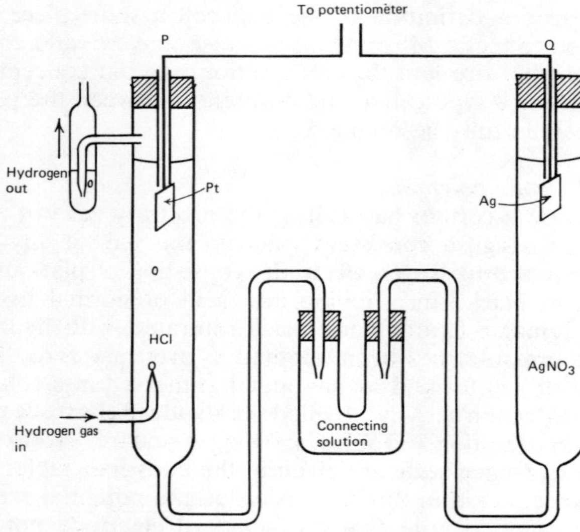

Fig. A.3 The determination of the electrode potential of silver

Effect of concentration

The standard hydrogen electrode consists of platinum immersed in acid of normal hydrogen-ion activity saturated with hydrogen at one atmosphere pressure. It takes up the potential at which the passage of elementary hydrogen from the molecular state to the ionic is balanced by the passage from the ionic state to the molecular. An increase in the hydrogen pressure of the atmosphere, producing an increase in the concentration of molecular hydrogen, will favour the change in the first direction, but not in the second, thus upsetting the balance. A fresh balance may, however, be obtained at a more negative potential. Thus an increase of hydrogen pressure above one atmosphere will render the potential negative, whilst a decrease will render it positive, the change being expressed approximately by the equation

$$E_p = E_1 - \frac{RT}{2F} \ln p$$

where E_p is the potential at pressure p atmospheres, E_1 the potential at one atmosphere, T the absolute temperature, R the gas constant, and F Faraday's number.

At constant gas pressure, an increase of hydrogen ion activity in the liquid (i.e. an increase of concentration) will make the potential positive and a decrease negative. The equation, of which a derivation will be found on p. 290, is

$$E_C = E_1 + \frac{RT}{F} \ln C = E_1 + \frac{2{\cdot}3RT}{F} \log_{10} C$$

Table A.1 Equilibrium potential of hydrogen at atmospheric pressure in different liquids

Activity of hydrogen ions in liquid	pH value	Example	Potential of hydrogen-saturated platinum, volts
Over 1·0M	Negative	HCl above 1·2M	Small positive values
1·000M	0·000	1·2M HCl	0·000 (arbitrary zero)
10^{-1}M	1	0·12M HCl	−0·058
10^{-2}M	2	0·011M HCl	−0·116
10^{-3}M	3	10^{-3}M HCl	−0·174
10^{-4}M	4	10^{-4}M HCl	−0·232
10^{-5}M	5	10^{-5}M HCl	−0·290
10^{-6}M	6		
10^{-7}M	7	Pure water	−0·406
10^{-8}M	8		
10^{-9}M	9	10^{-5}M KOH	−0·522
10^{-10}M	10	10^{-4}M KOH	−0·580
10^{-11}M	11	10^{-3}M KOH	−0·638
10^{-12}M	12	0·011M KOH	−0·696
10^{-13}M	13	0·12M KOH	−0·754
10^{-14}M	14	1·2M KOH	−0·812

where E_C is the potential at activity C, and E_1 the potential at normal activity of hydrogen ions. At 18° C, 2·3 RT/F is equal to 0·058 volts. The equation means, therefore, that, at a temperature of 18°C, in solutions where the activity can be regarded as equal to the concentration, the potential of the hydrogen electrode at one atmosphere pressure drops 0·058 volts every time the concentration is diminished to one-tenth of its previous value, as shown in Table A.1.

The potential of the hydrogen electrode constitutes a convenient way of determining the hydrogen-ion activity of the liquid surrounding it. The so-called pH scale is based on this. If the activity of hydrogen ions in a given liquid is 10^{-n}M the pH value of that liquid is said to be n. A neutral solution has a pH value of 7·0, whereas acid solutions show lower values, and alkaline liquids higher ones. M/1000 hydrochloric acid with a hydrogen-ion activity of about 10^{-3}M will have a pH value of about 3·0; M/1000 sodium hydroxide with a hydrogen-ion activity of about 10^{-11}M will have a value of about 11·0.

Amongst the numerous applications of this scale, one possessing special importance for corrosion problems concerns the character of soils and natural waters. Soils are usually acid in moorland districts and bog pools may have pH values of 5·0 or even lower. Rain water also is frequently acid. In London (Battersea Park) the pH of rain was 5·0 in July 1936 but lower (3·9) in the following January, since coal (which contains sulphur) is burnt more freely in winter, so that the atmosphere contains more sulphur acids. In the country, higher values are obtained, especially in summer; at Godalming the value was 7·6 in July but 5·3 in January. In industrial towns where fumes are allowed to escape from chemical works, it may be much lower;

3·0 has twice been recorded at Burnley. The pH value of the dew film deposited each night on metal in towns may have still lower values.

The potentials of the metals set out in the endpaper table are also altered if the activity (effective concentration) is not molar. Clearly, if the solution is diluted, the passage from the ionic to the metallic state will be slower, whereas the passage from the metal into solution remains unaltered. Thus the balance will be upset, but a fresh balance may be obtained at a more negative potential. The correction which must be applied is given by $\frac{RT}{nF}\ln C$, where n is the valency. This means that at 18°C, if activity and concentration are regarded as identical, every ten-fold dilution (say from M to M/10, or from M/10 to M/100) will shift the potential in a negative direction by about 0·058 volt for a monovalent ion, or 0·029 volt for a divalent ion. The theoretical shift of potential is approximately realized in the case of the more noble metals; but on some of the less noble metals the potential actually measured is found to be almost independent of the concentration of the metallic salt in the original solution; this occurs if the metal is capable of reacting with the solution, so that the concentration in the layer next to the metal becomes different from that in the body of the liquid.

Effect of films

The potentials given in the endpaper table refer to metal free from an oxide film. Aluminium, which covers itself with an oxide film, highly protective and difficult to remove, commonly gives values much less negative than that indicated, so that aluminium is frequently 'nobler' than zinc (i.e. it acts as the cathode of the cell Zn | Liquid | Al).

Effect of complex-forming substances

Whilst oxide films on the metal tend to shift the potential in the positive (noble) direction, the presence of salts which form complex ions containing the metal in question renders the potential abnormally negative. Thus copper placed in a solution containing cyanides may show values negative to hydrogen instead of the usual positive values; the cyanide ions combine with copper ions, forming complex ions such as $[Cu(CN)_2]^-$ and locking up the bulk of the metal in positions of low potential energy. The complex need not be an anion; in ammoniacal solutions, copper shows abnormally negative potentials, due to complex cations such as $[Cu(NH_3)_2]^+$. If the complex provides a specially stable position for the copper atoms within the liquid, the rate at which copper will be deposited on the metal will be slowed down, and equilibrium will only be obtained if the potential is shifted to a more negative value, so as to slow down the opposing passage from metal to liquid.

Potential in a solution originally free from ions of the metal under study

If a heavy metal is placed in a solution of some such salt as potassium chloride or sodium sulphate, the potential will tend to vary with time. Clearly, since the solution is initially free from ions of the metal in question, there can be no immediate equilibrium. Soon, however, owing to incipient corrosion, ions of that metal will begin to appear in the adjacent layers of the liquid. In general, a constant value of the potential will arise when some solid substance begins to separate out on the metallic surface, since the metal-ion concentration will then be kept constant at the saturation value. The solid may be either a salt or a hydroxide, and since the solubility of the solid body will be affected by the presence of a common anion in the solution, the value of the potential finally reached will be altered by the addition of a sodium or potassium salt providing such an anion. If the solid is a hydroxide, the potential will depend on the $(OH)^-$ concentration, that is on the pH value (since the OH^- concentration rises as the H^+ concentration falls). Gatty and Spooner[1040] found that the potential of lead in alkaline potassium chloride solution depended on the Cl^- concentration if the ratio Cl^-/OH^- exceeded $10^{5 \cdot 2}$, but on the pH value if the ratio was less than that value. The potential of iron immersed in an aqueous solution originally free from iron ions depends on the hydrogen-ion concentration, becoming steadily more negative as the pH value rises.

If solid silver chloride is stirred with potassium chloride of a certain concentration, the concentration of silver ions in the liquid will depend on the concentration of potassium chloride used. Thus a convenient reference electrode consists of silver surrounded by potassium chloride solution of known strength saturated with silver chloride. Similar reference electrodes can be based on mercury covered with various solutions saturated with mercurous salts. The well-known Calomel electrode is one of these. The values of some reference electrodes on the hydrogen scale are given in Table A.2.

Table A.2 Potentials of some practical reference electrodes at 25°C*

Hg	Solid Hg_2Cl_2	Solid KCl ('saturated calomel electrode')	+0·2415
Hg	Solid Hg_2Cl_2	M KCl ('molar calomel electrode')	+0·2800
Hg	Solid Hg_2Cl_2	M/10 KCl ('decimolar calomel electrode')	+0·3337
Ag	Solid AgCl	M/10 KCl	+0·2881
Cu	Saturated $CuSO_4.5H_2O$		+0·316

*Based on *Corrosion Handbook* edited by H. H. Uhlig, Wiley.

Determination of electrode potentials

It would be possible to determine the standard electrode potential of any metal (say silver) by placing it in a solution of a salt (say the nitrate) of standard activity, joining it to a standard hydrogen electrode through a connecting solution chosen to suppress liquid junction potentials (Fig. A.3), and measuring on a potentiometer the potential needed to balance the e.m.f. of the combination. The value (0·799 volt for silver) will represent the standard electrode potential of silver on the standard hydrogen scale.

In such cases, however it is often easier to substitute for the hydrogen electrode a standard half cell such as a calomel electrode, and measure the balance potential of the combination.

$$\underset{\text{(normal activity)}}{Ag \mid AgNO_3} \mid \underset{\text{(say concentrated } NH_4NO_3)}{\text{Connecting solution}} \mid \text{Molar KCl} \mid Hg_2Cl_2 \mid Hg$$

By 'adding' to the value obtained experimentally (say +0·519 volt) the value of the calomel half cell on the standard hydrogen scale (+0·280 volt), we get the value for silver on the hydrogen scale

$$0{\cdot}519 + 0{\cdot}280 = +0{\cdot}799 \text{ volt.}$$

It will be noted that for a base metal, which would have a potential strongly negative to calomel, the effect of this 'addition' would be to *decrease* the numerical value.

The calomel (or silver chloride) electrodes are useful in other ways. If (Fig. A.4) the open tip of a tubulus leading to such an electrode is brought up against a selected point (P) on a metal surface immersed in a solution, the value of the reading between the standard electrode and the metal will be affected by 'corrosion currents' passing to or from the metal at the point in question; the reading is displaced in one direction if the point is cathodic, and in the other if it is anodic, by an amount depending on the current density. If the relation between the potential reading and the cathodic or anodic current density has previously been determined by independent measurements on specimens to which known currents have been applied, it is possible

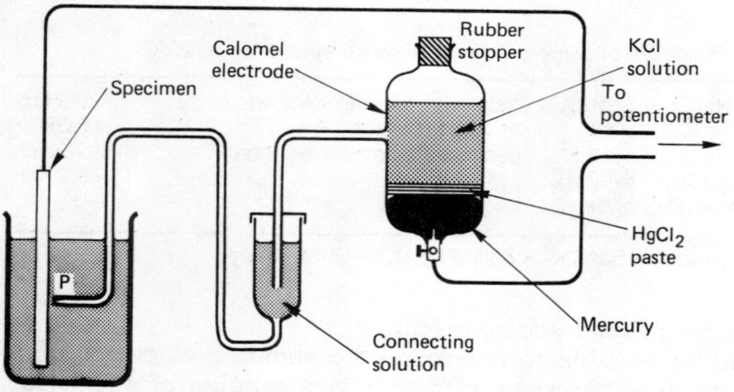

Fig. A.4 The calomel electrode used for measuring a local p.d.

to obtain a measure of the current density at any point on the corroding specimen, without actually tapping the current. Alternatively by moving the tip of the tubulus in turn to different points in the liquid near the metallic surface, and by taking readings at each point, the distribution of potential in the liquid may be studied, which will also serve to furnish a measure of the currents flowing. These two methods have been used at Cambridge by Hoar and Agar respectively to measure corrosion currents; the current strength was found to be equivalent in the sense of Faraday's Law to the corrosion velocity (p. 41).

Non-equilibrium conditions

Polarization

The potentials recorded in the endpaper table represent equilibrium values. In order that a metal shall suffer anodic attack, the potential must be moved slightly in the positive direction; if cathodic deposition is to take place upon it, the potential must be moved in a negative direction. These shifts of potential are known as anodic and cathodic 'polarization' respectively.

Overpotential of metallic deposition

In a few cases the changes required are quite large. On iron and nickel, cathodic deposition only becomes appreciable if the potential is made considerably more negative than the equilibrium value, whilst dissolution only becomes important if it is moved considerably in the opposite direction. This irreversibility of iron stands in contrast to the behaviour of silver, where a very small increase of potential from the equilibrium value will produce anodic attack, and a very small shift in the negative direction will produce cathodic deposition. The difference between the 'normal' metals (like silver) and the 'abnormal' ones (like iron) has been discussed by Piontelli; the abnormal metals have compact structures, the interatomic distances being small and the amplitude of thermal vibration small; thus more energy must be provided to help the passage of cations into the liquid.

Effect of complex salts

A solution of a complex cyanide, such as $KAg(CN)_2$, $KCu(CN)_2$ or $K_2Zn(CN)_4$, contains complex anions $[Ag(CN)_2]^-$, $[Cu(CN)_2]^-$ or $[Zn(CN)_4]^{2-}$, and, as explained on p. 284, metals show abnormal potentials in such solutions. For instance, when the ordinary Daniell cell

$$Cu \mid CuSO_4 \text{ solution} \mid ZnSO_4 \text{ solution} \mid Zn$$

furnishes current, the zinc is the anode and passes into solution, whilst the copper is cathode, fresh metal being deposited on it. On the other hand, the cell

Cu | KCN solution | $ZnSO_4$ solution | Zn

furnishes a current in the opposite direction, the copper being the anode. Similarly, although tin is cathodic to iron in most salt solutions, it becomes anodic to iron in certain organic acids, because the tin becomes locked up as complex anions (see p. 215).

Complex salt solutions are often used for electroplating, especially for coating articles of complicated geometrical shape, which, if plated in a simple salt solution, receive much deposit on the prominent parts, the recesses being neglected. If a complex salt solution is used, with most of the metal locked up as complexes of great stability, it is impossible to cause rapid deposition at the prominent parts, because polarization sets in, the potential being altered in such a manner that deposition is diverted to the recesses; these solutions are said to have good **throwing power**.

The same complex-ion solutions often provide smooth coats where simple salt solutions would give rugged macrocrystalline deposits, the atoms being deposited on existing crystals instead of starting new nuclei. The explanation may be similar to that of the improved throwing power, but it is generally supposed that the main cause of the smooth deposits obtained from cyanide baths is the adsorption of the cyanide as a 'poison' on the tips of the crystals, where growth would otherwise proceed; this favours the starting of new small crystals instead of the growth of the original ones. Organic addition agents, as well as metallic hydroxides, are also useful in obtaining smooth bright deposits (see p. 214). The 'abnormal' metals (iron and nickel) often give smooth deposits without complexing agents or other additives.

Types of polarization

The departure of the potential of an electrode from the equilibrium (or reversible) value when external current flows arises from three causes:

(1) **Concentration polarization**. If an external e.m.f. is applied to the cell

Silver | $AgNO_3$ solution | Silver

silver will be dissolved from the anode and deposited on the cathode. If the solution is vigorously stirred, a small e.m.f. suffices to force a large current through the cell, and the reading of a voltmeter joined in parallel with the cell represents mainly the ohmic drop IR, where I is the current and R the liquid resistance. If stirring is stopped, the current flowing will decline and the voltmeter reading will increase, since now there is a *back* e.m.f. connected with the fact that the solution is becoming stronger near the anode and weaker near the cathode; the potential changes at each electrode will be such as to oppose current flow. If the source of external e.m.f. be disconnected, the voltmeter will continue to show a reading, representing the back e.m.f.

In corrosion, the most important cases of concentration polarization are connected with oxygen exhaustion. Often corrosion velocity is controlled by the supply of oxygen, which is needed for the cathodic part of the corrosive reaction. If an e.m.f. is applied to a cell such as

Platinum	Aqueous solution containing oxygen	Platinum

the current connected with the cathodic reduction of oxygen can never exceed that corresponding to the rate at which oxygen reaches the cathode – whether by convection or diffusion. If the e.m.f. applied exceeds that needed to force this limiting amount of current through the cell, the potential at the cathode will fall to the level at which some other reaction, generally the evolution of hydrogen, becomes possible.

(2) **Approach resistance polarization**. If the anodic or cathodic reaction can take place only at certain points on a metallic surface, the ohmic resistance of the bottleneck approaches to these points represents a measurable potential drop. In this type of polarization, the shift of potential is a rectilinear function of the current. In the research described on p. 41, Hoar found iron partly immersed in a salt solution suffered potential shifts which were rectilinear functions of the current I (not of log I as would be expected if activation polarization were involved). This suggests that, at any moment, the anodic and/or cathodic reactions are proceeding at a limited number of small points – for which there is other evidence.

(3) **Activation potential**. When a metal electrode stands in equilibrium with a solution of a salt of the same metal, the potential difference V is such that cations are passing from metal to liquid at exactly the same rate as they are passing from liquid to metal. The current representing the electrical transfer is known as the *exchange current*, I_0, (the corresponding **exchange current density**, I_0/a, where a is the area, is written i_0). Only ions possessing sufficient energy can pass the hurdles separating sites of low energy in the metal and those in the solution. If we wish to produce say, anodic dissolution, we must apply a small potential, ΔV, from an external source so as to *help* passage *into* the liquid and *hinder* that *from* the liquid. The anodic current will now become

$$K_1 e^{-[W_1 - n\epsilon\alpha(V + \Delta V)]/kT}$$

and the cathodic current $K_2 C e^{-[W_2 + n\epsilon(1 - \alpha)(V + \Delta V)]/Kt}$ where ϵ is the electronic charge, W_1, W_2 are the energies needed to jump the hurdles from the two sides, α is the proportion of the potential drop available to help the anodic reaction ($(1 - \alpha)$ being available to hinder the cathodic reaction), C is the ionic concentration, K_1 and K_2 being constants. Near V, the net current passing is the difference between these two exponentials, but at a distance from equilibrium conditions ΔV is large and the second exponential becomes negligibly small, so that i will be roughly

$$K_1 e^{-[W_1 - n\epsilon\alpha(V + \Delta V)]/kT}$$

thus $i = i_0 e^{n\epsilon\alpha\Delta V/kT}$, since i_0 is the value of i when $\Delta V = 0$.

Thus

$$\ln (i/i_0) = n\epsilon\alpha\Delta V/kT$$

or

$$\Delta V = 2{\cdot}303 \frac{kT}{n\epsilon\alpha} \log_{10} \frac{i}{i_0} = b \log_{10} \frac{i}{i_0}$$

generally written $\Delta V = a + b \log_{10} i$, where a represents $-b \log_{10} i_0$. This is known as **Tafel's equation**. The proof given is an over-simplification. Often the reactions pass through stages, and the discussion becomes complicated, but the case presented serves to show the character of the arguments involved.

For the deposition or dissolution of normal metals, the activation energy is low, and activation polarization is therefore small – often being masked by concentration polarization. It is much higher for the 'abnormal' elements, like iron and nickel, which fall in the centre of the periodic table, but it becomes most important in connection with reactions involving oxygen or hydrogen.

Concentration correction

The influence of concentration on *equilibrium* potential (p. 282) can similarly be derived. At equilibrium

$$K_1 e^{-(W_1 - n\epsilon\alpha V)/kT} = C K_2 e^{-(W_2 + n\epsilon(1 - \alpha)V)/kT}$$

or

$$C = \frac{K_1}{K_2} e^{-(W_1 + W_2 + n\epsilon V)/kT}$$

Whence

$$\ln C = \ln \frac{K_1}{K_2} + \frac{- W_1 + W_2}{kT} + \frac{n\epsilon V}{kT}$$

or

$$V = K + \frac{kT}{n\epsilon} \ln C$$

where K is a constant independent of C. When C is unity, $V = K$. Therefore if ΔV is the alteration of potential when C departs from unity

$$\Delta V = \frac{kT}{n\epsilon} \ln C = \frac{RT}{nF} \ln C = 2{\cdot}3 \frac{RT}{nF} \log_{10} C.$$

At 18°C, $2{\cdot}3 \frac{RT}{nF} = 0{\cdot}058$ volt, explaining why a ten-fold shift of concentration (or more strictly activity) shifts the equilibrium potential by 58, 29 or 19 mV according to the valency – as shown in the endpaper table.

Hydrogen potential

If a platinum electrode is placed in acid of unit H^+ activity (e.g. 1·2M HCl) saturated with hydrogen at 1 atmosphere, equilibrium should be established at exactly 0·00 volts, and if the potential is slightly raised, molecular hydrogen should be converted to hydrogen ions; conversely if it is slightly lowered hydrogen bubbles might be expected to appear. On platinum (particularly if the effective area has been increased by depositing 'platinum black' on the electrode), bubbles can be seen at potentials not far below the equilibrium point, but with any other metal the *overpotential* which must be applied is quite high; with mercury or lead, it may be necessary to apply 0·5 volt or more before bubbles become visible, although measurable current will flow at much smaller polarization values; the hydrogen formed then passes into solution. In general, the relation between overpotential and current is given by Tafel's equation $\Delta V = a + b \log_{10} i$ and some accurate values for *a* and *b*, extracted by J. N. Agar[1025] from Russian work, are shown in Table A.3.

Table A.3 Overpotential constants (in volts) (A. N. Frumkin and others)

Metal	Pb†	Hg†	Cd‡	Zn†	Sn*	Cu†	Ag*	Fe*	Ni§	Co*	Pd#	Pt*
a	1·56	1·415	1·40	1·24	1·24	0·80	0·95	0·70	0·64	0·62	0·53	0·10
b	0·110	0·113	0·120	0·118	0·116	0·115	0·116	0·125	0·100	0·140	0·130	0·13

† in 0·5M H_2SO_4 ‡ in 0·65M H_2SO_4 * in M HCl § in 0·11M NaOH # in 1·1M KOH

The fact that different solutions have been used for different metals does not greatly complicate the situation. Iron, which shows the values $a = 0{\cdot}70$, $b = 0{\cdot}125$ in M HCl, shows $a = 0{\cdot}76$, $b = 0{\cdot}112$ in 2M NaOH.

Oxygen potential

Still more irreversible is the oxygen potential. The cathodic reaction $O_2 + 2H_2O + 4e^- \rightarrow 4OH^-$ is highly important in corrosion reactions, but proceeds in stages (hydrogen peroxide is an intermediate product), and only becomes rapid at potentials far below the theoretical equilibrium value, which should be 1·23 volts above the value of hydrogen in the same liquid (any shift in pH value alters the hydrogen and oxygen electrodes by the same amount). Perhaps this is fortunate, since the irreversibility of the reaction lowers the effective e.m.f. of the corrosion cells. Early attempts to establish the reversible potential experimentally were unsuccessful, since if there is the smallest trace of oxidizable impurity in the liquid (e.g. 10^{-5} mole dm^{-3} of SO_2, often present in 'pure' sulphuric acid), its oxidation will compete with the reaction $OH^- \rightarrow O_2$ and we shall fail to measure the potential at which $OH^- \rightarrow O_2$ balances $O_2 \rightarrow OH^-$. Only in 1956 did Bockris and Huq[1031], after careful removal of impurities by preliminary electrolysis, obtain a value of 1·24 volt, very close to the theoretical 1·23 volt; to achieve this, they had to reduce the impurity content below 10^{-11} mole dm^{-3}.

Author Index

Subject Index